Lecture Notes in Artific85

Subseries of Lecture Notes in Co

Edited by J. G. Carbonell, and J. S

Lecture Notes in Computer Science

Edited by G. Goos, J. Hartmanis, and J. van Leeuwen

Springer
Berlin
Heidelberg
New York
Barcelona
Hong Kong
London
Milan
Paris
Tokyo

Jacques Calmet   Belaid Benhamou
Olga Caprotti   Laurent Henocque
Volker Sorge (Eds.)

Artificial Intelligence, Automated Reasoning, and Symbolic Computation

Joint International Conferences
AISC 2002 and Calculemus 2002
Marseille, France, July 1-5, 2002
Proceedings

 Springer

Volume Editors

Jacques Calmet
University of Karlsruhe (TH)
Am Fasanengarten 5, Postfach 6980, D-76128 Karlsruhe, Germany
E-mail: calmet@ira.uka.de

Belaid Benhamou
Université de Provence, CMI
39 rue F. Juliot-Curie, 13453 Marseille Cedex 13, France
E-mail: Belaid.Benhamou@cmi.univ-mrs.fr

Olga Caprotti
Johannes Kepler University
Research Institute for Symbolic Computation (RISC-Linz)
A-4040 Linz, Austria
E-mail: ocaprott@risc.uni-linz.ac.at

Laurent Henocque
Université de la Méditerannée, ESIL
163 Avenue de Luminy, Marseille Cedex 09, France
E-mail: henocque@esil.univ-mrs.fr

Volker Sorge
University of Birmingham, School of Computer Science
Birmingham B15 2TT, United Kingdom
E-mail: V.Sorge@cs.bham.ac.uk

Cataloging-in-Publication Data applied for

Die Deutsche Bibliothek - CIP-Einheitsaufnahme

Artificial intelligence, automated reasoning, and symbolic computation :
joint international conferences ; proceedings / AISC 2002 and Calculemus
2002, Marseille, France, July 1 -5, 2002. Jacques Calmet ... (ed.). - Berlin
; Heidelberg ; New York ; Barcelona ; Hong Kong ; London ; Milan ; Paris ;
Tokyo : Springer, 2002
    (Lecture notes in computer science ; Vol. 2385 : Lecture notes in
    artificial intelligence)
    ISBN 3-540-43865-3

CR Subject Classification (1998): I.2.1-4, I.1, G.1-2, F.4.1

ISSN 0302-9743
ISBN 3-540-43865-3 Springer-Verlag Berlin Heidelberg New York

Springer-Verlag Berlin Heidelberg New York
a member of BertelsmannSpringer Science+Business Media GmbH

http://www.springer.de

© Springer-Verlag Berlin Heidelberg 2002
Printed in Germany

Typesetting: Camera-ready by author, data conversion by PTP-Berlin, Stefan Sossna
Printed on acid-free paper   SPIN 10870512   06/3142   5 4 3 2 1 0

Preface

AISC 2002, the 6th international conference on Artificial Intelligence and Symbolic Computation, and Calculemus 2002, the 10th symposium on the Integration of Symbolic Computation and Mechanized Reasoning, were held jointly in Marseille, France on July 1–5, 2002. This event was organized by the three universities in Marseille together with the LSIS (Laboratoire des Sciences de l'Information et des Systèmes).

AISC 2002 was the latest in a series of specialized conferences founded by John Campbell and Jacques Calmet with the initial title "Artificial Intelligence and Symbolic Mathematical Computation" (AISMC) and later denoted "Artificial Intelligence and Symbolic Computation" (AISC). The scope is well defined by its successive titles.

AISMC-1 (1992), AISMC-2 (1994), AISMC-3 (1996), AISC'98, and AISC 2000 took place in Karlsruhe, Cambridge, Steyr, Plattsburgh (NY), and Madrid respectively. The proceedings were published by Springer-Verlag as LNCS 737, LNCS 958, LNCS 1138, LNAI 1476, and LNAI 1930 respectively.

Calculemus 2002 was the 10th symposium in a series which started with three meetings in 1996, two meetings in 1997, and then turned into a yearly event in 1998. Since then, it has become a tradition to hold the meeting jointly with an event in either symbolic computation or automated deduction.

Both events share common interests in looking at Symbolic Computation, each from a different point of view: Artificial Intelligence in the more general case of AISC and Automated Deduction in the more specific case of Calculemus. Holding the two conferences jointly should trigger interdisciplinary research, with the first results expected at AISC 2004 (Austria) and at Calculemus 2003.

This volume includes papers accepted for presentation at both AISC 2002 and Calculemus 2002. From the 52 contributions submitted, 17 full papers were accepted for AISC, plus 7 full papers and 2 system descriptions for the Calculemus program. In addition, the invited speakers' abstracts are included as part of this volume. Several work-in-progress contributions were accepted for presentation at Calculemus 2002 but are not published in these proceedings.

We would like to express our thanks to the members of the two program committees, to the AISC steering committee, to the Calculemus trustees, to the referees, and to the organizing committee. Finally, we gratefully thank the sponsors for their financial support. Names are listed on the following pages.

April 2002

Jacques Calmet
Belaid Benhamou
Olga Caprotti
Laurent Henocque
Volker Sorge

Organization

Both AISC 2002 and CALCULEMUS 2002 were organized by the three universities of Marseille: L'Université de Provence (Aix-Marseille I), L'Université de la Méditerranée (Aix-Marseille II), La Faculté des Sciences de Saint-Jerôme (Aix-Marseille III), and the LSIS (Laboratoire des Sciences de l'Information et des Systèmes).

Conference Organization (AISC)

Conference and local chair:	Belaid Benhamou (Univ. de Provence, Aix-Marseille I)
Program chair:	Laurent Henocque (Univ. de la Méditerranée, Aix-Marseille II)
Steering committee:	Jacques Calmet (Univ. Karlsruhe, Germany)
	John Campbell (Univ. College London, UK)
	Eugenio Roanes-Lozano (Univ. Complutense de Madrid, Spain)

Symposium Organization (CALCULEMUS)

Program chairs:	Olga Caprotti (RISC, Hagenberg, Austria)
	Volker Sorge (Univ. of Birmingham, UK)
Local chair:	Belaid Benhamou (Univ. de Provence, Aix-Marseille I)

Local Committee (AISC / CALCULEMUS)

Gilles Audemard	(Univ. de Provence, Aix-Marseille I)
Belaid Benhamou	(Univ. de Provence, Aix-Marseille I)
Philippe Jegou	(Fac. de Saint Jerôme, Aix-Marseille III)
Laurent Henocque	(Univ. de la Méditerranée, Aix-Marseille II)
Pierre Siegel	(Univ. de Provence,Aix-Marseille I)

Program Committee (AISC)

Luigia C. Aiello	(Univ. La Sapienza, Roma, Italy)
Jose A. Alonso	(Univ. de Sevilla, Spain)
Michael Beeson	(San Jose State Univ., USA)
Belaid Benhamou	(Univ. de Provence, France)
Greg Butler	(Univ. Concordia, Montreal, Canada)
Simon Colton	(Univ. of York, UK)
Jim Cunningham	(Imperial College London, UK)
James Davenport	(Univ. of Bath, UK)
Carl van Geem	(LAAS-CNRS, Toulouse, France)
Reiner Haehnle	(Univ. of Technology, Chalmers, Sweden)
Deepak Kapur	(Univ. New Mexico, USA)
Luis M. Laita	(Univ. Politecnica de Madrid, Spain)
Luis de Ledesma	(Univ. Politecnica de Madrid, Spain)
Eric Monfroy	(Univ. de Nantes, France)
Jose Mira	(UNED, Spain)
Ewa Orlowska	(Inst. Telecommunications, Warsaw, Poland)
Jochen Pfalzgraf	(Univ. Salzburg, Austria)
Jan Plaza	(Univ. Plattsburgh, USA)
Zbigniew W. Ras	(Univ. North Carolina, Charlotte, USA)
Tomas Recio	(Univ. de Santander, Spain)
Peder Thusgaard Ruhoff	(MDS, Proteomics, Denmark)
Pierre Siegel	(Univ. de Provence, France)
Andrzej Skowron	(Warsaw Univ., Poland)
John Slaney	(ANU, Canberra, Australia)
Viorica Sofronie-Stokkermans	(Max Planck Institut, Germany)
Karel Stokkermans	(Univ. Salzburg, Austria)
Carolyn Talcott	(Stanford Univ., USA)
Rich Thomason	(Univ. of Pittsburgh, USA)
Dongming Wang	(Univ. Paris VI, France)

Additional Referees (AISC)

W. Ahrendt	M. Ceberio	A. C. Norman
F. Benhamou	M. Damsbo	H. Wang
R. Bernhaupt	L. Bordeaux	
B. Buchberger	M. Jaeger	
O. Caprotti	P.A. Madsen	

Program Committee (CALCULEMUS)

Alessandro Armando	(Univ. of Genova, Italy)
Christoph Benzmüller	(Univ. of Saarbrücken, Germany)
Jacques Calmet	(Univ. of Karlsruhe, Germany)
Alessandro Coglio	(Kestrel Institute, Palo Alto, USA)
Arjeh Cohen	(Univ. of Eindhoven, The Netherlands)
Simon Colton	(Univ. of Edinburgh, Scotland, UK)
James Davenport	(Univ. of Bath, UK)
William M. Farmer	(McMaster Univ., Hamilton, Canada)
Thérèse Hardin	(Univ. de Paris VI, France)
Hoon Hong	(Univ. of North Carolina State, Raleigh, USA)
Manfred Kerber	(Univ. of Birmingham, UK)
Michael Kohlhase	(Carnegie Mellon Univ., Pittsburgh, USA)
Steve Linton	(Univ. of St. Andrews, Scotland, UK)
Ursula Martin	(Univ. of St. Andrews, Scotland, UK)
Julian Richardson	(Heriot-Watt Univ., Edinburgh, Scotland,UK)
Renaud Rioboo	(Univ. de Paris VI, France)
Roberto Sebastiani	(Univ. di Trento, Italy)
Andrew Solomon	(Univ. of Technology, Sydney, Australia)
Andrzej Trybulec	(Bialystok Univ., Poland)
Volker Weisspfenning	(Univ. of Passau, Germany)
Wolfgang Windsteiger	(RISC, Hagenberg, Austria)

Additional Referees (CALCULEMUS)

M. Benerecetti	M. Jaume	G. Norman
D. Doligez	V. Ménissier-Morain	S. Ranise
A. Fiedler	M. Moschner	

Sponsoring Institutions

L'Ecole d'Ingénieur en Informatique de Luminy, ESIL
L'Université de Provence, Aix-Marseille I
Le Conseil Général de Marseille
La Mairie de Marseille
Calculemus Project
CologNet, European Network of Excellence

Table of Contents

Invited Talks

Constraint Acquisition 1
  Eugene C. Freuder

Expressiveness and Complexity of Full First-Order Constraints in
the Algebra of Trees 2
  Alain Colmerauer

Deduction versus Computation: The Case of Induction 4
  Eric Deplagne, Claude Kirchner

Integration of Quantifier Elimination with Constraint Logic
Programming 7
  Thomas Sturm

AISC Regular Talks

Towards a Hybrid Symbolic/Numeric Computational Approach in
Controller Design 12
  Madhu Chetty

Inductive Synthesis of Functional Programs 26
  Emanuel Kitzelmann, Ute Schmid, Martin Mühlpfordt, Fritz Wysotzki

A Symbolic Computation-Based Expert System for Alzheimer's
Disease Diagnosis 38
  *Begoña Herrero, Luis M. Laita, Eugenio Roanes-Lozano, Víctor Maojo,
  Luis de Ledesma, José Crespo, Laura Laita*

On a Generalised Logicality Theorem 51
  Marc Aiguier, Diane Bahrami, Catherine Dubois

Using Symbolic Computation in an Automated Sequent Derivation
System for Multi-valued Logic 64
  Elena Smirnova

The Wright ω Function 76
  Robert M. Corless, D.J. Jeffrey

Multicontext Logic for Semigroups of Contexts 90
  Rolf Nossum, Luciano Serafini

Indefinite Integration as a Testbed for Developments
in Multi-agent Systems 102
  J.A. Campbell

Expression Inference – Genetic Symbolic Classification
Integrated with Non-linear Coefficient Optimisation 117
  Andrew Hunter

A Novel Face Recognition Method 128
  Li Bai, Yihui Liu

Non-commutative Logic for Hand-Written Character Modeling........... 136
  Jacqueline Castaing

From Numerical to Symbolic Data during the Recognition of Scenarii..... 154
  S. Loriette-Rougegrez

On Mathematical Modeling of Networks and Implementation Aspects 168
  Regina Bernhaupt, Jochen Pfalzgraf

Continuous First-Order Constraint Satisfaction 181
  Stefan Ratschan

Coloring Algorithms for Tolerance Graphs: Reasoning and
Scheduling with Interval Constraints 196
  Martin Charles Golumbic, Assaf Siani

A Genetic-Based Approach for Satisfiability Problems 208
  Mohamed Tounsi

On Identifying Simple and Quantified Lattice Points in the 2SAT
Polytope 217
  K. Subramani

Calculemus Regular Talks

Integrating Boolean and Mathematical Solving: Foundations, Basic
Algorithms, and Requirements.. 231
  Gilles Audemard, Piergiorgio Bertoli, Alessandro Cimatti,
  Artur Korniłowicz, Roberto Sebastiani

The Meaning of Infinity in Calculus and Computer Algebra Systems...... 246
  Michael Beeson, Freek Wiedijk

Making Conjectures about Maple Functions 259
  Simon Colton

Employing Theory Formation to Guide Proof Planning 275
  Andreas Meier, Volker Sorge, Simon Colton

Unification with Sequence Variables and Flexible Arity Symbols
and Its Extension with Pattern-Terms 290
    Temur Kutsia

Combining Generic and Domain Specific Reasoning by Using Contexts.... 305
    Silvio Ranise

Inductive Theorem Proving and Computer Algebra in the MathWeb
Software Bus 319
    Jürgen Zimmer, Louise A. Dennis

YACAS: A Do-It-Yourself Symbolic Algebra Environment 332
    Ayal Z. Pinkus, Serge Winitzki

Focus Windows: A New Technique for Proof
Presentation... 337
    Florina Piroi, Bruno Buchberger

Author Index 343

Table of Contents

Constraint Acquisition*

Eugene C. Freuder

Cork Constraint Computation Centre
University College Cork, Cork, Ireland
`e.freuder@4c.ucc.ie`; `www.4c.ucc.ie`

Abstract. Many problems may be viewed as constraint satisfaction problems. Application domains range from construction scheduling to bioinformatics. Constraint satisfaction problems involve finding values for problem variables subject to restrictions on which combinations of values are allowed. For example, in scheduling professors to teach classes, we cannot schedule the same professor to teach two different classes at the same time. There are many powerful methods for solving constraint satisfaction problems (though in general, of course, they are NP-hard). However, before we can solve a problem, we must describe it, and we want to do so in an appropriate form for efficient processing. The Cork Constraint Computation Centre is applying artificial intelligence techniques to assist or automate this modelling process. In doing so, we address a classic dilemma, common to most any problem solving methodology. The problem domain experts may not be expert in the problem solving methodology and the experts in the problem solving methodology may not be domain experts.

---

* The author is supported by a Principal Investigator Award from Science Foundation Ireland.

Expressiveness and Complexity of Full First-Order Constraints in the Algebra of Trees

Alain Colmerauer

Laboratoire d'Informatique Fondamentale de Marseille, Université Aix-Marseille II,
France, alain.colmerauer@lim.univ-mrs.fr

Extended Abstract

What can be expressed by constraints, in the algebra of trees, if quantifiers and all the logical connectors are allowed? What is the complexity of algorithms for solving such general first-order constraints? This talk answers these two questions.

Preliminaries. Let F be a set of function symbols. The algebra of trees consists of the set of trees, whose nodes are labelled by elements of F, together with the construction operations linked to the elements f of F. Such an operation, with f of arity n, is the mapping $(a_1, \ldots, a_n) \mapsto a$, where a is the tree, whose initial node is labelled f and whose sequence of daughters is $a_1, \ldots, a_n$. A general first-order constraint is a formula made from: variables, elements of F, the equality symbol $=$, the logical constants and connectors $true, false, \neg, \wedge, \vee$ and the usual quantifiers $\exists, \forall$.

Expressiveness. With respect to expressiveness, we show how to express a constraint of the form

$$\varphi^n(x,y) \overset{\text{def}}{=} \begin{bmatrix} \exists u_0 \ldots \exists u_n \\ x = u_0 \wedge \\ \varphi(u_0, u_1) \wedge \\ \varphi(u_1, u_2) \wedge \\ \cdots \\ \varphi(u_{n-1}, u_n) \wedge \\ u_n = y \end{bmatrix}$$

by an equivalent constraint $\psi_n(x,y)$, of size almost proportional to the size of the constraint $\varphi(x,y)$. More precisely, we show that there exists a constant c such that, for any n,

$$|\psi_n(x,y)^{\alpha(n)}| \leq c\,|\varphi(x,y)|,$$

where $\alpha(n)$ is the huge integer

$$\alpha(n) \overset{\text{def}}{=} \underbrace{2^{\cdot^{\cdot^{\cdot^{\left(2^{\left(2^{2^2}\right)}\right)}}}}}_{n}.$$

For $n = 5$, this integer is already larger than the number of atoms of the universe.

J. Calmet et al. (Eds.): AISC-Calculemus 2002, LNAI 2385, pp. 2–3, 2002.

Complexity. With respect to complexity, by making use of the previous result, we show that there exists a constant d and an integer n_0 such that:

For any $n \geq n_0$, there exists a constraint ψ of size n, without free variables, such that any algorithm, which decides whether ψ is satisfied in the algebra of trees, executes at least $\alpha(\lfloor dn \rfloor)$ instructions.

Deduction versus Computation: The Case of Induction

Eric Deplagne and Claude Kirchner

LORIA & INRIA,
615 rue du Jardin Botanique, BP 101,
54602 Villers-lès-Nancy Cedex, Nancy, France.
{Eric.Deplagne,Claude.Kirchner}@loria.fr

Abstract. The fundamental difference and the essential complementarity between computation and deduction are central in computer algebra, automated deduction, proof assistants and in frameworks making them cooperating. In this work we show that the fundamental proof method of induction can be understood and implemented as either computation or deduction.

Inductive proofs can be built either explicitly by making use of an induction principle or implicitly by using the so-called induction by rewriting and inductionless induction methods. When mechanizing proof construction, explicit induction is used in proof assistants and implicit induction is used in rewrite based automated theorem provers. The two approaches are clearly complementary but up to now there was no framework able to encompass and to understand uniformly the two methods. In this work, we propose such an approach based on the general notion of deduction modulo. We extend slightly the original version of the deduction modulo framework and we provide modularity properties for it. We show how this applies to a uniform understanding of the so called induction by rewriting method and how this relates directly to the general use of an induction principle.

Summary

Induction is a fundamental proof method in mathematics. Since the emergence of computer science, it has been studied and used as one of the fundamental concepts to build mathematical proofs in a mechanized way. In the rising era of proved softwares and systems it plays a fundamental role in frameworks allowing to search for formal proofs. Therefore proofs by induction have a critical role in proof assistants and automated theorem provers. Of course these two complementary approaches of proof building use induction in very different ways. In proof assistants like COQ, ELF, HOL, Isabelle, Larch, NQTHM, PVS, induction is used explicitly since the induction axiom is applied in an explicit way: the human user or a clever tactics should find the right induction hypothesis as well as the right induction variables and patterns to conduct the induction steps. In automated theorem provers specific methods have been developed to

J. Calmet et al. (Eds.): AISC-Calculemus 2002, LNAI 2385, pp. 4–6, 2002.

automatically prove inductive properties. The most elaborated ones are based on term rewriting and saturation techniques. They are respectively called induction by rewriting and inductionless induction or proof by consistency. Systems that implement these ideas are Spike, RRL or INKA.

The latter automated methods have been studied since the end of the seventies and have shown their strengths on many practical examples from simple algebraic specifications to more complicated ones like the Gilbreath card trick. But what was intriguing from the conceptual point of view was the relationship between explicit and implicit induction: implicit induction was shown to prove inductive theorems, but the relationship with the explicit use of the induction principle was open.

In this work, we provide a framework to understand *both* approaches in a *unified* way. One important consequence is that it allows us to combine in a well-understood way automated and assisted proof search methods. This reconciliation of the two approaches will allow automated theorem provers and proof assistants to collaborate in a safe way. It will also allow proof assistants to embark powerful proof search tactics corresponding to implicit induction techniques. This corresponds to the deduction versus computation scheme advocated in [1] under the name of deduction modulo: we want some computations to be made blindly i.e. without the user interaction and in this case this corresponds to implicit induction; but one also needs to explicitly control deduction, just because we know this is unavoidable and this can also be more efficient.

It is thus not surprising to have our framework based on deduction modulo. This presentation of first-order logic relies on the sequent calculus modulo a congruence defined on terms and propositions. But since we need to formalize the induction axiom which is by essence a second-order proposition, we need to use the first-order representation of higher-order logic designed in [2]. In this formalism, switching from explicit induction to implicit one becomes clear and amounts to push into the congruence some of the inductive reasoning, then to apply standard automated reasoning methods to simplify the goal to be proved and possibly get a better representation of the congruence.

This work relies on the notions and notations of deduction modulo [1] as well as on the first-order presentation of higher-order logic presented in [2]. We refer to these two papers for full definitions, details and motivations of the framework. Using this new framework, we uniformly review the induction by rewriting method and show how it directly relates to the induction principle, thus providing proof theoretic instead of model theoretic proofs of this rewrite based method.

Consequently, since the proof method is completely proof theoretic, to any rewrite based inductive proof we can canonically associate an explicit proof in the sequent calculus, thus providing a proof assistant with all the necessary informations to replay the proof as needed.

References

1. Gilles Dowek, Thérèse Hardin, and Claude Kirchner. Theorem proving modulo. Rapport de Recherche 3400, Institut National de Recherche en Informatique et en Automatique, April 1998.
   ftp://ftp.inria.fr/INRIA/publication/RR/RR-3400.ps.gz.
2. Gilles Dowek, Thérèse Hardin, and Claude Kirchner. HOL-λσ an intentional first-order expression of higher-order logic. *Mathematical Structures in Computer Science*, 11(1):21–45, 2001.

Integration of Quantifier Elimination with Constraint Logic Programming

Thomas Sturm

University of Passau, Germany
sturm@uni-passau.de
http://www.fmi.uni-passau.de/~sturm/

Abstract. We examine the potential of an extension of constraint logic programming, where the admissible constraints are arbitrary first-order formulas over some domain. Constraint solving is realized by effective quantifier elimination. The arithmetic is always exact. We describe the conceptual advantages of our approach and the capabilities of the current implementation CLP(RL). Supported domains are currently $\mathbb{R}$, $\mathbb{C}$, and $\mathbb{Q}_p$. For our discussion here we restrict to $\mathbb{R}$.

1 Constraint Logic Programming

Logic programming languages have emerged during the early seventies with Prolog by Colmerauer and Kowalski being the by far most prominent example. The major conceptual contribution was disconnecting *logic* from *control* [Kow79]. The programmer should not longer be concerned with specifying and coding algorithmic control structures but instead *declaratively* specify a problem within some formal logical framework (Horn clauses). The system would then provide a universal control algorithm (resolution) for solving the specified problem. Prolog became surprisingly successful during the eighties. This pure approach of declarative specification, however, turned out to be not sufficiently efficient. On the basis of the observation that the arithmetic capabilities of the processing machine had remained unused, there had then been numbers added to Prolog and *built-in predicates* on these numbers. This approach, however, was not compatible with the original idea of separating logic and control.

This dilemma has been resolved with the step from logic programming (LP) to *constraint logic programming* (CLP) around the mid of the eighties. CLP combines logic programming languages with *constraint solvers*. Constraint solving was another established declarative programming paradigm that had come into existence already in the early sixties in connection with graphics systems. A *constraint solving problem* is given by a finite set of *constraints*. A constraint is a relational dependence between several objects, variables, and certain functions on these numbers and variables. The type of objects and the admitted functions and relational dependences make up the *domain* of the constraint solver. One example are linear programming problems (the target function can be coded as an extra constraint). A *solution* of a constraint system is *one* binding of all variables such that all constraints are simultaneously satisfied. A constraint solver

J. Calmet et al. (Eds.): AISC-Calculemus 2002, LNAI 2385, pp. 7–11, 2002.

computes such a solution if possible, or otherwise states that the system is not *feasible*. In CLP, constraints may appear, besides regular atoms, within the bodies of program clauses and within queries. Constraint solvers are supposed to admit at least *equations* as valid constraints such that the unification within resolution can be replaced by a constraint solving step.

The initial step towards this type of systems was Colmerauer's Prolog II introducing negated equality "$\neq$" and an extended unification that could handle infinite cyclic terms, also known as rational trees. Around 1988, three constraint logic systems of high influence appeared independently: CHIP, CLP(R), and Prolog III. CHIP includes constraint solvers for arithmetic over finite domains, finite Boolean algebra, and linear rational arithmetic. CLP(R) supports real linear arithmetic using floating point numbers. Colmerauer's Prolog III supports Boolean constraints, finite lists, and exact linear rational arithmetic.

2   Effective Quantifier Elimination

The historical development of effective quantifier elimination (QE) is independent from that of CLP. It origins from model theory, where the existence of quantifier-free equivalents over some model class has important theoretical consequences. Given a first-order formula over a domain in the above sense, a quantifier elimination procedure computes an equivalent formula that does *not* involve any quantifiers. For parameter-free formulas this amounts to solving the corresponding decision problem for the domain provided that variable-free atomic formulas can be evaluated to truth values, which is most often the case.

QE is an extremely powerful tool, and for many domains it can be shown that such procedures do not exist. One famous example for this is the ordered ring of integers. Luckily, they exist for a variety of very interesting domains, such as for instance the following:

- Real closed fields (real numbers $\mathbb{R}$ with ordering).
- Algebraically closed fields (complex numbers $\mathbb{C}$).
- Linear formulas in discretely valued fields (p-adic numbers $\mathbb{Q}_p$ for primes p with p-adic valuation). QE for general formulas can be obtained here by adding infinitely many unary relations that state the existence of n-th roots.

For a thorough survey of implemented real QE methods and application examples, we refer the reader to [DSW98]. Besides QEPCAD as the currently most advanced complete implementation of partial cylindrical algebraic decomposition for the reals, we mention here the computer logic system REDLOG [DS97] based on REDUCE, which implements the domains listed above. In addition, REDLOG provides interfaces to other QE implementations including QEPCAD. REDLOG is the platform for the implementation of CLP(RL), which is our connection between CLP and QE discussed in the sequel.

Before turning to this, we wish to mention that CLP(RL) is not the first system combining LP with QE. Around 1992, Hong and others have started to develop CLP(RISC) which combines LP with QEPCAD as a constraint solver [Hon93]. In contrast to CLP(RL) the use of QE in CLP(RISC) is not visible to the user, who just gains a solver for pure relational constraints of arbitrary degree over the reals. In other words, CLP(RISC) completely lives in the CLP world.

3   QE-Based CLP

Neglecting the programming language aspect of CLP a little bit, it can be considered an approach to modeling certain scenarios and automatically deriving knowledge about these scenarios. This is done by combining methods of mathematical logic in the LP part with algebraic methods in the constraint solving part.

This very characterization as well applies to the application of effective quantifier elimination procedures. Within this framework, a domain is chosen in the same way as in CLP. Then scenarios are modeled by first-oder formulas over this domain. Information is derived by possibly adding further first-order formulas, possibly adding further quantifiers, and then automatically computing a quantifier-free equivalent of the obtained input system.

It is a considerable restriction of CLP that logical constructions are restricted to Horn formulas. In spite of extensive research, there is no satisfying solution for handling arbitrary Boolean combinations of constraints or quantification. Vice versa, it is a considerable restriction of QE that it operates exclusively over the chosen domain. There is no facility for defining extra functions or predicates. Also there is no formalism for connecting quantifier elimination procedures over several domains to cooperatively solve a problem.

These observations, together with the fact that all services required from a constraint solver for use within CLP can be formulated as QE problems, gave rise to the idea for CLP(RL), which uses quantifier elimination procedures in the same way as constraint solvers are used in CLP [Stu02]. There are at least two possible ways to look at this:

1. In terms of CLP, the gain is a sophisticated constraint solver, which can handle not only relational dependences but arbitrary first-order formulas over the corresponding domain. Over the reals, constraints of arbitrary degree are supported with exact arithmetic.
2. From a QE point of view, the definition of extra predicates and functions in the LP part becomes possible. Furthermore, there is a perspective for smoothly integrating QE procedures over several domains by means of a type concept.

For both sides, the extensions are conservative in the following sense: CLP programmers can restrict to relational dependences, or use the extended concepts only to such an extent that they feel comfortable about it. Vice versa, every QE problem can be stated by means of an empty program with a query consisting of a single quantified "constraint."

We are going to give an impression of CLP(RL) over the reals by means of some simple examples. All computations have been performed on a 667 MHz Pentium III.

Arbitrary Degrees and Exact Arithmetic

Our first example is taken from Hong [Hon93]. The program describes the Wilkinson polynomial equation:

$$\text{wilkinson}(X, E) \leftarrow \prod_{i=1}^{20}(X + i) + EX^{19} = 0.$$

Mind that the left hand side polynomial of the equation occurs in the program in expanded form. On the query $\leftarrow$ wilkinson$(X, 0)$, $-20 \leqslant X \leqslant -10$ we obtain after 0.3 s the answer $\bigvee_{i=1}^{20} X + i = 0$.

For the query $\leftarrow$ wilkinson$(X, 2^{-23})$, $-20 \leqslant X \leqslant -10$ with a slight perturbation, we obtain after 0.9 s the following answer (in expanded form):

$$8388608 \cdot \left(\prod_{i=1}^{20}(X + i) + 2^{-23}X^{19} \right) = 0 \wedge X + 20 \geqslant 0 \wedge X + 10 \leqslant 0.$$

This answer is contradictory, which could be tested, e.g., by applying QE to its existential closure. QEPCAD immediately yields "false" on this. Since our CLP(RL) lives inside a computer algebra system, we leave the responsibility of how to proceed with such an answer to the user. They might alternatively apply the partly numerical function realroots of REDUCE to the left hand side polynomial of the equation. This yields after 0.5 s the result

$$X \in \{-20.8469, -8.91725, -8.00727, -6.9997, -6.00001, -5, -4, -3, -2, -1\}.$$

We have just been reminded how sensitive the root behavior of polynomials and thus algebraic equations and inequalities are even to smallest rounding errors. Within CLP(RL) all arithmetic is exact. We have learned that the price to be payed is possibly obtaining only implicit solutions. Then one has the choice either to remain exact, or to apply approximate methods.

Disjunction

In traditional CLP, constraints are finite sets of relational dependences regarded as conjunctions. All suggested approaches to introducing disjunction in the CLP literature are based on extending the resolution procedure to handle these constructs. This leads to further restrictions of completeness of the resolution procedure while the true declarative semantics of disjunction has never been obtained. Within our framework, we leave the treatment of disjunction to QE. Consider the following program for the minimum:

$$\min(X, Y, Z) \leftarrow (X \leqslant Y \wedge Z = X) \vee (Y \leqslant X \wedge Z = Y).$$

The answers that can be derived from this program are as complete and concise as the definition itself. For the query $\leftarrow \min(3, 4, Z)$ we obtain $Z - 3 = 0$. For $\leftarrow \min(X, Y, 3)$ the answer is

$$(X - 3 = 0 \wedge Y - 3 \geqslant 0) \vee (X - 3 \geqslant 0 \wedge Y - 3 = 0).$$

Asking for $\leftarrow \min(X, Y, Z)$, we obviously get the definition itself. All these computations take no measurable time.

Quantified Constraints

Since our admissible constraints are first-order formulas, they may also contain quantification. This does not increase the expressiveness but supports the concise formulation of programs. The following program describes that in real 2-space the point (u_1, u_2) is the image of the point (x_1, x_2) under central projection from the punctual light source (c_1, c_2):

$$\mathrm{pr}(C_1, C_2, X_1, X_2, U_1, U_2) \leftarrow \exists T \left(T > 0 \wedge \bigwedge_{i=1}^{2} U_i = T(X_i - C_i) \right).$$

Notice that this description covers all degenerate cases that arise when some of the points or coordinates coincide. The following is a possible quantifier-free description with 10 atomic formulas:

$$(C_1 = 0 \wedge C_2 = 0 \wedge U_1 = X_1 \wedge U_2 = X_2) \vee$$
$$(C_2 \neq 0 \wedge C_2 U_2 > C_2 X_2 \wedge C_1 U_2 - C_1 X_2 - C_2 U_1 + C_2 X_1 = 0) \vee$$
$$(C_1 \neq 0 \wedge C_1 U_1 > C_1 X_1 \wedge C_1 U_2 - C_1 X_2 - C_2 U_1 + C_2 X_1 = 0).$$

In higher dimensions the effect becomes more dramatic. The corresponding quantifier-free description in 3-space has 18 atomic formulas, the one in 4-space has 28.

4 Conclusions

We have introduced an extension of CLP, where the constraints are arbitrary first-order formulas over some domain. Constraint solving then consists in various applications of QE. The advantages of our approach include constraints of arbitrary degree, exact arithmetic, absolutely clean treatment of disjunction and other Boolean operators, and quantified constraints. Alternatively, our approach can be considered an extension of QE by LP facilities. Our concept is implemented in our system CLP(RL) based on REDLOG.

References

[DS97]   Andreas Dolzmann and Thomas Sturm. Redlog: Computer algebra meets computer logic. *ACM SIGSAM Bulletin*, 31(2):2–9, June 1997.

[DSW98]   Andreas Dolzmann, Thomas Sturm, and Volker Weispfenning. Real quantifier elimination in practice. In B. H. Matzat, G.-M. Greuel, and G. Hiss, editors, *Algorithmic Algebra and Number Theory*, pages 221–247. Springer, Berlin, 1998.

[Hon93]   Hoon Hong. RISC-CLP(Real): Constraint logic programming over real numbers. In Frederic Benhamou and Alain Colmerauer, editors, *Constraint Logic Programming: Selected Research*. MIT Press, 1993.

[Kow79]   Robert A. Kowalski. Algorithm = Logic + Control. *Communications of the ACM*, 22(7):424–435, July 1979.

[Stu02]   Thomas Sturm. Quantifier elimination-based constraint logic programming. Technical Report MIP-0202, FMI, Universität Passau, D-94030 Passau, Germany, January 2002.

Towards a Hybrid Symbolic/Numeric Computational Approach in Controller Design

Madhu Chetty

Gippsland School of Computing and Information Technology
Monash University, Churchill-3842, Australia
madhu.chetty@infotech.monash.edu.au

Abstract. Application of general computer algebra systems like MAPLE V® can prove advantageous over conventional 'numerical' simulation approach for controller design. In this paper, an approach for the application of hybrid symbolic/numeric computations to obtain explicit equations leading to the design of an output feedback controller is presented. The methodology for controller design using symbolic algebra is exemplified by considering the design of an excitation controller for a simplified model of the synchronous generator connected to an infinite bus. The output feedback controller is obtained from a symbolic full-state feedback controller by eliminating feedback from unmeasurable states using the free parameters in the symbolic feedback gain expressions. The entire analysis is carried out using the MATLAB® symbolic algebra toolbox that supports MAPLE V®.

1 Introduction

Due to the phenomenal developments in computer hardware and software, the approach to the solution of control and other problems is also undergoing fundamental changes. Implementation of new tools and techniques, such as the new generation of mathematical computation systems like MAPLE V® or MATHEMATICA®, have also increased the mathematical facilities available to investigators. Designers can now hopefully conceive techniques that were traditionally considered too complex or tedious. Such situations can arise in the modeling, sensitivity analysis and design or optimisation aspects of control systems. The recent development of linkages to application software such as MATLAB®, familiar to control engineers, will open further interesting possibilities for investigations in this area.

Recently, in the field of control engineering, symbolic computation has been used for studying zero dynamics, which plays an important role in the areas of modelling, analysis and control of linear and non-linear systems [1]. Furthermore, calculations of controllability and observability grammians and the determination of balanced realizations have also been considered in a symbolic framework [2]. The use of symbolic algebra in the analysis and design of robust control systems and the necessity

J. Calmet et al. (Eds.): AISC-Calculemus 2002, LNAI 2385, pp. 12–25, 2002.

of symbolic environment for the development of the software implementation of the algorithms has also been demonstrated [3]. In [4], usefulness of the complex symbolic computations for the analysis of systems with parametric uncertainty has been highlighted. Further, it has been shown in [5] that the mapping approach, which is considered to be numerically unstable, performs better than the other known pole assignment methods in a symbolic algebra environment. Stability investigations using symbolic Hurwitz determinants and eigenvalue sensitivity analysis were reported by the present authors in [6]. Other problems in control system design, e.g. designing a sub-optimal controller using only the available state measurements [7], or parameter sensitivity and robustness analyses (some of which may have *explicit and even closed form solutions* [8]), could now be revisited by using the power of symbolic algebra packages available.

This paper highlights the salient features of hybrid symbolic/numeric computation technique. This is illustrated using a simplified configuration of the simple power system consisting of a synchronous generator connected to an infinite bus. For the sake of simplicity, the analysis is carried out with two simplifications in power system representation, although a realistic generator representation is considered. The first simplification is that the excitation system of the generator is representted by a first order system instead of the traditional third order IEEE Type-1 excitation sytem. The second simplification is that a direct feedback of speed is considered in the excitation system, instead of the application of a conventional third order power system stabiliser. The full-state controller for the excitation control of synchronous generator is obtained using the Bass-Gura algorithm [9]. The elements of feedback gain matrix *K* are obtained in symbolic form i.e. expressed in terms of the real and imaginary parts of the desired closed loop poles specified also in symbolic form. Use of the symbolic representation of the desired closed-loop poles allows choice or imposition of appropriate relations among them (e.g. between the real and imaginary parts of the poles) to achieve other desirable goals. One such obvious goal is to eliminate the unmeasurable states from feedback law by requiring the corresponding gains to be set to zero. Plots are presented for the variation of the feedback gains of the output feedback controller and the parameters of one complex pair of desired eigenvalue as a function of the real part of the other assigned complex eigenvalue. For the investigations reported in this paper, the MATLAB® symbolic algebra toolbox was chosen to carry out investigations.

2   Pole Placement Controller Using Full State Feedback

Consider the state-space model of a single input single output linear system described by the following state and output equations.

$$\dot{x} = Ax + Bu$$
$$y = Cx + Du \tag{1}$$

x is the column vector of state variables. The matrices A, B, C and D are the system matrices. y is the plant output vector.

The open loop characteristic polynomial of the plant matrix *A* can be written as

$$s^n + a_1 s^{n-1} + a_2 s^{n-2} \cdots + a_{n-1}s + a_n \tag{2}$$

Here n is the order of the system and a_1, a_2 ....a_n are the constants of polynomial.

Let the desired closed loop polynomial of the plant be given as

$$s^n + \alpha_1 s^{n-1} + \alpha_2 s^{n-2} \cdots + \alpha_{n-1}s + \alpha_n \tag{3}$$

α_1, α_2, ..., α_n are the constants. These values fix the location of the desired closed loop poles of the plant effectively deciding the plant behaviour.

The coefficients of the two polynomials given in eqn. (1) and eqn. (2), can be arranged as row vectors as follows

$$a = \begin{bmatrix} a_1 & a_2 & \cdots & a_{n-1} & a_n \end{bmatrix} \tag{4}$$

$$\alpha = \begin{bmatrix} \alpha_1 & \alpha_2 & \cdots & \alpha_{n-1} & \alpha_n \end{bmatrix} \tag{5}$$

The full-state feedback vector *K* can be obtained by the Bass-Gura formula [9] as

$$K = (\alpha - a)(\beta^{-1})^T \gamma^{-1} \tag{6}$$

where β is the lower-triangular Toeplitz matrix given by eqn. (7),

$$\begin{bmatrix} 1 & 0 & 0 & \cdots & 0 & 0 \\ a_1 & 1 & 0 & \cdots & 0 & 0 \\ a_2 & a_1 & 0 & \cdots & 0 & 0 \\ \vdots & a_2 & 1 & \cdots & \vdots & 0 \\ a_{n-2} & \vdots & \vdots & \cdots & 1 & \vdots \\ a_{n-1} & a_{n-2} & a_{n-3} & \cdots & a_1 & 1 \end{bmatrix} \tag{7}$$

γ is the controllability matrix defined by

$$\gamma = \begin{bmatrix} B & AB & A^2 B & \cdots & A^{n-1}B \end{bmatrix}$$

and Z^T denotes the transpose of a matrix Z.

3   Partial State Feedback

The application of Bass–Gura formula discussed above gives the full-state feedback matrix K_s as

$$\mathbf{K}_s = \begin{bmatrix} \mathbf{K}_{s1} & \mathbf{K}_{s2} & \cdots & \mathbf{K}_{(n-1)s} & \mathbf{K}_{ns} \end{bmatrix} \tag{8}$$

where all the system states are fed back. With full state feedback, the matrix elements,

$$\mathbf{K}_{si} \neq 0; \; i = 1, 2, \cdots, n;$$

In practice, some of the states are not measurable and the actual feedback law used is a partial state feedback or an output feedback for the reasons discussed above.

The output feedback matrix K_o is then written as

$$\mathbf{K}_o = \begin{bmatrix} \mathbf{K}_{o1} & \mathbf{K}_{o2} & \cdots & \mathbf{K}_{(n-1)o} & \mathbf{K}_{no} \end{bmatrix} \tag{9}$$

where some of the elements of $\mathbf{K}_o$ are equated to zero reflecting the unmeasured states, i.e.

If the K_s matrix can be obtained in symbolic form, it may then prove very useful in implementing the output feedback controller. The gains of the state feedback then will be available in terms of the parameters of the desired closed loop poles. The output feedback matrix K_o can be formulated by proper choice of the desired closed-loop poles, based on the available measurements of the system states or outputs.

4   The Problem Formulation

The objective of the work reported here being mainly to illustrate a hybrid symbolic/numeric approach for the design, a simplification in power system representation is sought to minimise the complexities and details. Following simplifications are carried out in modelling the excitation system and also the transfer function for the supplementary feedback. In spite of these assumptions and simplifications, a realistic representation of the generator and the power system is maintained. Any increase in the plant order is avoided without affecting the design procedure. Without these assumptions, the order of the system would have been nine instead of the current fourth order system that will merely complicate the presentation of essential ideas of the paper.

4.1   Assumptions and Simplifications in Plant Modelling

1. In power system stability studies, an IEEE Type-1 excitation system is considered for investigations. This representation of excitation system requires additional 3 states. Hence, a much simpler first order representation with a single time constant is considered in this work.
2. In practice, an auxiliary signal such as the speed signal is fed back via a power system stabiliser comprising a cascaded lead-lag network and a wash out circuit. This representation requires three additional states. For simplification, a direct feedback of speed is considered in this paper.

With these assumptions, the closed loop power system under investigation with the supplementary feedback incorporated is shown in Figure 1.

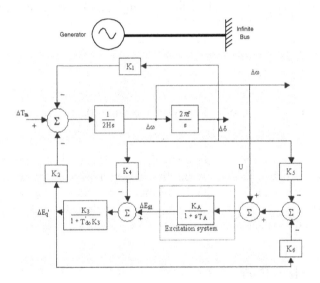

Fig. 1. Synchronous generator connected to an Infinite Busbar

4.2   The Plant Model

The constants K_1, K_2, ... K_6 depend on the operating conditions and system parameters. The relations to evaluate the constants, operating conditions and parameters of the system considered for analysis are the same as those of [12-13].

The dynamic power system model in state space form with the supplementary feedback of $\Delta\omega$ is given as

$$\Delta \dot{\omega} = \frac{1}{2H}\left(K_1 \Delta \delta - K_2 \Delta E_q'\right) + \left(\frac{1}{2H}\right)\Delta T_m$$

$$\Delta \dot{\delta} = 2\pi f\, \Delta\omega$$

$$\Delta \dot{E_q'} = \frac{1}{T_{do}' K_3}\left(-\Delta E_q' - K_3 K_4 \Delta\delta + K_3 \Delta E_{fd}\right)$$

$$\Delta \dot{E}_{fd} = \frac{1}{T_A}\left(-\Delta E_{fd} - K_A K_6 \Delta E_q' - K_A K_5 \Delta\delta + K_A U\right)$$

The synchronous generator is represented as a simplified third order system [14]. The state variables are the increments in rotor speed $\Delta\omega$, rotor angle $\Delta\delta$ and voltage behind the transient reactance $\Delta E_q'$. In this paper, the generator excitation system is represented by a first order system and the associated state of the exciter is ΔE_{fd}. A supplementary feedback of speed deviation $\Delta\omega$ to the excitation system is necessary to ensure an improved transient behavior.

For the system under consideration, the control signal U is same as $\Delta\omega$. The state vector Δx of the combined generator-exciter system is then given by

$$\Delta x = \begin{bmatrix} \Delta\omega & \Delta\delta & \Delta E_q' & \Delta E_{fd} \end{bmatrix} \tag{10}$$

The state-space model of the linearised power system incorporating the inner-loop feedbacks incorporated can be described by the following state and output equations.

$$\Delta\dot{x} = A\,\Delta x + \Gamma\,\Delta T_m$$
$$y = C\,\Delta x \tag{11}$$

where Γ is the disturbance matrix and ΔT_m is the step disturbance.

5   Full-State Feedback Controller

For the given operating condition $(P_o = 0.8 \text{ and } Q_o = 0.6)$, the four eigenvalues of the open-loop system (without incorporating the feedback U, i.e. $\Delta\omega$, of the speed signal) are

+0.10279 + j 5.5008
+0.10279 − j 5.5008
-6.3709
-14.298

Due to the location of a complex pair of eigenvalues in the right half of s-plane, it can be seen that the system is unstable. The open loop characteristic polynomial is then obtained from the plant matrix A. Let the desired closed loop poles be defined as

$$\text{real1} \pm j(\text{img1}) \quad \text{and} \quad \text{real2} \pm j(\text{img2}) \tag{12}$$

With the desired closed-loop poles described symbolically in this manner, the following symbolic K_s matrix is obtained using eqn. (6)

$$K_s = \left[K_\omega \ \ K_\delta \ \ K_{E_q'} \ \ K_{E_{fd}} \right] \tag{13}$$

where the elements of the K_s matrix are given in symbolic form as follows

Feedback gain from the speed signal $\Delta \omega$ is

$K_\omega =$

$-.3796 - 3.805 \ \text{real1} - 3.805 \ \text{real2} - .5434 \times 10^{-16} \ \text{real1}^2 + .1271 \ \text{real1}^2$
$\text{real2} + .1271 \ \text{img1}^2 \ \text{real2} - .3526 \times 10^{-16} \ \text{real1}^2 \ \text{real2} - .5434 \times 10^{-16} \ \text{img1}^2$
$- .5434 \times 10^{-16} \ \text{img2}^2 - .5434 \times 10^{-16} \ \text{real2}^2 - .3526 \times 10^{-17} \ \text{img1}^2 \ \text{real}_2 -$
$.3526 \times 10^{-17} \ \text{real1}^2 \ \text{img2}^2 - .3526 \times 10^{-17} \ \text{img1}^2 \ \text{img2}^2 - .2174 \times 10^{-15}$
$\text{real1 real2} + .1271 \ \text{real1 img2}^2 + .1271 \ \text{real1 real2}^2$

$$\tag{14}$$

Feedback gain from rotor angle $\Delta \delta$ is

$K_\delta =$

$0.002418 \ \text{real1} + .002418 \ \text{real2} + .006059 \ \text{real1}^2 + .7726 \times 10^{-18} \ \text{real1}^2 \ \text{real2}$
$+ .7726 \times 10^{-18} \ \text{img1}^2 \ \text{real2} - .0002023 \ \text{real1}^2 \ \text{real2}^2 + .006059 \ \text{img1}^2 +$
$.006059 \ \text{img2}^2 + .006059 \ \text{real2}^2 - .0002023 \ \text{img1}^2 \ \text{real2}^2 - .0002023 \ \text{real1}^2$
$\text{img2}^2 - .0002023 \ \text{img1}^2 \ \text{img2}^2 + .02424 \ \text{real}_1 \ \text{real}_2 + .7726 \times 10^{-18} \ \text{real}_1$
$\text{img2}^2 + .7726 \times 10^{-18} \ \text{real1 real2}^2 - .1271$

$$\tag{15}$$

Feedback gain from the voltage behind the transient reactance, $K_{E_q'}$ is

$K_{E_q'} =$

$0.005556 \ \text{real1} + .005556 \ \text{real2} - .6458 + .006000 \ \text{real2}^2 + .02400 \ \text{real1}$
$\text{real2} + .006000 \ \text{img1}^2 + .006000 \ \text{img2}^2 + .006000 \ \text{real1}^2 - .4996 \times 10^{-18}$

$$img1^2 \ real2 - .4996 \times 10^{-18} \ real1^2 \ real2 - .4996 \times 10^{-18} \ real1 \ real2^2 - .4996$$
$$\times 10^{-18} \ real1 \ img2^2$$

$$(16)$$

Feedback gain from the excitation system state variable $K_{E_{fd}}$ is

$$K_{E_{fd}} =$$
$$-.002000 \ real1 - .002000 \ real2 - .02046$$

$$(17)$$

6   Output-Feedback Controller

As seen from eqn. (13), there are four state feedback signals. However, the synchronous generator is usually stabilised by an output feedback. This is because the speed signal (ω) is easily measurable while rotor angle (δ) and the transient voltage (E_q) signals are difficult to measure. The excitation system voltage is also easily measurable and can be fed back if required. For numerical analysis, proposed algorithms [14], [15], [16] are available for computing output feedback law. However, for a hybrid symbolic/numeric computations approach, we propose to implement the output feedback law by setting or forcing the gain from an unavailable signal to zero. The information contained in the symbolic expressions given by eqn. (14) to eqn. (17) can be used for obtaining the output feedback controller K_o from the state feedback controller K_s defined by eqn. (3).

6.1   Determining Range of Values for Imaginary Parts

As a first step, the range of values of img_1 and img_2 corresponding to the desired stable region (negative values of real parts of the roots $real_1$ and $real_2$) of the system was determined. This range of values for img_1 and img_2 is obtained by varying the values of $real_1$ (or $real2$) between $-3 \leq (real1 \ or \ real2) \leq -0.5$. It is observed that the corresponding range of values for the absolute values of the imaginary parts img_1 or img_2 lie between *3.68* and *10.408*.

6.2   Identifying Signals to Be Eliminated

For the reasons given above, it is desired to stabilise the plant with the feedback of only speed signal. Hence, the next step is to identify which of the remaining three state feedback signal gains from δ, E_{fd}, E_q (retaining the feedback from speed, ω) can be eliminated. This elimination process is achieved as follows.

Verifying the scope for elimination of E_fd feedback. From eqn. (18), it can be observed that the feedback of $K_{E_{fd}}$ depends directly on the values of *real1* and *real2* only.

For $K_{E_{fd}}$ to be eliminated, following relationship is obtained from eqn. (17) by forcing $K_{E_{fd}} = 0$.

$$real1 = -10.23 + real2 \qquad (18)$$

This is a linear relationship between $real_1$ and $real_2$.

It is verified by time simulations that for the system to have a satisfactory response, both complex roots must lie beyond -1.0. Thus, by applying the constraint on *real1* ≤ -1, a corresponding value of *real2 is* obtained from eqn. (18). These values of $real_1$ and $real_1$ are then substituted in eqn. (15) and eqn. (16) and the two equations solved for img_1 and img_2.

It is observed that this computation does not result in real values for img_1 and img_2. Hence, it can be concluded that feedback from $K_{E_{fd}}$ cannot be made zero and the output feedback matrix, K_o must essentially have feedback from E_{fd} to guarantee good system stability and both the poles located appropriately, say to the left of -1.0. This restriction, however, does not pose any practical difficulty since the signal E_{fd} *is easily measurable.*

Verifying the scope for elimination of δ and Eq' signals. Next, it is necessary to check whether the gains from remaining two signals, i.e. δ and E'_q, can be eliminated. For this purpose, from the range of noted values of *img2* between 3.68 and 10.408, the value of *img2* is chosen as *img2* = ±7.5 rad/sec (corresponding to *real2* approximately equal to -1.2). Fixing *img2* at this value, it is observed that the two symbolic equations i.e. $K_\delta = 0$ and $K_{E'_q} = 0$ can be solved simultaneously for various values of *real2*, implying that the signals δ and E'_q can be eliminated from feedback and still system stability maintained. To further investigate system behaviour, the value of *real2* is varied between -1.0 and -2.0 for the variables *img1* and *real1* for each of the values of *real2*. The corresponding numerical values for K_ω and $K_{E_{fd}}$ are then computed. Plots of K_ω, $K_{E_{fd}}$, *real1* and *img1* as a function of *real2* are then obtained. These are shown in Figures 2(a)-2(d).

It may be noted that the second and third elements of the feedback matrix, corresponding to feedback gains of δ and E_q, are negligible (thus eliminated from the feedback), effectively resulting in an output feedback controller using only $\Delta\omega$ (gain

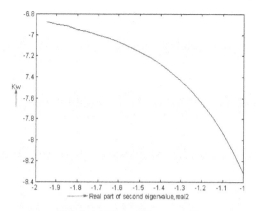

Fig. 2. (a) Plot of K_ω vs. real2

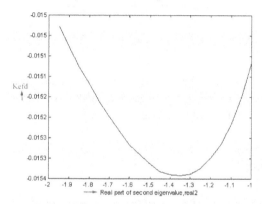

Fig. 2. (b) Plot of K_{efd} vs. real2

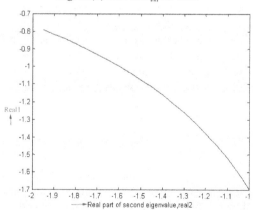

Fig. 2. (c) Plot of real1 vs. real2

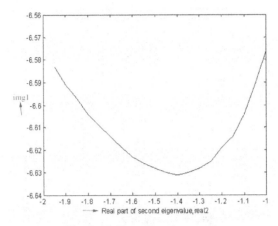

Fig. 2. (d) Plot of img1 vs. real2

Fig. 2. Effects of variation of real part of the second eigenvalue, real2

of –7.38) and ΔE_{fd} (gain of –1.53e-002). Further, the fourth element of the matrix, corresponding to the gain of E_{fd} is also very small and thus the effective control is thus reduced to the feedback from of speed signal ω alone.

7 Time Responses

The time responses of the speed signal of the generator rotor of both the open loop and closed loop power system are shown in Figure 3. It can be clearly seen that the third order open loop power system, which is highly unstable, has been stabilised by the symbolically designed output feedback controller with significant improvement in transient behaviour. It is observed that the system settles down in less than 4 seconds to produce a satisfactory transient behaviour. As expected, the performance of the output feedback controller does not compare well with a conventional power system stabiliser [12], [13] primarily due to the assumptions mentioned in sec. 4.1.

8 Conclusions

A novel approach to controller design using mixed symbolic/numeric computation is presented. The simplified power system consists of a generator represented by a third order model. Since the objective is basically to illustrate the hybrid symbolic-numeric design approach, complexities are avoided by making assumptions in the excitation system and in the feedback path of the speed signal from their usual representation reported in the literature. A symbolic full-state feedback controller is designed using

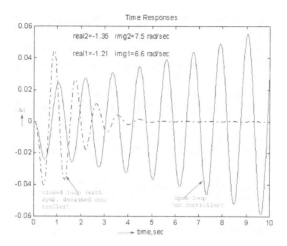

Fig. 3. Time responses of speed signal

the Bass-Gura algorithm. A simplified output controller is then obtained from the full-state controller by eliminating the states, which are not easily measurable due to the physical constraints of the power system. For the power system (with the simplifications) under consideration, it was found that the feedback gain of excitation system voltage depends on the real parts of the desired closed-loop poles and its feedback is essential for system stability. Numerical analysis is carried out and plots of the gains K_ω and $K_{E_{fd}}$ are obtained as a function of the real parts of the desired closed-loop poles. Similarly, the real and imaginary parts of the closed-loop poles are obtained for the variation in the real part of the other complex pole selected as the desired closed-loop pole. The time responses indicate that the system, which was open loop unstable, has been stabilised with a significant improvement in its transient performance.

Acknowledgement. The author wishes to thank Dr. Kishor Dabke, honorary research associate, for his help, discussions and valuable suggestions.

References

1.   A. G.(Bram) de Jager, Applications of zero dynamics with symbolic computation, *UKACC international conference on control'98*, September 1998, pp. 1311-1316.
2.   H. Baki and N.Munro, Implementation of balanced realisation algorithms in a symbolic environment, *UKACC International conference on control'98*, pp. 1317-1321.

3. E. Kontogiannis and N. Munro, The use of symbolic algebra in robust control, *UKACC International conference on control'98*, September 1998. pp. 1328-1332.
4. D.J. Balance and W. Chen, Symbolic computation in value sets of plants with uncertain parameters, *UKACC International Conference on Control'98*, September 98, pp. 1322-1327.
5. M.T. Soylemez and N.Munro, Pole assignment and symbolic algebra: A new way of thinking, *UKACC International Conference on Control'98*, September 1998, pp. 1306-1310.
6. M. Chetty and K.P Dabke, Symbolic computations: An overview and application to controller design, *proceedings of international conference on Information, Decision and Control*, Adelaide, 8th-10th February, 1999, pp. 451-456.
7. K. P. Dabke, Linear control with incomplete state feedback and known initial-state statistics, *Int. J. Control*, vol. 11, No. 1, pp. 133-141, 1970.
8. G. Zames, Input-Output Feedback Stability and Robustness, 1959-85, *IEEE Control Systems*, vol. 16, No. 3, pp. 61-66, June 1996.
9. T. Kailath, Linear systems, *Prentice Hall, Englewood cliffs, New Jersey*, 1980.
10. B. Porter and M.A. Woodhead, Performance of optimal control systems when some of the state variables are not measurable, *Int. Jr. of Control*, vol. 8, pp. 191-195, 1968.
11. The Mathworks Inc., Symbolic math tool for use with MATLAB, user's guide, Version 2.
12. M.L. Kothari, J. Nanda and K. Bhattacharya, Discrete-mode power system stabilisers, *Proc. IEE part-C, Generation Transmission and Distribution*, November, 1993, pp. 523-531.
13. M. Aldeen and M. Chetty, A dynamic output power system stabiliser, *Control'95*, October 1997, vol. 2, pp. 575-579.
14. D.D. Moerder and A.J Calise, Convergence of a numerical algorithm for calculating optimal output feedback gains, *IEEE Trans. On Automatic Control*, Vol. AC-30, No. 9, pp. 900-903, September 1985.
15. W.S. Levine and M. Athans, On the determination of optimal output feedback gains for linear multivariable systems, *IEEE Trans on Automatic Control*, Vol. AC-15, pp.44-48, 1970.
16. S.S Choi and H.R. Sirisena, Computation of optimal output feedback controls for unstable linear multivariable systems, IEEE Trans on Automatic Control, Vol. AC-22, pp.134-136, Feb 1977.
17. P.M. Anderson and A.A. Fouad, *Power system control and stability*, New York, IEEE press, 1994.

APPENDIX-A: System Parameters

The nominal parameters of the system under investigation are as follows. All data is in pu except the inertia constant, H and the time constants which are given in seconds.

Operating conditions

$P_o = 0.8$, $Q_o = 0.6$, $f = 50\,\text{Hz}$, $V_{to} = 1.0$

Generator

$H = 5.0\,\text{s}$, $T'_{do} = 6.0\,\text{s}$, $x_d = 1.6$, $x'_d = 0.32$, $x_q = 1.55$

First order excitation system

$K_A = 50.0$, $T_A = 0.05\,\text{s}$

Transmission line

$x_e = 0.4$

APPENDIX-B: Model Constants

The constants $K_1, \ldots\ldots, K_6$ are evaluated using the relations given below. The subscript o means steady state value. The steady state values of d-q axis voltage and current components (such as I_{qo}, V_{do} etc.) for the single machine infinite bus system are calculated using the phasor diagram relations as given in [14].

$$K_1 = \frac{x_q - x'_d}{x_e + x'_d} I_{qo} V_\infty \sin \delta_0 + \frac{E_{qo} V_\infty}{x_e + x_q} \cos \delta_0$$

$$K_2 = \frac{V_\infty}{x_e + x'_d} \sin \delta_0$$

$$K_3 = \frac{x'_d + x_e}{x_d + x_e}$$

$$K_4 = \frac{x_d - x'_d}{x_e + x'_d} V_\infty \sin \delta_0$$

$$K_5 = \frac{x_q}{x_e + x_q} \frac{V_{do}}{V_{to}} V_\infty \cos \delta_0 - \frac{x'_d}{x_e + x'_d} \frac{V_{qo}}{V_{to}} V_\infty \sin \delta_0$$

$$K_6 = \frac{x_e}{x_e + x'_d} \frac{V_{qo}}{V_{to}}$$

Inductive Synthesis of Functional Programs

Emanuel Kitzelmann[1], Ute Schmid[2], Martin Mühlpfordt[3], and Fritz Wysotzki[1]

[1] Department of Computer Science, Technical University Berlin, Germany,
{jemanuel,wysotzki}@cs.tu-berlin.de,
[2] Institute of Computer Science, University of Osnabrück, Germany,
schmid@informatik.uni-osnabrueck.de
[3] German National Research Center for Information Technology (GMD), Darmstadt, Germany,
mamue@ipsi.fhg.de

Abstract. We present an approach to folding of finite program terms based on the detection of recurrence relations in a single given term which is considered as the kth unfolding of an unknown recursive program. Our approach goes beyond Summers' classical approach in several aspects: It is language independent and works for terms belonging to an arbitrary term algebra; it allows induction of *sets* of recursive equations which are in some arbitrary 'calls' relation; induced equations can be dependent on more than one input parameters and we can detect interdependencies of variable substitutions in recursive calls; the given input terms can represent incomplete unfoldings of an hypothetical recursive program.

Keywords: Inductive program synthesis, folding, recursive program schemes
Topics: Symbolic computations and machine learning, term rewriting

1 Introduction

Automatic induction of programs from I/O examples is an active area of research since the sixties and of interest for AI research as well as for software engineering [LM91]. We present an approach which is based on the recurrence-detection method of Summers [Sum77]. Induction of a recursive program is performed in two steps: First, input/output examples are rewritten into a finite program term, and second, the finite term is checked for recurrence. If a recurrence relation is found, the finite program is folded into a recursive function which generalizes over the given examples. The first step of Summers' approach is knowledge dependent: In general, there are infinitely many possibilities to represent input/output examples as terms. Summers deals with that problem by restricting his approach to structural list problems. Alternatively, the finite program can be generated by the user [SE99] or constructed by AI planning [SW00].

In the following, we are only concerned with the second step of program synthesis – folding a finite program term in a recursive program. This corresponds to program synthesis from traces and is an interesting problem in its own right. Providing a powerful approach to folding is crucial for developing synthesis tools for practical applications. Furthermore, because recursive programs correspond to a subset of context-free tree grammers, our approach to folding can be applied to artificial and natural grammer inference problems. Finally, combining program synthesis and AI planning makes it

J. Calmet et al. (Eds.): AISC-Calculemus 2002, LNAI 2385, pp. 26–37, 2002.

possible to infer general control policies for planning domains with recursive domains, such as the Tower of Hanoi [SW00].

Our approach extends Summers' approach in several aspects: First, it is language independent and works for terms belonging to an arbitrary term algebra, while Summers was restricted to Lisp programs; second, it allows induction of *sets* of recursive equations which are in some arbitrary 'calls' relation, while Summers was restricted to induction of a single, linear recursive equation; third, induced equations can be dependent on more than one input parameters and we can detect interdependencies of variable substitutions in recursive calls, while Summers was restricted to a single input list; finally, the given input terms can represent incomplete unfoldings of an hypothetical recursive program – which is important if the program terms to be folded are obtained by other sources, such as a user or an AI tool.

In the following, we first present our basic terminology, then we formulate the induction problem. We will give a simple example of inductive program synthesis to illustrate our approach and afterwards present the approach formally. Because the approach is rather complex, we will only present the central definitions and theorems and do not give the algorithms. A complete description can be found in [Sch01].

2 Basic Terminology

2.1 Terms and Term Rewriting

Terms. A signature Σ is a set of (function) symbols with $\alpha : \Sigma \to \mathcal{N}$ giving the arity of a symbol. With X ($X \cap \Sigma = \emptyset$) we denote the set of variables, with $\mathcal{T}_\Sigma(X)$ the terms over Σ and X, and with $\mathcal{T}_\Sigma$ the ground terms (terms without variables). $\mathbf{var}(t)$ is the set of all variables in term t. We use tree and term as synonyms. If $\{x_1, \dots, x_n\}$ are the variables of a term t, then $t[x_1 \leftarrow t_1, \dots, x_n \leftarrow t_n]$ (or $t[t_1, \dots, t_n]$ for short) denotes the tree obtained by simultaneously substituting the terms t_i for each occurence of the variables x_i in t, $i = 1, \dots, n$.

A position in t is defined in the usual way as a sequence of natural numbers: (a) λ is the root position of t, (b) if $t = f(t_1, \dots, t_n)$ and u is a position in t_i then $i.u$ is a position in t. The composition $v \circ w$ of two positions is defined as $u.w$, if $v = u.\lambda$, the composition $v \circ k$ of a position v and a natural number k as $u.k.\lambda$, if $v = u.\lambda$. A partial order over positions is defined by $v \leq w$ iff $v = w$ or it exists a position u with $v \circ u = w$. A set of positions U is called *segmentation set* if for all $u \in U$ it holds: $\not\exists u' \in U$ with $u' < u$.

A subterm of a term t at a position u (written $t|_u$) is defined as: (a) $t|_\lambda = t$, (b) if $t = f(t_1, \dots, t_n)$ and u a position in t_i, then $t|_{i.u} = t_i|_u$, $i = 1, \dots, n$. For a term t and a position u, function $\mathbf{node}(t, u)$ returns the fixed symbol $f \in \Sigma$, if $t|_u = f(t_1, \dots, t_n)$. The set of all positions at which a fixed symbol f appears in a term is denoted by $\mathbf{pos}(t, f)$. With $\mathbf{pos}(t)$ we refer to the set of all positions in term t. Obviously it holds that $v \in \mathbf{pos}(t)$, if $w \in \mathbf{pos}(t)$ and $v \leq w$. The replacement of a sub-term $t|_w$ by a term s in a term t is written as $t[w \leftarrow s]$.

A term $p \in \mathcal{T}_\Sigma(\{y_1, \dots, y_n\})$ is called first order pattern of a term $t \in \mathcal{T}_\Sigma(X)$, $\{y_1, \dots, y_n\} \cap X = \emptyset$, if there exist subtrees t_i, such that $t = p[y_1 \leftarrow t_1, \dots, y_n \leftarrow t_n]$, $i = 1, \dots, n$. A pattern p of a term t is called *trivial*, if p is a variable and *non-trivial* otherwise. We write $p \leq t$ if p is a pattern of t and $p < t$ if p and t can not be unified by variable renaming only. p is called maximal (first order) pattern of t, if $p \leq t$ and there exists no term p' with $p < p'$ and $p' \leq t$.

Term rewriting. A term rewrite system over Σ is a set of pairs of terms $\mathcal{R} \subseteq \mathcal{T}_\Sigma(X) \times \mathcal{T}_\Sigma(X)$. The elements (l, r) of $\mathcal{R}$ are called rewrite rules and are written as $l \to r$. A term t' can be derived in one rewrite step from a term t using $\mathcal{R}$ ($t \to_\mathcal{R} t'$), if there exists a position u in t, a rule $l \to r \in \mathcal{R}$, and a substitution $\sigma : X \to \mathcal{T}_\Sigma(X)$, such that (a) $t|_u = \sigma(l)$ and (b) $t' = t[u \leftarrow \sigma(r)]$. $\mathcal{R}$ implies a rewrite relation $\to_\mathcal{R} \subseteq \mathcal{T}_\Sigma(X) \times \mathcal{T}_\Sigma(X)$ with $(t, t') \in \to_\mathcal{R}$ if $t \to_\mathcal{R} t'$. The reflexive and transitive closure of $\to_\mathcal{R}$ is $\stackrel{*}{\to}_\mathcal{R}$.

2.2 Recursive Program Schemes

Let Σ be a signature and $\Phi = \{G_1, \ldots, G_n\}$ a set of function variables with $\Sigma \cap \Phi = \emptyset$ and arity $\alpha(G_i) = m_i > 0$. A recursive program scheme (RPS) $\mathcal{S}$ on a signature Σ, variables X and a set of function variables $\Phi = \{G_1, \ldots, G_n\}$ with $\Sigma \cap \Phi = \emptyset$ is a pair $(\mathcal{G}, t_0)$ with a calling "main program" $t_0 \in \mathcal{T}_{\Sigma \cup \Phi}(X)$ and $\mathcal{G}$ as a system of n equations (recursive "subprograms"): $\mathcal{G} = \langle G_i(x_1, \ldots, x_{m_i}) = t_i \rangle$ with $t_i \in \mathcal{T}_{\Sigma \cup \Phi}(X), i = 1, \ldots, n$.

Language of an RPS. For an RPS $\mathcal{S} = (\mathcal{G}, t_0)$ and a special symbol Ω, the equations in $\mathcal{G}$ constitute rules $\mathcal{R}_\mathcal{S} = \{G_i(x_1, \ldots, x_{m_i}) \to t_i \mid i \in \{1, \ldots, n\}\}$ of a term rewrite system. The system additionally contains rules $\mathcal{R}_\Omega = \{G_i(x_1, \ldots, x_{m_i}) \to \Omega \mid i \in \{1, \ldots, n\}\}$. We write $\stackrel{*}{\to}_{\Sigma, \Omega}$ for the reflexive and transitive closure of the rewrite relation implied by $\mathcal{R}_\mathcal{S} \cup \mathcal{R}_\Omega$.

For a substitution with ground terms $\beta : X \to \mathcal{T}_\Sigma$, called *initial instantiation of variables* X, the set of all terms $\mathcal{L}(\mathcal{S}, \beta) = \{t \mid t \in \mathcal{T}_{\Sigma \cup \{\Omega\}}, \beta(t_0) \stackrel{*}{\to}_{\Sigma, \Omega} t\}$ is called the *language* generated by $\mathcal{S}$ with initial instantiation β. For a particular equation $G_i \in \mathcal{G}$ the rewrite relation $\to_{G_i, \Omega}$ is implied by the rules $\mathcal{R}_{G_i, \Omega} = \{G_i(x_1, \ldots x_{m_i}) \to t_i, G_i(x_1, \ldots x_{m_i}) \to \Omega\}$ and for an instantiation $\beta : \{x_1, \ldots x_{m_i}\} \to \mathcal{T}_\Sigma$ of parameters of G_i, the language generated by subprogram G_i is the set of all terms $\mathcal{L}(G_i, \beta) = \{t \mid t \in \mathcal{T}_{\Sigma \cup \{\Omega\} \cup \Phi \setminus \{G_i\}}, \beta(G_i(x_1, \ldots, x_{m_i})) \stackrel{*}{\to}_{G_i, \Omega} t\}$.

Relations between subprograms. For an RPS $\mathcal{S}$ let $H \notin \Phi$ be a function variable (for the unnamed main program). A relation $\mathbf{calls}_\mathcal{S} \subseteq \{H\} \cup \Phi \times \Phi$ between subprograms and main program of $\mathcal{S}$ is defined as: $\mathbf{calls}_\mathcal{S} = \{(H, G_i) \mid G_i \in \Phi, \mathbf{pos}(t_0, G_i) \neq \emptyset\} \cup \{(G_i, G_j) \mid G_i, G_j \in \Phi, \mathbf{pos}(t_i, G_j) \neq \emptyset\}$. The transitive closure $\mathbf{calls}^*_\mathcal{S}$ of $\mathbf{calls}_\mathcal{S}$ is the smallest set $\mathbf{calls}^*_\mathcal{S} \subseteq \{H\} \cup \Phi \times \Phi$ for which holds: (a) $\mathbf{calls}_\mathcal{S} \subseteq \mathbf{calls}^*_\mathcal{S}$, (b) for all $P \in \{H\} \cup \Phi$ and $G_i, G_j \in \Phi$: If $P \ \mathbf{calls}^*_\mathcal{S} \ G_i$ and $G_i \ \mathbf{calls}^*_\mathcal{S} \ G_j$ then $P \ \mathbf{calls}^*_\mathcal{S} \ G_j$.

For a recursive equation $G_i(x_1, \ldots, x_{m_i}) = t_i$ in $\mathcal{G}$, the *set of transitively called (tc) equations* $\mathcal{G}_i$ with initial equation G_i is defined as $\mathcal{G}_i = \{G_j(x_1, \ldots, x_{m_j}) = t_j \mid G_i \ \mathbf{calls}^*_\mathcal{S} \ G_j\}$. A set of tc equations for an initial equation G_i is the set of all equations in an RPS called by G_i, directly or indirectly by means of further equations.

For a set of tc equations $\mathcal{G}_i$ and its initial equation G_i the extended rewrite relation $\to_{\mathcal{G}_i, \Omega}$ is implied by the rules $\mathcal{R}_{\mathcal{G}_i} = \{G_i(x_1, \ldots, x_{m_i}) \to t_i \mid G_i \in \mathcal{G}_i\}$, $\mathcal{R}_\Omega = \{G_i(x_1, \ldots, x_{m_i}) \to \Omega \mid G_i \in \mathcal{G}_i\}$ and for an instantiation $\beta : \{x_1, \ldots x_{m_i}\} \to \mathcal{T}_\Sigma$ the set of all terms $\mathcal{L}_{ext}(G_i, \beta) = \{t \mid t \in \mathcal{T}_{\Sigma \cup \{\Omega\}}, \beta(G_i(x_1, \ldots, x_{m_i})) \stackrel{*}{\to}_{\mathcal{G}_i, \Omega} t\}$ is the *extended language* generated by subprogram G_i with instantiation β.

2.3   Unfolding of RPSs

For a recursive equation $G_i(x_1, \ldots, x_{m_i}) = t_i$ with parameters $X_i = \{x_1, \ldots x_{m_i}\}$ the set of *recursion points* is given by $U_{rec} = \mathbf{pos}(t_i, G_i)$ with indices $R = \{1, \ldots, |U_{rec}|\}$. Each recursive call of G_i at position $u_r \in U_{rec}, r \in R$ in t_i implies substitutions $\sigma_r :$ $X_i \rightarrow \mathcal{T}_\Sigma(X_i)$ of the parameters in G_i. The *substitution terms*, denoted by $\mathbf{sub}(x_j, r)$, are defined by $\mathbf{sub}(x_j, r) = \sigma_r(x_j)$ for all $x_j \in X_i$ and it holds $\mathbf{sub}(x_j, r) = t_i|_{u_r \circ j}$.

The set of *unfolding points*, denoted by U_{unf}, is constructed over U_{rec} and inductively defined as the smallest set for which holds: (a) $\lambda \in U_{unf}$, (b) if $u_{unf} \in U_{unf}$ and $u_{rec} \in U_{rec}$, then $u_{unf} \circ u_{rec} \in U_{unf}$. Unfolding points imply *compositions of substitutions*, denoted by $\mathbf{sub}^*(x_j, u_{unf})$, which are inductively defined by: (a) $\mathbf{sub}^*(x_j, \lambda) = x_j$, (b) $\mathbf{sub}^*(x_j, u_{unf} \circ u_r) = \mathbf{sub}(x_j, r)\{x_1 \leftarrow \mathbf{sub}^*(x_1, u_{unf}), \ldots, x_{m_i} \leftarrow \mathbf{sub}^*(x_{m_i}, u_{unf})\}, j = 1, \ldots, m_i$.

Unfoldings. For an initial instantiation $\beta : X_i \rightarrow \mathcal{T}_\Sigma$, the instantiations of parameters in unfoldings of an equation G_i are indexed by U_{unf} and defined as: $\beta_{u_{unf}}(x_j) = \beta(\mathbf{sub}^*(x_j, u_{unf}))$. The set of all *unfoldings* Υ_i of equation G_i over instantiation β is indexed by U_{unf} and defined as $\Upsilon_i = \{v_{u_{unf}} | v_{u_{unf}} = \beta_{u_{unf}}(t_i)\}$.

The following lemma states the relation between the concepts introduced, that is, between an RPS and its unfoldings:

Lemma 1   (Inductive Structure of $\mathcal{L}$). *Let* $t \in \mathcal{L}(G_i, \beta)$ *be an element of the language of G_i over an initial instantiation* $\beta : X_i \rightarrow \mathcal{T}_\Sigma$. *Then for all* $u_{unf} \in U_{unf} \cap \mathbf{pos}(t)$ *it holds:* $t|_{u_{unf}} = \Omega$ *or* $t|_{u_{unf}}[U_{rec} \leftarrow \Omega] = v_{u_{unf}}[U_{rec} \leftarrow \Omega]$. *(Proof by induction over U_{unf}.)*

The relations existing between a given RPS and its unfoldings can be exploited for the reverse process – the induction of an unknown RPS from a given term which is considered as some unfolding of a set of recursive equations.

3   Induction of Recursive Program Schemes

3.1   Initial Programs

An *initial program* is a ground term $t \in \mathcal{T}_{\Sigma \cup \{\Omega\}}$ which contains at least one Ω. A t which contains function variables, i.e. $t \in \mathcal{T}_{\Sigma \cup \Phi \cup \{\Omega\}}$, is called *reduced initial program*. We use initial program and initial tree as synonyms. For terms which contain Ωs, an order relation is defined by: (a) $\Omega \leq_\Omega t'$, if $\mathbf{pos}(t', \Omega) \neq \emptyset$, (b) $x \leq_\Omega t'$, if $x \in X$ and $\mathbf{pos}(t', \Omega) = \emptyset$, (c) $f(t_1, \ldots, t_n) \leq_\Omega f(t'_1, \ldots t'_n)$, if $\forall i \in \{1, \ldots, n\}$ it holds $t_i \leq_\Omega t'_i$.

Initial programs correspond to the elements of the language of an RPS or a set of tc equations, whereas reduced initial programs correspond to elements of the language of a particular recursive equation. We say that an RPS $\mathcal{S} = (\mathcal{G}, t_0)$ *explains* an initial program t_{init}, if there exists an instantiation $\beta : \mathbf{var}(t_0) \rightarrow \mathcal{T}_\Sigma$ of the parameters of the main program t_0 and a term $t \in \mathcal{L}(\mathcal{S}, \beta)$, such that $t_{init} \leq_\Omega t$. $\mathcal{S}$ is a *recurrent explanation* of t_{init}, and if exists a term $t' \in \mathcal{L}(\mathcal{S}, \beta)$ which can be derived by at least two applications of rules $\mathcal{R}_\mathcal{S}$, such that $t' \leq_\Omega t_{init}$. Analogously we say that a set of tc equations (or a particular recursive equation) explains a (reduced) initial program. An equation/RPS explains a set of initial programs T_{init}, if it explains all terms $t_{init} \in T_{init}$ and if there is a recurrent explanation for at least one of them.

3.2   Characteristics of Program Schemes

There are some restrictions of RPSs which can be folded using our approach:

No Nested Program Calls: Calls of recursive equations within subtstitution terms of
another call of an equation are not allowed, that is $U_{rec} \cup U_{sub}$ is a segmentation
set for each equation in $\mathcal{G}$.

No Mutual Recursion: There are no recursive equations $G_i, G_j, i \neq j$, with
G_i **calls**$_S^*$ G_j and G_j **calls**$_S^*$ G_i, that is, the relation **calls**$_S^*$ is antisymmetric.

The first restriction is semantical, that is, it reduces the class of calculable functions
which can be folded. The second restriction is only syntactical since each pair of mutually
recursive functions can be transformed into semantically equivalent functions which are
not mutually recursive.

The appearance of RPSs which will be folded are characterized in the following way:
For each program body t_i of an equation in $\mathcal{G}$ it holds

- **var**$(t_i) = \{x_1, \ldots, x_{m_i}\}$, that is, all variables in the program head are used in the
  program body, and
- **pos**$(t_i, G_i) \neq \emptyset$, that is, each equation is recursive.

These characteristics do not restrict the class of RPSs which can be folded.
Furthermore the folded RPSs are minimal according to the following conditions:

No unused subprograms: H **calls**$_S^*$ G_i.

No unused parameters: For each instantiation $\beta : X_i \to \mathcal{T}_\Sigma$ and instantiations $\beta_j :
X_i \to \mathcal{T}_\Sigma, j = 1, \ldots, m_i$ constructed as $\beta_j(x_k) = t$, if $k = j, t \in \mathcal{T}_\Sigma, t \neq \beta(x_j)$,
$\beta_j(x_k) = \beta(x_k)$, if $k \neq j$ holds $\mathcal{L}(G_i, \beta) \neq \mathcal{L}(G_i, \beta_j)$.

No identical parameters: For all $\beta : X_i \to \mathcal{T}_\Sigma$ and all $x_i, x_j \in X_i$ holds: For all
unfoldings $\upsilon_{u_{unf}} \in \Upsilon_i, u_{unf} \in U_{unf}$ with instantiation $\beta_{u_{unf}}$ of variables follows
$i = j$ from $\beta_{u_{unf}}(x_i) = \beta_{u_{unf}}(x_j)$.

Definition 1 (Substitution Uniqueness of an RPS). *An RPS $\mathcal{S} = (\mathcal{G}, t_0)$ over Σ and
Φ which explains recursively a set of initial trees T_{init} is called* substitution unique *wrt
T_{init} if there exists no $\mathcal{S}'$ over Σ and Φ which explains T_{init} recursively and for which
it holds: (a) $t_0' = t_0$, (b) for all $G_i \in \Phi$ holds $\mathbf{pos}(t_i, G_i) = \mathbf{pos}(t_i', G_i) = U_{rec}$,
$t_i'[U_{rec} \leftarrow \Omega] = t_i[U_{rec} \leftarrow \Omega]$, and it exists an $r \in R$ with $\mathbf{sub}(x_j, r) \neq \mathbf{sub}'(x_j, r)$.*

Substitution uniqueness guarantees that it is not possible to replace a substitution term in
$\mathcal{S}$, such that the resulting RPS $\mathcal{S}'$ still explains a given set of initial programs recursively.
It can be shown that each RPS satisfying the given characteristics is substitution unique
wrt the set of all terms which can be derived by it.

3.3   The Synthesis Problem

Now all preliminaries are given to state the synthesis problem:

Definition 2 (Synthesis Problem). *Let $T_{init} \subset \mathcal{T}_{\Sigma \cup \{\Omega\}}$ be a set of initial programs.
The synthesis problem is to induce*

- *a signature Σ,*
- *a set of function variables $\Phi = \{G_1, \ldots, G_n\}$,*
- *a minimal RPS $\mathcal{S} = (\mathcal{G}, t_0)$ with a main program $t_0 \in \mathcal{T}_{\Sigma \cup \Phi}(X)$ and a set of recursive equations $\mathcal{G} = \langle G_1(x_1, \ldots, x_{m_1}) = t_1, \ldots, G_n(x_1, \ldots, x_{m_n}) = t_n \rangle$*

such that

- *$\mathcal{S}$ recursively explains T_{init}, and*
- *$\mathcal{S}$ is substitution unique (Def. 1).*

4 A Simple Example

We will illustrate our method, described in the following sections, by means of a (very) simple example. Consider the following RPS $\mathcal{S} = (\mathcal{G}, t_0)$:

$$\mathcal{G} = \langle\, G(x) = if(eq0(x), 1, *(x, G(pred(x)))) \,\rangle, \quad t_0 = G(x).$$

If the symbols are interpreted in the usual way, this RPS calculates the factorial function. An example for an initial program which can be generated by this RPS with initial instantiation $\beta(x) = succ(succ(0))$ is $t_{init} =$

if(eq0(succ(succ(0))),   1,(succ(succ(0)),*
  if(eq0(pred(succ(succ(0))))),   1,(pred(succ(succ(0))),*
    if(eq0(pred(pred(succ(succ(0)))))),1,(pred(pred(succ(succ(0)))),Ω)))))).*

In a first step (Sect. 5.1), the recursion points (positions, at which a recursive call appeared) will be infered. Only those positions are possible which lie on a path leading to an Ω (only one in this example). Furthermore, it must hold that the positions between the root and the recursion point reiterate itself up to arriving the Ω. This results in $U_{rec} = \{2.3.\lambda\}$ in this example. The minimal pattern which includes the mentioned symbols is $t_{skel} = if(x_1, x_2, *(x_3, \Omega))$. The recursion point devides the initial term into three *segments*:

$$if(eq0(succ(succ(0))), 1, *(succ(succ(0)), G))$$
$$if(eq0(pred(succ(succ(0)))), 1, *(pred(succ(succ(0))), G))$$
$$if(eq0(pred(pred(succ(succ(0))))), 1, *(pred(pred(succ(succ(0)))), G))$$

Since there are no further Ωs in the initial tree, the searched for equation can only contain the found position as recursion point. In more complex RPSs it might happen that not all Ωs can be explained in the described way. In this case, we assume that another equation is called by the searched for equation. An RPS for the resulting subtrees will be induced seperately.

In a second step (Sect. 5.2), the program body will be constructed by extending the term t_{skel} by all positions which remain constant over the segments. This method results in the term $\hat{t}_G = if(eq0(x_1), 1, *(x_1))$. Identical subtrees in one segment are represented by the same variable (x_1 in our example).

At last (Sect. 5.3), the substitution terms for the variables will be infered. The subtrees which differ over the segments represented by the variables in $\hat{t}_G$ are instantiations of the parameters in the recursive equation. For the single variable in our example we obtain the trees:

$$succ(succ(0)) \tag{1}$$
$$pred(succ(succ(0))) \tag{2}$$
$$pred(pred(succ(succ(0)))) \tag{3}$$

Now a recurrence relation will be searched such that the instantiations in one segment can be generated by instantiations in the preceding segment in a recurrent way. This results in the substitution term $\mathbf{sub}(x_1, 1) = pred(x_1)$, since for the instantiations in the three segments it holds $(2) = pred((1))$, and $(3) = pred((2))$.

With our method we are able to fold initial trees for RPSs of substantially more complexity, namely with a constant part in the main programs, with more than one recursive equation, and with interdependent, switching and hidden variables in the equations. In the following, we present the theory and methodology formally, concluding each section with a theorem which backs up our approach.

5 Inducing a Subprogram

5.1 Segmentation

When we start with segmentation, the hypothesis is that there exists a set of tc equations $\mathcal{G}_i$ with initial equation $G_i(x_1, \ldots, x_{m_i}) = t_i$ which explains a set of given initial trees T_{init}. The goal is to find a set of possible recursion points U_{rec} for such an equation G_i which divides the initial trees into *segments* which correspond to the unfoldings of G_i.

Each equation of an RPS occurs in some context. For instance, the main program might be a term $t_0 = G_1([1, 2, 3])$, that is, G_1 is called in an "empty" context. Within G_1, another equation G_2 might be called with in a term $t' = cons(42, G_2(x))$. In this case, $cons(42, \, . \,)$ is the context of G_2. In the following, we call a recursive equation together with its context a *sub-scheme*. Consequently, for each recursive equation G_i, a set of hypothetical recursion points U_{rec} and a (possibly empty) set of sub-scheme positions U_{sub} must be determined.

Sub-scheme positions. The set of *sub-scheme positions* of an equation G_i is defined as $U_{sub} = \{u \circ k \in t_i \mid \exists u_{rec} \in U_{rec} : u < u_{rec}, \not\exists u_{rec} \in U_{rec} : u \circ k \leq u_{rec}, \exists u_{sp} \in \mathbf{pos}(t_i, G_j) : i \neq j \text{ and } u \circ k \leq u_{sp}\}$. Sub-scheme positions in a program term indicate the call of another subprogram in the subterm at a sub-scheme position. Moreover it must hold, that the sub-scheme positions are the "deepest" positions in a program term at which a call of a subprogram is possible wrt the condition of no nested program calls.

The Ωs in an initial tree give us a first restriction for the choice of possible recursion points, since each recursion point (and moreover each unfolding point constructed over the recursion points) and also each sub-scheme position composed to an unfolding point must lie on a path leading to an Ω in the initial tree. We say that an Ω is explained by the recursion points and sub-scheme positions. It must hold that *all* Ωs in t_{init} are explained in this way.

Method. The method is to search in the initial trees for possible recursion points. If a position was found, the set of sub-scheme positions is determined by the yet unexplained

Ωs. Such a pair (U_{rec}, U_{sub}) is called a *valid hypothesis of recursion points and sub-scheme positions*. Up to now, we only regarded the structure of the initial trees. We now regard additionally the symbols (i. e., the node labels) which lie between the root of the initial trees and the so far infered recursion points. These symbols imply a special minimal pattern of the body of the searched for subprogram which is defined by means of the function **skeleton**.

Definition 3 (Skeleton). *The* skeleton *of a term* $t \in \mathcal{T}_{\Sigma \cup \{\Omega\}}(X)$, *written* **skeleton**$(t)$, *is the minimal pattern of* t *for which holds* **pos**$(t, \Omega) = $ **pos**(**skeleton**$(t), \Omega)$.

Obviously it holds: $t \leq_\Omega$ **skeleton**(t). The mentioned pattern is named t_{skel} and defined as **skeleton**$(t_{init}[U_{rec} \cup U_{sub} \leftarrow \Omega])$. As stated in the following definition, t_{skel} must reiterate itself at each unfolding point constructed over the calculated recursion points, since each unfolding point indicates an unfolding of the same recursive equation G_i (see Lemma 1).

Definition 4 (Valid Segmentation). *Let* U_{unf} *be the unfolding points constructed over a set of possible recursion points* U_{rec}. *The hypothesis* (U_{rec}, U_{sub}) *together with* t_{skel} *is called a* valid segmentation *of an initial tree* t_{init}, *if for all* $u_{unf} \in U_{unf} \cap$ **pos**(t_{init}) *holds: Given the set of all positions of* Ωs *in* $t_{init}|_{u_{unf}}$ *which are above the recursion points as* $U^{cut}_{u_{unf}} = \{u \mid$ **pos**$(t_{init}|_{u_{unf} \circ u}) = \Omega, \exists u_{rec} \in U_{rec}$ *with* $u < u_{rec}\}$, *it holds* $t_{skel}[U^{cut}_{u_{unf}} \leftarrow \Omega] \leq_\Omega t_{init}|_{u_{unf}}$. (U_{rec}, U_{sub}) *with* t_{skel} *is called a* valid recurrent segmentation *of* t_{init}, *if additionally it holds that:* $\exists u_{rec} \in U_{rec} : U_{rec} \subseteq t_{init}|_{u_{rec}}$.

If a hypothesis can be verified by means of t_{skel}, then the trees will be searched for further possible recursion points, because the algorithm should detect a set of recursion points as large as possible. This sequence of steps for inferring a valid segmentation – (a) search for a further possible recursion point, (b) calculate the set of sub-scheme positions, (c) calculate t_{skel} and verify the hypothesis – will be repeated, until no further possible recursion point can be found.

It is possible, that a valid segmentation doesn't lead to an equation which explains the initial trees together with a set of tc equations. Therefore, the presented method results in a backtracking algorithm.

Theorem 1 (Segmentation Theorem). *If a set of tc equations* $\mathcal{G}_i$ *with initial equation* $G_i(x_1, \ldots x_{m_i}) = t_i$ *with recursion points* U_{rec} *and sub-scheme positions* U_{sub} *explains an initial tree* t_{init} *for an initial instantiation* $\beta : \{x_1, \ldots x_{m_i}\} \to \mathcal{T}_\Sigma$, *then* (U_{rec}, U_{sub}) *together with* t_{skel} *is a valid segmentation for* t_{init}.

If $\mathcal{G}_i$ *explains a set of initial trees* T_{init}, *then* (U_{rec}, U_{sub}) *together with* t_{skel} *is a valid segmentation for all trees in* T_{init} *and a valid recurrent segmentation for at least one tree in* T_{init}.

Proof. Let be Υ_i the set of all unfoldings of equation G_i and $u_{unf} \in U_{unf} \cap$ **pos**(t_{init}) an arbitrary unfolding point in t_{init}. If (case 1) $U_{rec} \subset t_{init}|_{u_{unf}}$, then follows $U^{cut}_{u_{unf}} = \emptyset$. From the definition of unfoldings (Sect.2.3) and Lemma 1 follows $t_{init}|_{u_{unf}}[U_{rec} \cup U_{sub} \leftarrow \Omega] = v_{u_{unf}}[U_{rec} \cup U_{sub} \leftarrow \Omega]$. From this and with definition of t_{skel} follows the assertion. For (case 2) $U_{rec} \not\subset t_{init}|_{u_{unf}}$ the proof is analogous to case 1.

The second paragraph of the theorem follows directly from the just proved first part, the definition of recursive explanation (Sect.2.3) and Definition 4.

If a valid recurrent segmentation (U_{rec}, U_{sub}) was found for a set of initial trees T_{init} and if $U_{sub} \neq \emptyset$, then induction of an RPS is performed recursively over the set of subtrees at the positions in U_{sub}. Therefore, for each $u \in U_{sub}$ a new set of initial trees will be constructed by including the subtrees at position u in each segment of each tree in T_{init} to T_{init}^u.

5.2 Inducing a Program Body

When we start with inferring the body of subprogram G_i, a valid segmentation (U_{rec}, U_{sub}) is given for T_{init}. Moreover, for each sub-scheme position $u \in U_{sub}$ a sub-scheme $\mathcal{S}^u = (\mathcal{G}^u, t_0^u)$ of G_i which explains the subtrees in T_{init}^u is already induced. That allows us to fold (the inverse process of unfolding) the initial trees to reduced initial trees, denoted by T_{red}, which can be explained by one recursive equation (the initial equation of the set of tc equations explaining T_{init}, Sect.3.1).

Method. A subprogram body is uniquely determined by a valid segmentation:

Definition 5 (Valid Subprogram Body). *For a valid segmentation (U_{rec}, U_{sub}) of T_{init} and the reduced initial trees $t_{red} \in T_{red}$, the term $\hat{t}_G \in \mathcal{T}_{\Sigma \cup \Phi^u \cup G}(X)$ is defined as the maximal pattern of all complete segments $\{t_{red}|_{u_{unf}}[U_{rec} \leftarrow G] \mid u_{unf} \in U_{unf} \cap \mathbf{pos}(t_{rec}), U_{rec} \subset t_{rec}|_{u_{unf}}\}$ and is called* valid subprogram body.

A segment is complete, if it includes all recursion points in U_{rec}. The following lemma states that this method results in a valid solution.

Lemma 2 (Maximization of the Body). *Let E be a finite index set with indices $e \in E$. Let $G(x_1, \ldots, x_n) = t_G$ be a recursive equation with $X = \{x_1, \ldots, x_n\}$, $t_G \in \mathcal{T}_{\Sigma \cup \Phi}(X)$ and initial instantiations $\beta_e : X \to \mathcal{T}_\Sigma$ for all $e \in E$. Then it exists a recursive equation $G'(x_1, \ldots, x_{n'}) = t'_G$ with $X' = \{x_1, \ldots, x_{n'}\}$, $t_{G'} \in \mathcal{T}_{\Sigma \cup \Phi}(X')$, unfolding points U_{unf} and initial instantiations $\beta'_e : X' \to \mathcal{T}_\Sigma$ for all $e \in E$, such that $\mathcal{L}(G, \beta_e) = \mathcal{L}(G', \beta'_e)$ for each $e \in E$. Additionally, for each $x \in X'$ it holds that the instantiations which can be generated by G' from β'_e, $\{\beta'_{(e, u_{unf})}(x) \mid e \in E, u_{unf} \in U_{unf}\}$ do not share a common not-trivial pattern.*

Proof. It can be shown that substitution terms exist which generate the instantiations in the unfoldings of elements in $\mathcal{L}(G', \beta'_e)$ from the initial instantiation β'_e. The idea is to extend the body of equation G by the common pattern of the instantiations in the unfoldings of the elements of $\mathcal{L}(G, \beta_e)$. Then the initial instantiation will be reduced by this pattern. It must be considered, that variables can be interdependent.

The maximal pattern of a set of terms can be calculated by first order anti-unification. Only *complete* segments are considered. For incomplete segments, it is in general not possible to obtain a consistent introduction of variables during generalization. The variables in the resulting subprogram body represent that subtrees which differ over the segments of the initial trees. Identical subtrees are represented by the same variable.

Equivalence of sub-schemes. Because we handle induction of sub-schemes $\mathcal{S}^u$ as independent problem, we must insure that if for each $u \in U_{sub}$ exists a sub-scheme $\mathcal{S}^u$ which explains the trees in T_{init}^u *and* it exists a recursive equation which explains the resulting reduced initial trees T_{red}, then for arbitrary other sub-schemes S'^u explaining the trees in each T_{init}^u, it exists a recursive equation which explains the resulting reduced initial trees T'_{red}. Such an "Equivalence of Sub-Schemes"-condition can be shown by considering the parameters with the initial instantiations of two different schemes which both explain the same initial trees. If they are constructed by maximizing the body, as described, it holds, that the parameters and instantiations are equal.

Theorem 2 (Subprogram Body Theorem). *If it exists a set of tc equations $\mathcal{G}'$ with initial equation $G'(x_1, \ldots, x_{n'}) = t_{G'}$ with recursion points U_{rec} and sub-scheme positions U_{sub} which explains T_{init}, then it exists a set of tc equations $\mathcal{G}$ with initial equation $G(x_1, \ldots, x_n) = t_G$ which explains T_{init} and such it holds: $t_G[U_{rec} \leftarrow G] = \hat{t}_G$. (Proof follows from Def. 5, Lemma 2 and the fact, that two independently infered sub-schemes are equivalent (see text).)*

5.3 Inducing Substitution Terms

The subtrees of the trees in T_{red} which differ over the segments represent instantiations of the variables $\mathbf{var}(\hat{t}_G)$. The goal is to infer a set of variables X with $\mathbf{var}(\hat{t}_G) \subset X$ and substitution terms for each variable and each recursion point in U_{rec}, such that the corresponding subtrees can be generated by an initial instantiation and compositions (Sect.2.3) of the infered substitution terms. We are able to deal with quite complex substitution terms. Consider the following examples:

$$f_1(x, y) = if(eq0(x), y, +(x, f_1(pred(x), +(x, y)))) \tag{4}$$
$$f_2(x, y, z) = if(eq0(x), +(y, z), +(x, f_2(pred(x), z, succ(y)))) \tag{5}$$
$$f_3(x, y, z) = if(eq0(x), y, +(x, f_3(pred(x), z, succ(y)))) \tag{6}$$

A variable might be substituted by an operation involving other program parameter (4). Additionally, variables can switch there positions (given in the head of the equation) in the recursive call (5). Finally, there might be "hidden" variables which only occur within the recursive call (6). If you look at the body of f_3, variable z occurs only within the recursive call. The existence of such a variable cannot be detected when the program body is constructed by anti-unification but only a step later, when substitutions for the recursive call are inferred.

Method. The method for inferring substitution terms is based on the fact that instantiations of variables in an unfolding (resp. a segment) can be generated by instantiating the variables in the substitution terms with the instantiations of the preceding unfolding, $\beta_{u_{unf} \circ u_r}(x_j) = \beta_{u_{unf}}(\mathbf{sub}(x_j, r))$ for an arbitrary variable and recursion point index.

Moreover – starting with $u = \lambda$ – it holds that the substitution terms are inductively characterized by:

$$\mathbf{sub}(x_j, r)|_u = \begin{cases} x_k & \forall u_{unf} \in U_{unf} : \\ & \beta_{u_{unf} \circ u_r}(x_j)|_u = \beta_{u_{unf}}(x_k) \\ f(\mathbf{sub}(x_j, r)|_{u \circ 1}, \ldots, & \forall u_{unf} \in U_{unf} : \\ \mathbf{sub}(x_j, r)|_{u \circ n}) & \mathbf{node}(\beta_{u_{unf} \circ u_r}(x_j), u) = f \in \Sigma \\ & with\ arity\ \alpha(f) = n. \end{cases}$$

This characterization can directly be transformed into a recursive algorithm which calculates the substitution terms, if we identify the instantiations in the unfoldings with the subtrees in T_{red} which differ over the segments. If for a variable in $\mathbf{var}(\hat{t}_G)$ no substitution term can be calculated, because at a position u it holds neither condition (a) nor (b), then it will be assumed that a "hidden" variable (that is $x \notin \mathbf{var}(\hat{t}_G)$) is on this position in the substitution term. The set of variables X will be extended by a new variable $x \notin X$ and then reversed application of condition (a) yields the instantiations of the new variable in the unfoldings. The substitution terms for the new variable will be generated as described.

Incomplete unfoldings. Because we allow for incomplete unfoldings in the initial trees, it can occur that for a particular variable no represented subtree in an unfolding can be found. In this case, the instantiation of the variable is undefined in this unfolding and will be set to $\perp$. This can result in substitution terms which generate subtrees (representing instantiations) of the subtrees in the preceding unfolding (as described above), but don't generate the same instantiations by applying the original definition (Sect.2.3). Furthermore, it can occur that a substitution term is not unique wrt Definition 1. By means of the following definition, the infered substitution terms will be verified wrt the mentioned problems.

Definition 6 (Valid Substitution Terms). *Let be X_h a set of hidden variables with $X_h \cap \mathbf{var}(\hat{t}_G) = \emptyset$, $X = X_h \cup \mathbf{var}(\hat{t}_G)$ the set of infered variables. The set of all substitution terms $\mathbf{sub}(x,r)$, $x \in X, r \in R$ is called valid, if it holds:*

Consistency: $\forall t_e \in T_{red} \, \exists \beta_e : X \to \mathcal{T}_\Sigma \, \forall x_j \in \mathbf{var}(\hat{t}_G), u_{unf} \in U_{unf} \cap t_e, u \in \mathbf{pos}(\hat{t}_G, x_j) : t_e|_{u_{unf} \circ u} =_\perp \beta_e(\mathbf{sub}^*(x_j, u_{unf}))$.

Uniqueness: *It not exists a variable $x \in X$ and a recursion point u_r such that another substitution term $\mathbf{sub}'(x,r) \neq \mathbf{sub}(x,r)$ exists which is consistent.*

Minimality: *Let be $X_h' \subset X_h$ a set of variables and $X' = X_h' \cup \mathbf{var}(\hat{t}_G)$: It doesn't exist a set of consistent and unique substitution terms for the variables X'.*

Theorem 3 (Substitution Theorem). *Let $\mathbf{sub} : X \times R \to \mathcal{T}_\Sigma(X)$ with $X = \{x_1, \ldots, x_n\}$ be valid substitution terms. Let $G(x_1, \ldots, x_n) = t_G$ be a recursive equation such that it holds: (a) $t_G[U_{rec} \leftarrow G] = \hat{t}_G$ and (b) $\forall r \in R, x_j \in X : t_G|_{u_r \circ j} = \mathbf{sub}(x_j, r)$. Then it holds: The set of tc equations $\mathcal{G} = \mathcal{G}^u \cup \{G(x_1, \ldots, x_n) = t_G\}$ explains T_{init} substitution unique.*

Proof. From the condition of consistence in Definition 6 and Theorem 2 follows directly that equation G explains T_{red} constructed over T_{init} and the sub-schemes $\mathcal{S}^u$. From the condition of uniqueness follows directly that this explanation is substitution unique (as postulated in Def. 3.3). Additionally, with Lemma 1 follows that $\mathcal{G} = \mathcal{G}^u \cup \{G(x_1, \ldots, x_n) = t_G\}$ explains T_{init} and from the substitution uniqueness of all sub-schemes $\mathcal{S}^u$ and equation G follows that the explanation of T_{init} by $\mathcal{G}$ is substitution unique too.

6   Inducing an RPS

If an RPS can be induced from a set of initial trees using the approach presented here, then induction can be seen as a proof of existence of a recursive explanation – given the restrictions presented in Sect.3.2.

Theorem 4 (Existence of an RPS). *Let T_{init} be a set of initial trees indexed over E. T_{init} can be explained recursively by an RPS iff:*

1. *T_{init} can be recursively explained by a set of tc equations, or*
2. *$\forall e \in E : \exists f \in \Sigma$ with $\alpha(f) = n$, $n > 0$, and $\mathbf{node}(t_e, \lambda) = f$, and $\forall T^k = \{t_e|_{k.\lambda} \mid k = 1, \ldots, n\}$ it holds:*
   a) *$\forall t \in T^k : \mathbf{pos}(t, \Omega) = \emptyset$, or*
   b) *$\forall t \in T^k : \mathbf{pos}(t, \Omega) \neq \emptyset$ and it exists an RPS $\mathcal{S}^k = (\mathcal{G}^k, t_0^k)$ which recursively explains the trees in T^k.*

(Proof in [Sch01].)

To inductively construct an RPS from a set of initial trees, first a valid segmentation for the initial trees is searched-for. If one is found, body and substitutions are constructed as described above; if segmentation fails, recursively sub-schemes for the subtrees will be induced, and at the end, the final RPS will be constructed. A consequence of inferring sub-schemes separately is that subprograms are introduced *locally* with unique names. It can happen that two such subprograms are identical. After folding is completed, the set of subprograms can be reduced.

7   Conclusion

We presented a new, powerful approach to folding of finite program terms, focussing on the formal framework. The synthesis algorithms and a variety of examples can be found in [Sch01]. The induction method described in this paper is able to deal with all (tail, linear, tree recursive) structures which can be generated by a set of recursive functions of arbitrary complexity. We already demonstrated the applicability of our approach to control-rule learning for planning [SW00]. In future we plan to investigate applicability to grammar learning and to enduser programming.

References

[LM91]   M. L. Lowry and R. D. McCarthy, editors. *Autmatic Software Design*. MIT Press, Cambridge, MA, 1991.

[Sch01]   Ute Schmid. Inductive synthesis of functional programs – Learning domain-specific control rules and abstract schemes. http://www.inf.uos.de/schmid/pub-ps/habil.ps.gz, Mai 2001. unpublished habilitation thesis.

[SE99]   S. Schrödl and S. Edelkamp. Inferring flow of control in program synthesis by example. In *Proc. Annual German Conference on Artificial Intelligence (KI'99), Bonn*, LNAI, pages 171–182. Springer, 1999.

[Sum77]   P. D. Summers. A methodology for LISP program construction from examples. *Journal ACM*, 24(1):162–175, 1977.

[SW00]   U. Schmid and F. Wysotzki. Applying inductive programm synthesis to macro learning. In *Proc. 5th International Conference on Artificial Intelligence Planning and Scheduling (AIPS 2000)*, pages 371–378. AAAI Press, 2000.

A Symbolic Computation-Based Expert System for Alzheimer's Disease Diagnosis*

Begoña Herrero[1], Luis M. Laita[1], Eugenio Roanes-Lozano[2], Víctor Maojo[1],
Luis de Ledesma[1], José Crespo[1], and Laura Laita[3]

[1] Depto. de Inteligencia Artificial, Facultad de Informática, Universidad Politécnica
de Madrid, Boadilla del Monte, 28660-Madrid, Spain
[2] Depto. de Álgebra, Universidad Complutense de Madrid, c/ Rector Royo Villanova
s/n, 28040-Madrid, Spain
[3] Escuela de Enfermería, Universidad Complutense de Madrid
{laita,eroanes}@fi.upm.es

Abstract. In this paper we summarize a method of construction of
a rule-based expert system (denoted RBES) for Alzheimer's disease
diagnosis. Once the RBES is constructed, Symbolic Computation
techniques are applied to automatically both verify (that is, check for
consistency) and extract new knowledge to produce a diagnosis.

Keywords: Rule-based expert systems, diagnosis of Alzheimer's disease,
ideal membership problem, Gröbner bases

1 Introduction

In this paper we summarize a method of construction of a rule-based expert
system (denoted as RBES) for Alzheimer's disease diagnosis. The full RBES is
described in [6][1]. After building the RBES, symbolic computation techniques are
applied to both automatically verify (that is, check for consistency) and extract
new knowledge to output a diagnosis.

RBES usually have two components: a "knowledge base"and an "inference
engine". In this article, the knowledge base and the inference engine are, respectively, symbolically expressed and symbolically implemented using the computer
algebra language CoCoA [4][2]. We use the 3.0b MSDOS version of CoCoA because, even though the Windows version [14] is more user friendly, it appears to
be less efficient.

---

* Partially supported by projects TIC2000-1368-C03-01 and TIC2000-1368-C03-03
(Ministry of Science and Technology, Spain).
[1] This Master thesis (in Spanish) contains a large amount of information about the
illness, an explanation of the 343 rules entered, a glossary and a diskette with the
entire program.
[2] CoCoA, a system for doing Computations in Commutative Algebra. Authors: A.
Capani, G. Niesi, L. Robbiano. Available via anonymous ftp from:
cocoa.dima.unige.it

J. Calmet et al. (Eds.): AISC-Calculemus 2002, LNAI 2385, pp. 38–50, 2002.
© Springer-Verlag Berlin Heidelberg 2002

Our particular knowledge base consists of production rules, ADDI (additional information given by the experts) and facts. Classic bi-valued logic will be used in this system.

The "production rules" are, in this article, logical implications of the following forms:

$$\circ x[i] \wedge \circ x[j] \wedge ... \wedge \circ x[k] \to \circ y[l]$$
or
$$\circ x[i] \vee \circ x[j] \vee ... \vee \circ x[k] \to \circ y[l]$$

(or, possibly, a combination of $\vee$ and $\wedge$ in the antecedent), where "$\circ$" represents the symbol "$\neg$" or no symbol at all.

For example the first rule of our system, R1, refers to memory disorder. It is:

$$x[1] \wedge x[2] \wedge x[3] \wedge [4] \to x[5] \ .$$

The most relevant "facts" are what are known as "potential facts", which are the literals (that is, variables or variables preceded by "$\neg$"). These are on the left-hand side of the production rules and never on the right-hand side.

For example the first fact of our system, x[1], is "difficulty in remembering what day it is" (the list of the potential facts can be found in section 4.2).

Potential facts allow "fire" the rules. A rule is fired when, given the literals of its left-hand side, its right-hand side is obtained by the logical rule of modus ponens (from α and $\alpha \to \beta$, to get β).

It is important to note that, in this paper, we consider the set of both potential facts and their negations. But, in each case study (that is, each class of patients w.r.t. the equivalence relation "his/her relatives have answered the questions about the symptoms the same way and the results of the neurological evaluation and laboratory tests are equal"), we consider one maximal consistent subset of such a set of facts. "Consistent", because a fact and its negation are never included in any case study; "maximal", because either each fact or its negation must be included. Note that, depending on which maximal consistent set of facts is used, different rules are fired, giving different results.

Regarding the additional auxiliary information, denoted "ADDI", these are formulae that the experts judge should or must be taken into account.

Let us say that CoCoA requires all logical formulae of the RBES to be written in prefix form. For instance, the prefix form of the first formula above is:

```
IMP(AND1(AND1(AND1...(AND1(ox[i],ox[j]),...,ox[k])...),oy[l]).
```

AND1 (also OR1 in other formulae) are used instead of AND and OR, because these last two words are reserved words in CoCoA.

Our inference engine proceeds as follows: first, logical formulae are automatically translated into polynomials and, second, "Gröbner bases" and "normal forms" are applied to these polynomials (using CoCoA). A substantial part of the background theory is the original work of a group of researchers to which the authors belong [2,5,9,15].

The information about Alzheimer's disease has been gathered from several sources, mainly [3,12] and several experts in the field. It has been coordinated by the fourth author (MD and director of the Medical Informatics Research Group at the Universidad Politécnica de Madrid). Public information can be found on the web.

2   The Alzheimer Knowledge Base

The knowledge base we have developed contains 343 production rules, grouped in the following subsets:

(1) Clinical history and neuropsychological tests (production rules 1 to 54). For example, production rule R6 (written in prefix form) is:

```
R6:=NF(IMP(AND1(AND1(AND1(AND1(NEG(x[12]),NEG(x[13])),
NEG(x[14])),NEG(x[15])),NEG(x[16])),NEG(x[17])),I);³
```

that translates the IF-THEN statement:

"If the patient experiences no problems in using machines or domestic appliances AND no problems in pursuing his/her preferred hobbies AND no problem in dressing AND no problem in gesturing using hands and face AND no problem in writing and drawing THEN the patient has no difficulty in making controlled movements".

(2) Neurological evaluation; subdivided in turn into:
  (2.1) Cranial pairs (production rules 55 to 80).
  (2.2) Reflexes (production rules 81 to 100).
  (2.3) Motor system (production rules 101 to 120). For example, production rule R119 (written in prefix form) is:

```
R119:=NF(IMP(OR1(OR1(OR1(OR1(x[160],x[163]),x[167]),x[169]),
x[172]),x[173]),I);
```

that translates the IF-THEN statement:

"If spinal column muscle injury exists OR extrapyramidal injury exists OR spinal cord injury exists OR cerebellum injury exists OR psychogenic disturbances exist THEN motor disturbance exists".

  (2.4) Sensitivity (production rules 121 to 122).
  (2.5) Coordination (production rules 123 and 124).
  (2.6) Normal or abnormal walking (production rules 125 and 126).
  (2.7) Neurological evaluation (production rules 127 and 128).
(3) Laboratory tests; subdivided in its turn into:
  (3.1) Nutritional deficit (production rules 129 and 130).
  (3.2) Hypothyroidism (production rules 131 and 132).
  (3.3) Neurosyphilis (production rules 133 and 134).
  (3.4) Hypercalcemia (production rules 135 and 136).
  (3.5) HIV (production rules 137 and 138).

---

[3] The notation NF(..., I); means "Normal Form, modulo the ideal I". The ideal I. will be specified later.

(3.6) Hematology (production rules 139 to 272).

(3.7) Biochemistry (production rules 273 to 318). For example, production rule R291 (written in prefix form) is:

```
R291:=NF(IMP(AND1(AND1(AND1(AND1(AND1(AND1(AND1(
NEG(x[266]),x[264]),x[268]),x[270]),x[273]),x[275]),x[277]),
x[279]),x[280]),I);
```

that translates the IF-THEN statement:

"IF the urea level is abnormal AND the glucose, cholesterol, uric acid, creatinine, HDL-cholesterol, LDL-cholesterol and triglycerides levels are all normal THEN Biochemistry results are normal".

(3.9) Systematic analysis of urine (production rules 319 to 343).

For reasons of simplicity and effectiveness, our inference engine studies each one of these subsets of production rules separately, together with their respective maximal consistent sets of facts and ADDIs.

The translation from the medical knowledge to logic is not one-to-one, as usually happens in the medical field. On the contrary, the translation of classes of equivalent propositional formulae into classes of equal polynomials is one-to-one [7].

3 Outline of the Algebraic Background of the Inference Engine

The basic concepts and results on which the inference engine is based are summarized next and explained in more detail in the Appendix. The full development can be found in [2,5,9,15].

The inference engine proceeds as follows.

- First step: it translates production rules into polynomials. The translation of the basic bi-valued logic formulae is:

$$\begin{aligned}
\neg x_1 &\rightsquigarrow 1 + x_1 \\
x_1 \vee x_2 &\rightsquigarrow x_1 x_2 + x_1 + x_2 \\
x_1 \wedge x_2 &\rightsquigarrow x_1 x_2 \\
x_1 \rightarrow x_2 &\rightsquigarrow x_1 x_2 + x_1 + 1
\end{aligned}$$

- Second step: it applies a mathematical result, which relates the fact that any piece of information, say α, follows from (more precisely, "is a tautological consequence of"[4]) the facts, production rules and any other additional information (ADDI) of a RBES, iff the polynomial that translates the (negation of the) above-mentioned piece of information belongs to a certain polynomial ideal. For the sake of clarity and for efficiency reasons, this ideal is actually the sum of three ideals $I + L + N$. I is the ideal generated by the polynomials of the form $x_i^p - x_i$, the role of which is to simplify the exponents of the

---

[4] A formula A_0 is a tautological consequence of other formulae $A_1, A_2, ..., A_m$ iff whenever $A_1, A_2, ..., A_m$ are true, A_0 is also true.

variables x_i. L is the ideal generated by the polynomials that translate the negations of the facts in some chosen maximal consistent set of facts. N is the ideal generated by the polynomials that translate the negations of both the production rules and the ADDIs.

The theorem says:

Theorem 1. *A formula A_0 of the language in which the RBES is expressed is a tautological consequence, in any p-valued logic, where p is a prime number, of a set of formulae $\{A_1, A_2, ..., A_m\}$ iff the polynomial translation of the (negation of) A_0 belongs to the ideal generated in the quotient ring $Z_p[x_1, x_2, ..., x_n]/I$ by the polynomial translation of the (negations of) $A_1, A_2, ..., A_m$, where I is the ideal generated by the polynomials $x_1^p - x_1, x_2^p - x_2, ..., x_n^p - x_n$.*

Corollary 1. *In the context of the RBES mentioned above, A_0 follows from a certain maximal consistent set of facts, which polynomial translation (of their negations) generate ideal L, and certain production rules and other additional information, which polynomial translation (of their negations) generate ideal N, iff*

$$pol(NEG(A_0)) \in I + L + N$$

in the ring $Z_p[x_1, x_2, ..., x_n]$.

The polynomial variables $x_1, ..., x_n$ each correspond to the propositional variables contained in the RBES.

Two important steps can be performed thanks to this theorem:

(1) look for inconsistencies in the expert system
(2) extract new information from the information contained in the expert system.

Let us detail them.

(1) Let us recall that a RBES is inconsistent if all formulae of the language of the RBES are (tautological) consequences of a maximal consistent set of facts and the information contained in the RBES (because, in particular, both a formula and its negation, that is, a contradiction, would be consequences of the RBES).

It can be proved from the main result above that inconsistency is expressed by the algebraic fact that polynomial 1 belongs to the ideal generated by the polynomials that translate the (negation of) facts, rules and ADDIs (the ideal is the whole ring in this case). That is, the RBES is inconsistent if and only if

$$1 \in I + L + N .$$

Therefore, step (1) merely involves typing the following CoCoA command:

```
GBasis(I+L+N);
```

(`GBasis` means "Gröbner basis"). If the output is [1], there is inconsistency in the expert system.

Let us observe that, to assure that there are no inconsistencies in the expert system this should be done for all maximal consistent set of facts.

(2) Let us just say that to extract information translated as, say, a formula α (of the language in which the RBES is built), from the rules, ADDI and facts, all that has to be done is to type:

```
NF(NEG(α), I+L+N);
```

(where `NF` means "Normal Form"). The output of this command is 0 iff α follows from the information provided in `I+L+N`.

An introduction to Gröbner bases (GB) can be found in [16,1].

Other methods could be used to perform these tasks. For instance SAT-solvers may be more efficient in the bi-valued case. Even though, GB-based methods also have advantages, e.g.:

- we wanted to choose a general tool for all systems (see section "Acknowledgements" at the end). GB-based methods can be applied to multi-valued logics (which is not the case for SAT-solvers)
- GB-based method allows to save time when adding a new polynomial. Adding new polynomials to a set that is already a GB is faster in general than starting the calculations from scratch.

4   CoCoA Implementation of the Expert System

4.1   Introduction

As advanced in Section 2, the 343 production rules of the Alzheimer RBES are divided into three subsets:

(1) clinical history and neuropsychological tests
(2) neurological evaluation
(3) laboratory tests

These subsets are, in turn, subdivided into the subsubsets already listed in Section 2.

Each of these subsubsets, when fired using appropriate maximal consistent sets of facts, gives rise to partial information on Alzheimer's disease.

As the RBES is quite large, for reasons of space, simplicity and clarity, we will deal in some detail with only the partial information that refers to clinical history. This means dealing with 49 production rules, 23 potential facts (and their negations) and one ADDI. This partial information leads to either a positive answer- "the patient has cognitive disturbance"- or a negative one- "the patient does not have cognitive disturbance". This existence or non-existence of cognitive disturbances is what we have called "diagnosis of the clinical history", and is expressed by variable $x[74]$.

As the full verification of such a RBES is a huge task, the program checks for any inconsistency that could appear in the current process, alerting the user if needed.

The information dealing with neuropsychological tests, neurological evaluation and laboratory test results is treated similarly.

The different partial information on clinical history, neuropsychological tests, neurological evaluation and laboratory test results, makes up in turn, the facts of a final production rule that returns the diagnosis of the patient's illness.

4.2   Clinical History

In the code below, comments are preceded by --.

The polynomial ring A with coefficients in Z_2 (that is, permitting coefficients 0 and 1) and 318 variables (for the whole RBES) is declared first:

```
A ::= Z/(2)[x[1..318]];
USE A;
```

Definition of the ideal I:

```
I:=Ideal(x[1]^2-x[1],x[2]^2-x[2],x[3]^2-x[3],....,x[74]^2-x[74],
...,x[318]^2-x[318]);
```

Translation of the basic bi-valued logic formulae to polynomials:

```
NEG(M):=NF(1+M,I);
OR1(M,N):=NF(M+N-M*N,I);
AND1(M,N):=NF(M*N,I);
IMP(M,N):=NF(1+M+M*N,I);
```

The following is the list of the potential facts that appear in the 49 production rules corresponding to clinical history (for the sake of space, only affirmative facts are included). Both these facts and their negations make up a list that is presented to family members of the patient who the doctors suspect has Alzheimer's disease. They should select one of the two alternatives for each symptom in the list. That is, they must provide a "maximal consistent set of facts".

```
F1:=x[1];   -- difficulty in remembering what day it is.
F2:=x[2];   -- difficulty in remembering what he/she has eaten.
F3:=x[3];   -- difficulty in remembering recent events.
F4:=x[4];   -- difficulty in learning new things.
F5:=x[6];   -- difficulty in finding the right words.
F6:=x[7];   -- mistakes when naming things.
F7:=x[8];   -- poor pronunciation.
F8:=x[9];   -- inability to read.
```

```
F9:=x[10];  -- problems with verbal expressions.
F10:=x[12]; -- problems with using machines.
F11:=x[13]; -- problems with pursuing hobbies.
F12:=x[14]; -- problems with getting dressed.
F13:=x[15]; -- problems with making the right gesture.
F14:=x[16]; -- problems with writing.
F15:=x[18]; -- problems with recognizing common objects.
F16:=x[19]; -- problems with recognizing common smells.
F17:=x[20]; -- problems with recognizing people.
F18:=x[21]; -- mixing up hands.
F19:=x[23]; -- problems with cooking simple meals.
F20:=x[24]; -- problems with money use.
F21:=x[25]; -- problems with very simple mathematical operations.
F22:=x2[6]; -- alteration of planning capabilities.
F23:=x[71]; -- MMSE test score of less than 24.
```

The production rules referring to clinical history are included below (there are 49 production rules and 74 propositional variables). In the following code each rule is assigned first (R:=...), and is then shown (R;) and remarked upon afterwards (--...):

```
R1:=NF(IMP(AND1(AND1(AND1(x[1],x[2]),x[3]),x[4]),x[5]),I);
R1;  -- Refers to memory disorder.
R2:=NF(IMP(OR1(OR1(OR1(NEG(x[1]),NEG(x[2])),NEG(x[3])),NEG(x[4])),
          NEG(x[5])),I);
R2;  -- Refers to memory disorder.
R3:=NF(IMP(AND1(AND1(AND1(AND1(x[6],x[7]),x[8]),x[9]),x[10]),
          x[11]),I);
R3;  -- Refers to aphasia (difficulty with verbal expression).
R4:=NF(IMP(OR1(OR1(OR1(OR1(NEG(x[6]),NEG(x[7])),NEG(x[8])),
          NEG(x[9])),NEG(x[10])),NEG(x[11])),I);
R4;  -- Refers to aphasia.
R5:=NF(IMP(AND1(AND1(AND1(AND1(x[12],x[13]),x[14]),x[15]),x[16]),
          x[17]),I);
R5;  -- Refers to apraxia (difficulty in coordinating movements)
...
...
R48:=NF(IMP(AND1(x[70],OR1(OR1(x[71],x[72]),x[73])),x[74]),I);
R48; -- Refers to disturbances of two or more cognitive functions.
R49:=NF(IMP(NEG(x[70]),NEG(x[74])),I); R49; -- Refers to
disturbances of two or more cognitive functions.
```

The ADDI says that cognitive disturbance needs to be corroborated by tests:

```
ADDI:=NF(NEG(Y(x[70],x[74])),I);
```

Suppose that the 23 facts have been affirmatively answered when referring to some patient. They form a maximal consistent set of facts, which gives rise to the ideal L:

```
L:=Ideal(NEG(F1),NEG(F2),NEG(F3),NEG(F4),NEG(F5),NEG(F6),
         NEG(F7),NEG(F8),NEG(F9),NEG(F10),NEG(F11),NEG(F12),
         NEG(F13),NEG(F14),NEG(F15),NEG(F16),NEG(F17),NEG(F18),
         NEG(F19),NEG(F20),NEG(F21),NEG(F22),NEG(H23));
```

The ideal for all rules and the ADDI of the clinical history is:

```
N:=Ideal(NEG(R1),NEG(R2),NEG(R3),...,NEG(R48),NEG(R49),NEG(ADDI));
```

Consistency checking:

```
G:=GBasis(I+L+N);
```

(remember that if the output is [1], which is not the case here, there would be inconsistency).

We have implemented a simple program named CONSIST that checks the ideal incrementally and underlines the first rule that, when added, produces inconsistency.

If the answer to NF(NEG(x[74]),I+L+N); is 0 (we assume that there are no inconsistencies), x[74] is tautological consequence of the information in the RBES, that is, cognitive disturbance supported by tests exists. Remember that the existence or non-existence of cognitive disturbances is what was called "diagnosis of clinical history" above.

4.3 Production Rule for Final Diagnosis

Once (1) the clinical history and neuropsychological tests, (2) the neurological evaluation and (3) the laboratory tests have been evaluated as shown in the previous section for clinical history, a production rule can be applied leading to the diagnosis. Its code is shown below. Note that this RBES will give Alzheimer's disease as a diagnosis only if there are no other simultaneous serious illnesses with similar symptoms.

```
If (Clinical History=0) And (Depression=0)
   And (Neurological-Evaluation=0) And (Nutritional-Evaluation<>0)
   And (Hypothyroidism<>0) And (Neurosyphilis<>0)
   And (Hypercalcemia<>0) And (HIV<>0) And (Hematology=0)
   And (Biochemistry=0) And (Urine=0) And (EAProbable=0)
 Then Alzheimer:=0
 Else Alzheimer:=1 ;
End ;
```

We must remark that the assessment that "Alzheimer disease occurs under the assumption that other simultaneous serious illnesses are not present", is not a trivial assertion. As in many cases in Medicine there are several illnesses with similar symptoms that are difficult to distinguish. The complete system includes production rules that translate precise knowledge and is able to perform this discarding of presence of other illnesses. These rules are based on complex tests: neurological tests, laboratory tests (for instance hematological: folic acid, TSH, T4L, T3, VDRL, treponemic, HIV, VCM, HCM, CHCM), radiological tests (ECG, CAT, NMR, SPECT, PET...).

5   Performance, Motivation, and Evaluation

The input to the expert system (i.e., which of the facts hold) are the clinical history (including the answers of the family of the patient to the questions asked by the system) and the results of the neurological tests, laboratory tests, radiological test... The extraction of knowledge for each block of rules (e.g. to obtain the partial diagnosis for each of the parts of the neurological evaluation -see section 2-) takes a few seconds in a standard 1.1MHz 384 MB of RAM PC. When all the blocks have been fired, the final diagnosis is obtained in a similar timing.

About the evaluation, the system is still a prototype. It has been reviewed by three medical doctors (two of them specialized in degenerative illnesses). Their opinion is positive, and they think that it could be a very useful tool for family doctors, as the early detection of this illness is not always obvious for a non-specialist. That was the main motivation to design this system.

Nevertheless a validation of the system through its use in medical practice is needed (and is planned as the next step of its development).

6   Conclusion

We have summarized the construction of a RBES for diagnosis of Alzheimer's disease. Once the RBES has been constructed, symbolic computation techniques were applied to automatically verify and extract new knowledge and output a diagnosis. Both the knowledge base and the inference engine have been symbolically expressed and implemented using the computer algebra language CoCoA. As the full RBES is quite large (it contains 343 production rules), we have detailed only the rules that lead to the particular partial information about the illness which we have called "clinical history".

This partial information, together with others (neurological evaluation and laboratory tests), are in turn the antecedents of a set of final production rules that lead to a diagnosis. As illustration, one such final production rule has been considered by way of an illustration.

Acknowledgements. This new work by our research team is another medical application of algebraic techniques. Our work in this field began by studying verification and knowledge extraction in Coronary By-Pass Surgery Medical Practice Guidelines. It was possible thanks to the support received from the Spanish Ministry of Health (FIS 95/1952) and the Ministry of Science and Technology (DGES PB96-0098-C04) [10,11]. This was followed by the development of two RBES for the detection of anorexia [13] and depression [8], respectively (both of which were developed within a Spanish Ministry of Science and Technology research project, TIC2000-1368-C03, which has supported this work too).

We would also like to thank the anonymous referees for their most valuable comments.

7 Appendix

7.1 Introduction

- Let us detail how a polynomial is assigned to each logical formula. This is achieved by assigning to each propositional variable X_i a monomial x_i and defining, for each connective c_j, a function:

$$f_j : (Z_p[x_1, x_2, ..., x_n]/I)^{s_j} \longrightarrow Z_p[x_1, x_2, ..., x_n]/I.$$

The symbol I represents the ideal generated by the polynomials $x_1^p - x_1, x_2^p - x_2, ..., x_n^p - x_n$:

$$I = <x_1^p - x_1, x_2^p - x_2, ..., x_n^p - x_n>$$

As the process has has been published by the authors elsewhere [9,15], we simply transcribe the final expressions, for example, for the functions f_j for Kleene's three-valued and modal logic. The letters q and r are variables that range over the elements (polynomials) of $Z_p[x_1, x_2, ..., x_n]/I$. In this case $p = 3$.

$f_\neg(q) = (2 + 2q) + I$
$f_\Diamond(q) = 2q^2 + I$
$f_\Box(q) = (q^2 + 2q) + I$
$f_\vee(q, r) = (q^2r^2 + q^2r + qr^2 + 2qr + q + r) + I$
$f_\wedge(q, r) = (2q^2r^2 + 2q^2r + 2qr^2 + qr) + I$
$f_\rightarrow(q, r) = (q^2r^2 + q^2r + qr^2 + 2q + 2) + I$

The complexity of the expressions of the functions $f_\rightarrow$ increase with the value of p. for instance f_j for $p = 7$ is:

$f_\rightarrow(q, r) = (3q^6r^2 + 6q^5r^3 + 4q^4r^4 + 6q^3r^5 + 3q^2r^6 + 3q^6r + 4q^5r^2 + 2q^4r^3 + 2q^3r^4 + 4q^2r^5 + 3qr^6 + 5q^5r + 6q^4r^2 + q^2r^3 + 6q^2r^4 + 5qr^5 + q^4r + 2q^3r^2 + 2q^2r^3 + qr^4 + 4q^3r + q^2r^2 + 4qr^3 + 4q^2r + 4qr^2 + 5qr + 6q + 6) + I$

- The functions f_j translate the basic propositional formulae $\neg X_i$, $\Diamond X_i$, $\Box X_i$, $X_i \wedge X_k$,..., $X_i \vee X_k$, $X_i \rightarrow X_k$ into (classes of) polynomials. The next definition determines a function θ that, interacting with the functions f_j, translates any propositional formula, in particular the rules and other items of any RBS, into (classes of) polynomials.
  $P_C(X_1, X_2, ..., X_n)$ represents the set of all well constructed propositional formulae with the propositional variables $X_1, X_2, ..., X_n$.
  $\theta : P_C(X_1, X_2, ..., X_n) \longrightarrow Z_p[x_1, x_2, ..., x_n]/I$ is a function from propositions to (classes of) polynomials, recursively defined as follows:
  $\theta(X_i) = x_i + I$, for all $i = 1, ..., n$.
  $\theta(A) = f_j(\theta(A_1), ..., \theta(A_{s_j}))$ if A is $c_j(A_1, ..., A_{s_j})$.

- Each logical truth valuation of propositional variables, v, can be extended recursively to a valuation of the propositional formulae in $P_C(X_1, X_2, ..., X_n)$, denoted v'. For each v we can define the homomorphism v^*:

$$v^* : Z_p[x_1, x_2, ..., x_n]/I \longrightarrow Z_p$$

such that $v^*(x_i + I) = v(X_i)$ for $i = 0, ..., n$. It can be proved that for any valuation v, $v' = v^* \cdot \theta$.

Lemma 1. $\{0\} + I = (\bigcap_{i=1,...,k} \ker(v_i^*)) \cap (\theta(P_C(X_1, X_2, ..., X_n)))$, where $k = p^n$ is the number of all valuations, the $v_i's$ range over all possible valuations.

Lemma 2. Let $A_1, A_2,..., A_m, A_0 \in P_C(X_1, X_2, ..., X_n)$. The following two assertions are equivalent:

(i) for all valuations $v_i (i = 1, ..., k)$ such that $v_i^*(\theta(A_1)) = v_i^*(\theta(A_2)) = ... = v_i^*(\theta(A_m)) = 0$ it follows that $v_i^*(\theta(A_0)) = 0$.
(ii) $\theta(A_0) \in< \theta(A_1), \theta(A_2), ..., \theta(A_m) >$.

Let us remember **Theorem 1**:

A formula A_0 is a tautological consequence of a set $\{A_1, A_2, ..., A_m\}$ of formulae, in any p-valued logic where p is a prime number, if and only if the polynomial translation of the negation of A_0 belongs to the ideal generated in $Z_p[x_1, x_2, ..., x_n]/I$ by the polynomial translation of the negations $A_1, A_2, ..., A_m$.

We can rewrite it in a more formal way:

Let $A_0, A_1, ..., A_m \in P_C(X_1, X_2, ..., X_n)$. The following assertions are equivalent:

(i) $\{A_1, A_2, ..., A_m\} \models A_0$,
(ii) $f_\neg(\theta(A_0)) \in< f_\neg(\theta(A_1)), ..., f_\neg(\theta((A_m)) >$ (in $Z_p[x_1, x_2, ..., x_n]/I$).

Proof. $\{A_1, A_2, ..., A_m\} \models A_0$ iff for any v, $v'(A_1) = p - 1, ..., v'(A_m) = p - 1$ implies $v'(A_0) = p - 1$ (remember that $p - 1$ is the value "true"). This is equivalent to the condition: $(p-1) - v'(A_1) = 0, ..., (p-1) - v'(A_m) = 0$ implies $(p-1) - v'(A_0) = 0$ which is equivalent to $v'(\neg A_1) = 0, ..., v'(\neg A_m) = 0$ implies $v'(\neg A_0) = 0$.
This implication is equivalent to that, for any v,

$$v_i^*(\theta(\neg A_1)) = 0, ..., v_i^*(\theta(\neg A_m)) = 0 \text{ implies } v_i^*(\theta(\neg A_0)) = 0$$

which is equivalent to

$$v_i^*(f_\neg(\theta(A_1))) = 0, ..., v_i^*(f_\neg(\theta(A_m))) = 0 \text{ implies } v_i^*(f_\neg(\theta(A_0)) = 0.$$

By the last lemma above, the last implication is equivalent to

$$f_\neg(\theta(A_0)) \in< f_\neg(\theta(A_1)), ..., f_\neg(\theta(A_m)) >.$$

References

1. V. Adams and P. Loustaunau: *An Introduction to Gröbner Bases*. Graduate Studies in Mathematics 3. American Mathematical Society Press, Providence, RI, 1994.
2. J.A. Alonso and E. Briales: *Lógicas polivalentes y bases de Gröbner*. In M. Vide (ed.): *Proceedings of the V Congress on Natural Languages and Formal Languages*. P.P.U. Press, Barcelona, 1989 (307-315).
3. American Psychiatric Association: *DSM III-R, Manual diagnóstico estadístico de los transtornos mentales*. Masson, Barcelona, 1988.
4. A. Capani and G. Niesi: *CoCoA User's Manual (v. 3.0b)*. Dept. of Mathematics, University of Genova, Genove, 1996.
5. J. Chazarain, A. Riscos, J.A. Alonso and E. Briales: Multivalued Logic and Gröbner Bases with Applications to Modal Logic. *Journal of Symbolic Computation*, **11** (1991) 181-194.
6. B. Herrero: *Desarrollo de un Sistema Experto para el Diagnóstico del Alzheimer usando CoCoA*. Master thesis (advisor: Luis M. Laita), Facultad de Informática, Universidad Politécnica de Madrid, 2002.
7. L.M. Laita, L. de Ledesma, E. Roanes-Lozano and E. Roanes-Macías: *An Interpretation of the Propositional Bolean ALgebra as a k-Algebra. IEffective Calculus*. In J. Calmet, J.A. Campbell (eds.): *Integrating Symbolic Mathematical Computation and Artificial Intelligence. Selected Papers of AIMSC-2*. Springer-Verlag LNAI-958, Berlin, 1995 (255-263).
8. L.M. Laita, G. González-Páez, E. Roanes-Lozano, V. Maojo, L. de Ledesma, L. Laita: *A Methodology for Constructing Expert Systems for Medical Diagnosis*. In: J. Crespo, V. Maojo, F. Martín (eds.): *Medical data Analysis. Procs. of IMSDA 2001*. Springer-Verlag LNCS-1933, 2001 (212-217).
9. L.M. Laita, E. Roanes-Lozano, L. de Ledesma, J.A. Alonso: A Computer Algebra Approach to Verification and Deduction in Many-valued Expert Systems. *Soft Computing*, **3/1** (1999) 7-19.
10. L.M. Laita, E. Roanes-Lozano and V. Maojo: *Inference and Verification in Medical Appropriateness Criteria*. In J. Calmet, J. Plaza (eds.): *Artificial Intelligence and Symbolic Computation, Proceedings of AISC'98*. Springer-Verlag LNAI-1476, Berlin, 1998 (183-194).
11. L. M. Laita, E. Roanes-Lozano, V. Maojo, L. de Ledesma, L. Laita: An Expert System for Managing Medical Appropriateness Criteria based on Computer Algebra Techniques. *Computers and Mathematics with Applications*, **42/12** (2001) 1505-1522.
12. J.J. López-Ibor (ed.): *CIE10. Transtornos mentales y del comportamiento. Descripciones clínicas y pautas para el diagnóstico*. Organización Mundial de la Salud - Ed. Meditor, 1992.
13. C. Pérez-Carretero, L.M. Laita, E. Roanes-Lozano, L. Lázaro, J. González-Cajal, L. Laita: A Logic and Computer Algebra Based Expert System for Diagnosis of Anorexia. *Mathematics and Computers in Simulation*, **58/3** (2002) 183-202.
14. D. Perkinson: *CoCoA 4.0 Online Help* (electronic file), 2000.
15. E. Roanes-Lozano, L.M. Laita and E. Roanes-Macías: A Polynomial Model for Multivalued Logics with a Touch of Algebraic Geometry and Computer Algebra. *Mathematics and Computers in Simulation*, **45/1** (1998) 83-99.
16. F. Winkler: *Polynomial Algorithms in Computer Algebra*. Springer, Wien, 1996.

On a Generalised Logicality Theorem

Marc Aiguier[1], Diane Bahrami[1], and Catherine Dubois[2]⋆

[1] Université d'Évry Val d'Essonne, CNRS UMR 8042, LaMI, F-91025 Évry, France
{aiguier, bahrami}@lami.univ-evry.fr
[2] Institut d'Informatique d'Entreprise, CEDRIC, Évry, France dubois@iie.cnam.fr

Abstract. In this paper, the correspondence between derivability (syntactic consequences obtained from ⊢) and convertibility in rewriting ($\overset{*}{\leftrightarrow}$), the so-called *logicality*, is studied in a generic way (i.e. logic-independent). This is achieved by giving simple conditions to characterise logics where (bidirectional) rewriting can be applied. These conditions are based on a property defined on proof trees, that we call *semi-commutation*. Then, we show that the convertibility relation obtained via semi-commutation is equivalent to the inference relation ⊢ of the logic under consideration.

Keywords: Formal system, semi-commutation, abstract rewrite tree, abstract convertibility relation, logicality
Topics: Term Rewriting, Reasoning , Integration of Logical Reasoning and Computer Algebra

1 Introduction

A classical result due to Birkhoff ensures that equational reasoning coincides with convertibility in rewriting for equational logic. This property, named *logicality*, is expressed by:

$$\Gamma \vdash t = t' \Longleftrightarrow t \overset{*}{\leftrightarrow}_\Gamma t' \qquad (\Gamma : \text{set of equations}).$$

V. van Oostrom [9] has shown that this result could be extended to any sub-equational logic (such as the rewrite logic [8]) and S. Kaplan [6] has shown that it also held for conditional equational logic (this last result as well as new ones about logicality of conditional rewrite systems can be found in [13]). We show here that logicality can be extended to a larger class of logics. To achieve this purpose, we study convertibility in rewriting in a generic way. We have a proof-theoretic approach, in the sense that we use the underlying formal system of logics to study rewriting. This comes from the observation that, for all logics where logicality holds, the convertibility relation $\overset{*}{\leftrightarrow}$ each time defines proof strategies which select proof trees equipped with a specific structure. Roughly

---

⋆ This work was partially supported by the European Commission under WGs Aspire (22704) and is partially supported by the French research program "GDR Algorithmique-Langage-Programmation (ALP)"

J. Calmet et al. (Eds.): AISC-Calculemus 2002, LNAI 2385, pp. 51–63, 2002.

speaking, these trees, that we will call *rewrite trees*, denote closures of basic rewriting relations. Thus, proof trees that specify basic rewriting relations are always above proof trees that specify closures (considering that the leaves are at the top and the root at the bottom of the tree). For instance, in the mono-sorted equational logic, a convertibility relation is obtained by closing any set of equations Γ under substitution and context and then under transitivity.

The logicality result will hold if we can ensure that all statements of the form $\Gamma \vdash \varphi$ accept some rewrite trees as proof trees. This can be checked thanks to basic proof tree transformations produced from the property of *semi-commutation* of some inference rules with other ones. For instance, in the mono-sorted equational logic, substitution semi-commutes (denoted by $\rightsquigarrow$) with transitivity:

$$
\text{Subst} \cfrac{\text{Trans} \cfrac{t=t' \quad t'=t''}{t=t''}}{\sigma(t)=\sigma(t'')} \quad \rightsquigarrow \quad \text{Trans} \cfrac{\text{Subst} \cfrac{t=t'}{\sigma(t)=\sigma(t')} \quad \text{Subst} \cfrac{t'=t''}{\sigma(t')=\sigma(t'')}}{\sigma(t)=\sigma(t'')}
$$

From the property of semi-commutation, we will then be able to divide the inference rules into two disjoint sets, Up and $Down$, such that Up semi-commutes with $Down$ and the converse is not true. Therefore, proof trees resulting from the composition of inference rules in Up will define basic rewriting steps, and those in $Down$ will be used to compose rewriting steps together.

We assume that the reader is conversant with the definitions of basic equational logic and abstract rewriting as found in the introductory chapters of a textbook on the subject such as [3]. The paper is organised as follows: in Section 2, we recall standard notations about formal systems, deductions and proof trees. In Section 3, we instantiate formal systems in order to deal with predicates. In Section 4, we introduce the concept of semi-commutation and, from this, we show how to define rewriting relations such that logicality holds. Section 5 proposes as an example to study rewriting in distributive lattices.

2   Preliminaries

A *formal system* (a so-called *calculus*) $\mathcal{S}$ over an alphabet Ω consists of a set F of strings over Ω, and a finite set R of n-ary relations on F. It is denoted by $\mathcal{S} = (F, R)$. The elements of F are usually called *formulae* and the relations of R are called *inference rules*. Thus, a rule with arity n $(n \geq 1)$ is a set of tuples $(\varphi_1, \ldots, \varphi_n)$ of strings in F. Each sequence $(\varphi_1, \ldots, \varphi_n)$ belonging to a rule r of R is called an *instance* of that rule with *premises* $\varphi_1, \ldots, \varphi_{n-1}$ and *conclusion* φ_n. It is usually written $\frac{\varphi_1 \cdots \varphi_{n-1}}{\varphi_n}$. A *deduction* in $\mathcal{S}$ from a set of formulae Γ of F is any finite sequence $(\psi_1, \ldots, \psi_m)$ of formulae such that $m \geq 1$ and, for all $i = 1, \ldots, m$, either ψ_i is an element of Γ or there is an instance $\frac{\varphi_1 \cdots \varphi_{n-1}}{\varphi_n}$ of a rule in $\mathcal{S}$ where $\varphi_n = \psi_i$ and $\{\varphi_1, \ldots, \varphi_{n-1}\} \subseteq \{\psi_1, \ldots, \psi_{i-1}\}$. A *theorem* from a set of formulae Γ in $\mathcal{S}$ is any formula φ such that there exists a deduction in $\mathcal{S}$ from Γ with last element φ. This is usually denoted by $\Gamma \vdash \varphi$. Instances can be composed to build *proof trees*. Thus, we obtain another way to denote deduction of theorems in formal systems. Formally, a *proof tree* π in a formal system $\mathcal{S}$ is a finite tree whose nodes are labelled with formulae of F in the following way: if a non-leaf node is labelled with φ_n and its predecessor nodes are labelled (from left to right) with $\varphi_1, \ldots, \varphi_{n-1}$, then $\frac{\varphi_1 \ \cdots \ \varphi_{n-1}}{\varphi_n}$ is an instance of a rule of $\mathcal{S}$. A proof tree π with root φ is denoted by $\pi : \varphi$ and $\mathcal{L}(\pi)$ denotes the set of leaves of π. Obviously, for any statement of the form $\Gamma \vdash \varphi$ in a formal system $\mathcal{S}$, there is an associated proof tree $\pi : \varphi$ whose leaves are instances of relations of arity 1 or formulae from Γ. We call any proof tree associated to the statement $\emptyset \vdash \varphi$ a *tautological proof tree*.

Using a standard numbering of the tree nodes by strings of natural numbers, we can refer to positions in a proof tree. Thus, given a proof tree π, a *position* of π is a string w on $I\!N$ which represents the path from the root of π to the subtree at that position. This subtree is denoted by $\pi|_w$. Given a position w in a proof tree π, $\pi[\pi']_w$ denotes the proof tree obtained from π by replacing the subtree at position w by π'. When w is a leaf position of π such that $\pi|_w = \varphi$ and $\pi' : \varphi$ is a proof tree, we use the expression $\pi \cdot_w \pi'$ to denote the proof tree $\pi[\pi']_w$. This operation is called *composition* of π and π' on (leaf) position w. Two proof trees $\pi : \varphi$ and $\pi' : \varphi$ are *equivalent w.r.t. a set of formulae Γ* if and only if they both are proof trees associated to the statement $\Gamma \vdash \varphi$.

3   Predicative Formal Systems

Rewriting necessarily deals with binary relations (e.g. equality [3], inclusion [7], more general transitive relations [4,12,10], the ideal membership problem, etc.). It can be constrained by other relations (such as the definedness predicate in the partial logic - see example below). This leads us to consider logics where formulas are of the form $p(t_1, \ldots, t_n)$ where p is a predicate name. Since we do not want to impose constraints that are useless for the establishment of the logicality result, the elements $t_1, \ldots, t_n$ do not necessarily belong to an inductively defined set.

Definition 1. (Predicative formal system). *A predicative formal system is a triple $\mathcal{P} = (T, P, R)$ such that T is a set, P is a set of n-ary relations on T and (F, R) is a formal system, with F the set of formulae defined as follows :*
$$F = \{p(t_1, \ldots t_n) \mid p \in P \wedge (t_1, \ldots t_n) \in p\}.$$

Example 1. **The Partial Conditional Logic**
Let us define the predicative formal system for the many-sorted partial conditional equational logic with definedness, total valuation and existential equality (see [2]) and then study rewriting within it. First, we recall the basic notions and notations of this logic. A signature $\Sigma = (S, F, V)$ contains a set S of sorts, a set F of partial function names with arity in S^+ and a S-indexed family V of sets of variables. For every $s \in S$, $W_\Sigma(V)_s$ denotes the standard set of terms of sort s over Σ, free with generating set V and $W_\Sigma(V)$ denotes the set $\coprod_{s \in S} W_\Sigma(V)_s$. Given a term t in $W_\Sigma(V)$, $Var(t)$ denotes the set of variables occurring in t. Atoms are either equations $t = t'$ where t and t' are terms of the same sort, or formulae $D(t)$ where t is a term. Semantically, $t = t'$ states that both sides of the equality are defined and denote the same value and $D(t)$ states that t is

54   M. Aiguier, D. Bahrami, and C. Dubois

defined, i.e., t necessarily yields a value when it is evaluated. A conjunction over Σ is a formula of the form $\alpha_1 \wedge \ldots \wedge \alpha_n$ where, for $1 \leq i \leq n$, α_i is an atom. The empty conjunction is denoted by $\emptyset$. Let us denote by $Conj_\Sigma$ the whole set of conjunctions over Σ. A conditional formula over Σ is then any formula of the form $c \Rightarrow \alpha$ where $c \in Conj_\Sigma$ and α is an atom. A substitution is an S-indexed family of functions $\sigma_s : V_s \rightarrow W_\Sigma(V)_s$. It is naturally extended to terms, conjunctions and conditional formulae. If c is a conjunction and Γ a set of conditional formulae, a substitution σ is c-defined with respect to Γ if and only if, for all $x \in V_s$, we have $\Gamma \vdash c \Rightarrow D(\sigma_s(x))$.

Given a signature $\Sigma = (S, F, V)$, we define the predicative formal system $\mathcal{P}_\Sigma = (T_\Sigma, P_\Sigma, R_\Sigma)$ as follows:

- $T_\Sigma = W_\Sigma(V)$;
- $P_\Sigma = E_\Sigma \cup D_\Sigma$ where E_Σ is the $(Conj_\Sigma \times S)$-indexed family $\{=_{c,s}\}_{(c,s) \in Conj_\Sigma \times S}$ of binary relations s.t. $(t, t') \in =_{(c,s)}$ iff t and t' are terms of sort s, and D_Σ is the $Conj_\Sigma$-indexed family $\{D_c\}_{c \in Conj_\Sigma}$ of unary relations on terms s.t. $D_c = T.^1$ Thus, an equation $t =_{(c,s)} t'$ (resp. $D_c(t)$) denotes the formula $c \Rightarrow t = t'$ (resp. $c \Rightarrow D(t)$), where $t, t' \in W_\Sigma(V)_s$. Afterwards, we will forget the sort s and simply write $=_c$ rather than $=_{(c,s)}$ when this does not bring about any ambiguity.
- F_Σ is the set of all formulae of the form $t =_c t'$ and $D_c(t)$.
- Finally, R_Σ is the set of inference rules generated from the following inference rule schemata, where α_c denotes either $t =_c t'$ or $D_c(t)$, C is a *context* (i.e., any term with a unique hole $\square$) and the notation $C[t]$ denotes the term obtained by substituting the symbol $\square$ by t in C :

Reflexivity: $\dfrac{}{x =_\emptyset x}$ **Symmetry:** $\dfrac{t =_c t'}{t' =_c t}$ **Transitivity:** $\dfrac{t =_c t' \quad t' =_c t''}{t =_c t''}$

Strictness: $\dfrac{D_c(f(t_1, \ldots, t_n))}{D_c(t_i)}$ **Replacement:** $\dfrac{t =_c t' \quad D_c(C[t])}{C[t] =_c C[t']}$

Existential equality: $\dfrac{t =_c t'}{D_c(t)}$ **Tautology:** $\dfrac{}{\alpha_c}$ if α occurs in c

Substitution: $\dfrac{\alpha_c}{\sigma(\alpha)_{\sigma(c)}}$ where σ is a c-defined substitution

Monotonicity: $\dfrac{\alpha_c}{\alpha_{c \wedge c'}}$ **Modus Ponens:** $\dfrac{\alpha_{c \wedge \alpha'} \quad \alpha'_c}{\alpha_c}$

4   Logicality

4.1   Predicates Appropriate for Rewriting

In a predicative formal system, statements of the form $\Gamma \vdash p(t, t')$ will be proved by rewriting if there is an associated *rewrite tree* $\pi : p(t, t')$. Any binary predicate having this property will be said *appropriate for rewriting*. These notions will be defined in this section.

Notation 1. We denote by $\pi = (\pi_1, \cdots, \pi_k, \varphi)$, with $k \geq 0$, the proof tree whose last inference rule is $\dfrac{\varphi_1, \ldots, \varphi_k}{\varphi}$ and such that, for all $i \in \{1, \ldots, k\}$, $\pi_i : \varphi_i$ is the subtree of π at position i (i.e. $\pi|_i = \pi_i : \varphi_i$).

---

^{1} Syntactically, we can write $c \Rightarrow D(t)$ for any term $t \in T$

Notation 2. Let S be a set of proof trees. We denote by $S^{\sharp}$ the closure of S under the composition operation. Formally, it is the least set (according to the set-theoretical inclusion) satisfying:

- $S \subseteq S^{\sharp}$;
- if $\pi \in S^{\sharp}$ with $\pi_{|w} = \varphi$ and $\pi' : \varphi \in S^{\sharp}$ then $\pi \cdot_w \pi' \in S^{\sharp}$.

Definition 2. (Structured proof trees). *Let $\mathcal{P} = (T, P, R)$ be a predicative formal system and $E \subseteq P$ a non-empty set of binary predicates. Let $Up_E = (Up_p)_{p \in E}$ and $Down_E = (Down_p)_{p \in E}$ be two families of sets of rule instances such that for all $p \in E$, every instance of R whose conclusion is of the form $p(t, t')$ belongs either to Up_p or to $Down_p$. The set $\mathcal{P}r_{Up_E > Down_E}$ is the least set of proof trees (according to the set-theoretical inclusion) such that:*

- $Up_E^{\sharp} \cup Down_E \subseteq \mathcal{P}r_{Up_E > Down_E}$;
- *Let $p \in E$ be a binary predicate, let $\frac{\varphi_1 \cdots \varphi_n}{p(t,t')}$ be an instance of $Down_p$ with $n > 0$, and let $(\pi_i : \varphi_i)_{1 \leq i \leq n}$ be n proof trees such that, if φ_i is of the form $p'(u, v)$ with $p' \in E$, then $\pi_i \in \mathcal{P}r_{Up_E > Down_E}$. Then, the tree $(\pi_1, \ldots, \pi_n, p(t, t'))$ belongs to $\mathcal{P}r_{Up_E > Down_E}$.*

The set $\mathcal{P}r_{Up_E > Down_E}$ contains the trees of $Up_E^{\sharp}$, the trees of $Down_E^{\sharp}$ and the trees described in figure 1.

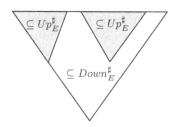

Fig. 1. Trees belonging to $\mathcal{P}r_{Up_E > Down_E}$

Definition 3. (Semi-commutation). *With the notations of Definition 2, Up_E semi-commutes with $Down_E$ if and only if for every $p, p' \in E$, and every proof tree π of the form:*

$$i' \frac{\psi_1 \ldots i \frac{\varphi_1 \ldots \varphi_n}{p(t,t')} \ldots \psi_m}{p'(u, v)}$$

where $i \in Down_p$, $i' \in Up_{p'}$ and $\psi_1 \ldots \psi_m$ are formulae, there exists an equivalent proof tree $\pi' \in \mathcal{P}r_{Up_E > Down_E}$ w.r.t. $\mathcal{L}(\pi)$.
When the semi-commutation property is satisfied, proof trees in $\mathcal{P}r_{Up_E > Down_E}$ are called rewrite trees.

Definition 3 calls for some comments:

- The sets Up_E and $Down_E$ will be used to build the convertibility relation (Up_E for the basic relation and $Down_E$ for its closure).
- Definition 3 does not ensure unicity for the couple $(Up_E, Down_E)$ (when it exists). For instance, in the equational logic framework, many couples $(Up_{\{=\}}, Down_{\{=\}})$ can be candidate. Let us suppose that $Up_{\{=\}}$ contains all rule instances except one instance of the transitivity rule $\frac{t_1=t_2 \quad t_2=t_3}{t_1=t_3}$ (which is thus the only instance of $Down_{\{=\}}$). Easily, we can show that all the instances of symmetry, replacement and substitution rules semi-commute with this instance (actually, they semi-commute with all instances of the transitivity rule - see examples below). Moreover, we have:
$$\frac{\frac{t_1=t_2 \quad t_2=t_3}{t_1=t_3} \quad t_3=t_4}{t_1=t_4} \quad \rightsquigarrow \quad \frac{t_1=t_2 \quad \frac{t_2=t_3 \quad t_3=t_4}{t_2=t_4}}{t_1=t_4}.$$
Consequently, $Up_{\{=\}}$ semi-commutes with $Down_{\{=\}}$.
Now, for many logics (anyway, all logics used in computer science or mathematics) the underlying inference relation $\vdash$ is generated from a finite set of inference rule schemata, that is, a single form with infinitely many instantiations. This allows to denote all the instances by a set of generic forms (up to meta-variable renaming). We take advantage of these forms to give a choice strategy for the two sets : we put all the instances of a same rule schema in the same set (either Up_E or $Down_E$). This strategy is natural : this is the way convertibility relations are built in the literature (even though the two sets are not mentioned).
This strategy also makes the study of semi-commutation easier. Indeed, the number of instances of a formal system is usually infinite. Therefore, checking the semi-commutation property between them can be a hard task. But this property can be checked only by reasoning on the generic forms. We will use this method in the two examples of the paper, but we have not formalised the rule schemata because the theory would have been too heavy.
- Usually, tautologies are simply recognised on their syntactical structure. For instance, in equational logic, tautologies are all equations of the form $t = t$. Regarding conditional equational logic, tautologies are either of the form $t =_c t$ or $t =_{c_1 \wedge t=t' \wedge c_2} t'$. In such logics, proof trees denoting tautologies are removed from the rewrite tree set. The underlying bidirectional rewriting will then be closed under tautologies.

The semi-commutation property between instances of Up_E and $Down_E$ gives rise to basic transformation rules. They transform any proof tree only composed of two instances (one from $Down_E$ and the other from Up_E) into an equivalent rewrite tree. Thus, if semi-commutation holds, there is one basic transformation rule for each one of these trees. When the set of basic transformation rules is terminating, it obviously allows to transform any proof tree into a rewrite tree.

Definition 4. (Appropriate for rewriting). *With the notations of Definition 3, E is said appropriate for rewriting if and only if there exists a couple*

$(Up_E, Down_E)$ such that Up_E semi-commutes with $Down_E$ and the application of the resulting transformation rules terminates.

As an immediate consequence of Definition 4, we obtain:

Proposition 5. *Let $\mathcal{P} = (T, P, R)$ be a predicative formal system and let $E \subseteq P$ be a set appropriate for rewriting. For any $p \in E$ and any proof tree $\pi : p(t, t')$, there exists a rewrite tree equivalent to π w.r.t. $\mathcal{L}(\pi)$.*

Proof As E is appropriate for rewriting, the proof tree transformation rules resulting of the semi-commutation property are terminating. Therefore, repeatedly applying these rules on any proof tree always terminates in a rewrite tree. □

However, termination cannot be ensured in general. New conditions must be added. In next section, we define a terminating sub-case from two simple easy-to-check conditions which generalise basic transformation rules obtained for usual logics (anyway all logics for which the logicality result holds).

4.1.1   A Terminating Sub-case

Here, we give two simple conditions which ensure that any set E satisfying the semi-commutation property is appropriate for rewriting (i.e. termination of basic proof tree transformation rules holds).

Definition 6. (Distributive). *With the notations of Definition 3, E is called distributive if and only if there exists a couple $(Up_E, Down_E)$ such that Up_E semi-commutes with $Down_E$ and the following conditions hold:*

1. *all instances $\frac{\varphi_1 \ldots \varphi_n}{\varphi} \in Up_E \cup Down_E$ satisfy: $\forall 1 \leq i \leq n,\ \varphi_i \neq \varphi$;*

2. *for every proof tree transformation rule $\dfrac{\psi_1 \ldots \frac{\varphi_1 \ldots \varphi_n}{p(t,t')} \ldots \psi_m}{p'(u,v)} \to \pi$ resulting from the semi-commutation property, π contains at least an instance of $Down_E$.*

Remark. The name "distributive" results from the fact that, for many logics, semi-commutation is recognised as "distribution" of some inference rules over other ones (see the two examples developed in this paper). But distribution rules are simple sub-cases of Definition 6, Point *2.*.

The interest of distributive sets is that the semi-commutation property gives rise to basic proof tree transformation rules which always terminate. This is expressed by the following lemma:

Lemma 7. *Any distributive set E is appropriate for rewriting.*

Proof   With every predicative formal system $\mathcal{P} = (T, P, R)$, we can associate the following multi-sorted signature $\Sigma_{\mathcal{P}} = (S, Op)$:

- $S = \{s_\varphi \mid \varphi \in F\}$;
- $Op = \{c_\varphi : \to s_\varphi \mid \varphi \in F\} \cup \{f_i : s_{\varphi_1} \ldots s_{\varphi_n} \to s_\varphi \mid i\ \frac{\varphi_1 \ldots \varphi_n}{\varphi} \in R\}$

Proof trees are then Σ_P-ground terms. Therefore, basic proof transformation rules resulting from the semi-commutation property can be transformed into the following term rewriting system: with every proof tree transformation rule $i' \dfrac{\psi_1 \ldots i \, \dfrac{\varphi_1 \cdots \varphi_n}{p(t,t')} \ldots \psi_m}{p'(u,v)} \to \pi$, we associate the rewriting rule:

$$f_{i'}(x_{\psi_1}, \ldots, f_i(x_{\varphi_1}, \ldots, x_{\varphi_n}), \ldots, x_{\psi_m}) \to t_\pi$$

where x_{ψ_j} (resp. x_{φ_k}) with $1 \leq j \leq m$ (resp. $1 \leq k \leq n$) is a variable of sort s_{ψ_j} (resp. s_{φ_k}), and t_π is the Σ_S-term obtained from π where any leaf φ has been replaced in t_π by the variable x_φ of sort s_φ.

Then, E is appropriate for rewriting if the rewriting system defined just above is terminating. Let $>$ be the partial order defined on Σ_P as follows: $\forall f_i : s_{\varphi_1} \ldots s_{\varphi_n} \to s_\varphi \in Up_E, \forall f_{i'} : s_{\varphi'_1} \ldots s_{\varphi'_n} \to s_\varphi \in Down_E$, $f_i > f_{i'}$. Obviously, $>$ is well-founded. Therefore, the induced lexicographic path order $>_{lpo}$ is a simplification order on $T_{\Sigma_S}(V)$. As E is distributive, t_π is of the form $f_j(t_1, \ldots, t_l)$ with $j \in Down_E$. Thus, $f_{i'}(x_{\psi_1}, \ldots, f_i(x_{\varphi_1}, \ldots, x_{\varphi_n}), \ldots, x_{\psi_m}) >_{lpo} f_j(t_1, \ldots, t_l)$ results of the fact that $f_i > f_j$ and Point 1. of Definition 6.   □

Example 2. **The Partial Conditional Logic** (continuation) Let us show that the set E_Σ is appropriate for rewriting. By Definition 4, this requires to study the semi-commutation property between all instances of inference rules of $\mathcal{P}_\Sigma$, and show that the resulting set of transformation rules is terminating. First, we can notice that the partial conditional logic has a predicative formal system defined from a finite set of rule schemata. Then, defining $Up_{=_c}$ and $Down_{=_c}$ for any $=_c \in E_\Sigma$ only requires to prove the following result :

Proposition 8. Replacement, Substitution, Monotonicity *and* **Symmetry** *semi-commute with* **Transitivity** *and* **Modus-Ponens**, *but not the converse. Moreover, all resulting transformation rules satisfy Definition 6, Point 2..*

Proof [(Sketch)] For lack of space, we cannot give the entire proof, which can be found in [1]. Here, let's only detail (as an example) the semi-commutation property between monotonicity and modus-ponens. To help the reading, we use the standard notation $c \Rightarrow t = t'$ (resp. $c \Rightarrow D(t)$) rather than $t =_c t'$ (resp. $D_c(t)$). The symbol $\rightsquigarrow$ denotes the transformation of a proof tree into an equivalent one.

$$\text{Mono} \cfrac{\text{MP} \cfrac{c \wedge t_1 = t_2 \wedge c' \Rightarrow t = t' \quad c \wedge c' \Rightarrow t_1 = t_2}{c \wedge c' \Rightarrow t = t'}}{c \wedge c'' \wedge c' \Rightarrow t = t'} \rightsquigarrow$$

$$\text{MP} \cfrac{\text{Mono} \cfrac{c \wedge t_1 = t_2 \wedge c' \Rightarrow t = t'}{c \wedge t_1 = t_2 \wedge c'' \wedge c' \Rightarrow t = t'} \quad \text{Mono} \cfrac{c \wedge c' \Rightarrow t_1 = t_2}{c \wedge c'' \wedge c' \Rightarrow t_1 = t_2}}{c \wedge c'' \wedge c' \Rightarrow t = t'}$$

Notice that this is distributivity of monotonicity over modus-ponens.   □

Modus-Ponens semi-commutes with **Transitivity** when it is juxtaposed on the left premise of the **Modus-Ponens** rule. This property is no longer verified on its right premise. **Transitivity** does not semi-commute with any other rule.

Thus, for any $=_c \in E_\Sigma$, $Up_{=_c}$ contains all the instances of **Replacement, Substitution, Monotonicity** and **Symmetry**, and $Down_{=_c}$ contains all instances of **Transitivity** and **Modus-Ponens** (with conclusions of the form $t =_c t'$).

4.2   Convertibility Relations

Here, we define the basic (bidirectional) rewriting relations associated with a set E appropriate for rewriting. Each relation is obtained by closing a set of equations Γ under the instances of Up.

Definition 9. (Basic rewriting relations).

Let $\mathcal{P} = (T, P, R)$ be a predicative formal system equipped with a set E appropriate for rewriting. Let Γ be a set of formulae of $\mathcal{P}$. For any $p \in E$, $\leftrightarrow_\Gamma^p$ is the least binary relation on T (according to the set-theoretical inclusion) satisfying :
$t \leftrightarrow_\Gamma^p t'$
if and only if either $p(t, t') \in \Gamma$ or there is an instance $\frac{\varphi_1 \ldots \varphi_n}{p(t,t')} \in Up_p$ such that

- *for all premises $p'(u, v) \in \{\varphi_1 \ldots \varphi_n\}$ with $p' \in E$ we have $u \leftrightarrow_\Gamma^{p'} v$,*
- *and for all other premises φ, we have $\Gamma \vdash \varphi$.*

Let us note $\leftrightarrow_\Gamma$ the E-indexed family $\leftrightarrow_\Gamma = \{\leftrightarrow_\Gamma^p\}_{p \in E}$.

Be careful, $\leftrightarrow_\Gamma^p$ does not mean at all that this relation is symmetric. It only means that operational aspects are not considered, and then no direction of rewriting is imposed. To deal with operational aspects of rewriting, one needs to provide each binary relation p with a reduction relation $>_p$ (not necessarily transitive) and two rewrite relations $\rightarrow_p = p \cap >_p$ *and* $\leftarrow_p = p \cap (>_p)^{-1}$. When p is symmetric we have: $\leftarrow_p = (\rightarrow_p)^{-1}$. However, this last point is not the concern of the logicality result where rewriting is studied as a deduction mechanism.

Example 3. **The partial conditional logic** (continuation)
Given a set of conditional formulae Γ and a conjunction $c \in Conj_\Sigma$, $\leftrightarrow_\Gamma^{=_c}$ is the least binary relation on T satisfying the following clauses:

- if $t =_c t' \in \Gamma$, then $t \leftrightarrow_\Gamma^{=_c} t'$;
- if $t \leftrightarrow_\Gamma^{=_c} t'$ and $\sigma : V \rightarrow W_\Sigma(V)$ is a c-defined substitution w.r.t. Γ, then $\sigma(t) \leftrightarrow_\Gamma^{=_{\sigma(c)}} \sigma(t')$;
- if $t \leftrightarrow_\Gamma^{=_c} t'$ and C is a context such that $\Gamma \vdash D_c(C[t])$, then $C[t] \leftrightarrow_\Gamma^{=_c} C[t']$;
- if $t \leftrightarrow_\Gamma^{=_c} t'$ and $c' \in Conj_\Sigma$, then $t \leftrightarrow_\Gamma^{=_{c \wedge c'}} t'$.

A more concise definition of each $\leftrightarrow_\Gamma^{=_c}$ can be given. It is the result of the study of semi-commutation between the rule schemata representing the instances of $Up_{=_c}$. Indeed, it is easy to show that **Substitution** semi-commutes with **Replacement** and **Monotonicity** (substitutions are homomorphisms) and that **Replacement** semi-commutes with **Monotonicity**.
Thus, $t \leftrightarrow_\Gamma^{=_c} t'$ if and only if there exists a conditional equation $u =_{c'} v$ in Γ, a substitution σ, a conjunction $c'' \in \mathcal{C}_\Sigma$ and a context C such that:

1. $\forall x \in Var(u) \cup Var(v),\ \Gamma \vdash D_{c'}(\sigma(x))$,   2. $c = \sigma(c') \wedge c''$,

3. $\Gamma \vdash D_c(C[t])$,   4. $t = C[\sigma(u)]$ and $t' = C[\sigma(v)]$.

Definition 10. (Closure of rewriting steps). *With the notations of Definition 9, and given a rewriting relation $\leftrightarrow_\Gamma$, we define the E-indexed family $\overset{*}{\leftrightarrow}_\Gamma = \{\overset{*}{\underset{\Gamma}{\leftrightarrow}}{}^{p}\}_{p \in E}$ of binary relations on T such that, for any $p \in E$, $\overset{*}{\underset{\Gamma}{\leftrightarrow}}{}^{p}$ is the closure of $\overset{p}{\underset{\Gamma}{\leftrightarrow}}$ under instances of $Down_p$.*
In other words, $\overset{p}{\underset{\Gamma}{\leftrightarrow}} \subseteq \overset{*}{\underset{\Gamma}{\leftrightarrow}}{}^{p}$ and, for any instance $\frac{\varphi_1 \ \cdots \ \varphi_n}{p(t,t')} \in Down_p$ such that

– *for each premise of the form $p'(u,v)$ with $p' \in E$, we have $u \overset{*}{\underset{\Gamma}{\leftrightarrow}}{}^{p'} v$,*
– *and for the other premises φ_i, we have $\Gamma \vdash \varphi_i$,*

then $t \overset{*}{\underset{\Gamma}{\leftrightarrow}}{}^{p} t'$.

As already explained above, tautologies are usually recognised on their syntactical structure, and then proof trees denoting them are removed from the rewrite tree set. In this case, closure of rewriting steps are decomposed in two closures. The first is under proof trees of $Down_E$ (e.g. transitivity in equational logic). The second closure is under tautologies (e.g. all equations of the form $t = t$ in equational logic). This last closure can be defined in this generic framework as follow: $\overset{0}{\leftrightarrow}_\Gamma = \{(u,v) \mid \exists p \in E,\ \emptyset \vdash p(u,v)\}$.

From Definition 10, the following result holds:

Proposition 11. *$t \overset{*}{\underset{\Gamma}{\leftrightarrow}}{}^{p} t'$ iff there exists a rewrite tree $\pi : p(t,t')$ with all non-tautology leaves in Γ.*

Proof

($\Rightarrow$) We can even build this rewrite tree thanks to the definition of $\overset{*}{\underset{\Gamma}{\leftrightarrow}}{}^{p}$. We prove it by induction on the structure of $\overset{*}{\underset{\Gamma}{\leftrightarrow}}{}^{p}$:

 – If $t \overset{p}{\underset{\Gamma}{\leftrightarrow}} t'$, there is a proof tree $\pi' : p(t,t')$ belonging to $Up_p^\sharp$. Since all proof trees of $Up_p^\sharp$ are rewrite trees (cf. Definition 2), so is $\pi' : p(t,t')$.

 – If $t \overset{*}{\underset{\Gamma}{\leftrightarrow}}{}^{p} t'$, then, by Definition 10, there is a proof tree $\frac{\varphi_1,\dots,\varphi_n}{p(t,t')}$ in $Down_p$ such that, for all φ_i of the form $p'(u_i,v_i)$ ($p' \in E$), we have $u_i \overset{*}{\underset{\Gamma}{\leftrightarrow}}{}^{p'} v_i$ (less than k steps) and for the other φ_j, we have $\Gamma \vdash \varphi_j$. By induction hypothesis, for all φ_i there is a single rewrite tree $\pi_i : \varphi_i$, corresponding to $u_i \overset{*}{\underset{\Gamma}{\leftrightarrow}}{}^{p'} v_i$. Moreover, for all φ_j, there is a proof tree $\pi_j : \varphi_j$. By Definition 2, the tree obtained by replacing every φ_i by the rewrite tree $\pi_i : \varphi_i$ and all φ_j by the tree $\pi_j : \varphi_j$ is still a rewrite tree.

($\Leftarrow$) Three cases can occur :

- $\pi \in Up_p^\sharp$.

  Let φ be a leaf of π. By hypothesis, φ is either a tautology or an element of Γ. Therefore, if $\varphi = p'(u,v)$ with $p' \in E$, then $u \leftrightarrow_\Gamma^{p'} v$ (Definition 9). If φ is any other leaf, then either $\emptyset \vdash \varphi$ or $\Gamma \vdash \varphi$.

  According to Definition 9, and since $\pi \in Up_p^\sharp$, we have $t \leftrightarrow_\Gamma^p t'$ and thus $t \overset{*}{\leftrightarrow}_\Gamma^p t'$ (Definition 10).

- $\pi \in Down_p$.

  Let φ be a premise of π. Such as in the previous case, if $\varphi = p'(u,v)$ with $p' \in E$, then $u \leftrightarrow_\Gamma^{p'} v$, and thus $u \overset{*}{\leftrightarrow}_\Gamma^{p'} v$ (Definition 10). If φ is any other leaf, then $\Gamma \vdash \varphi$. Thus, from Definition 10, $t \overset{*}{\leftrightarrow}_\Gamma^p t'$.

- $\pi = (\pi_1 : \varphi_1, \ldots, \pi_n : \varphi_n, p(t,t'))$ such that:

  - the instance $\frac{\varphi_1 \cdots \varphi_n}{p(t,t')}$ belongs to $Down_p$,
  - $\forall i$ such that $\varphi_i = p_i(u,v)$ with $p_i \in E$, $\pi_i : \varphi_i$ is a rewrite tree.

  Since $\forall i$ $\pi_i : \varphi_i$ is a subtree of π, the non-tautology leaves of π_i also belong to Γ. Thus, by induction hypothesis, $\forall \pi_i : p_i(u,v)$ such that $p_i \in E$, we have $u \overset{*}{\leftrightarrow}_\Gamma^{p_i} v$. For the other premises φ_i, we have $\Gamma \vdash \varphi_i$. Therefore, according to Definition 10, $t \overset{*}{\leftrightarrow}_\Gamma^p t'$. $\qquad\square$

Theorem 12. (Logicality.) *Let $\mathcal{P} = (T, P, R)$ be a predicative formal system equipped with a set E appropriate for rewriting. Let Γ be a set of formulae. For any predicate $p \in E$ and any formula $p(t,t')$, we have:*

$$t \overset{*}{\leftrightarrow}_\Gamma^p t' \Leftrightarrow \Gamma \vdash p(t,t')$$

Proof Follows from Proposition 5 and Proposition 11. $\qquad\square$

5   Example: The Distributive Lattice Theory

Herein, we give the predicative formal system for distributive lattice [12]. This example of logic, unlike the previous one, allows us to study bidirectional rewriting on a binary predicate which is not symmetric.

Signatures for this theory are any set $\Sigma \cup \{\wedge; \vee; , ; (;)\}$ where Σ is a set of function names and $\wedge$ (resp. $\vee$) is the function join (resp. the function meet) to denote infimum (resp. supremum). We suppose here that every function of Σ is monotonic in all its variable positions, i.e. Σ is compatible (cf. [11])[2]. Let X be a set of variables. Then, T is the set of lattice terms free with generating set X in the class of all $\Sigma \cup \{\wedge; \vee\}$-algebras. Formulae are strings of the form $t \preccurlyeq t'$ or $t = t'$ where t and t' are lattice terms. Thus, P only contains the two binary relations $=$ and $\preccurlyeq$. Finally, R is the set of instances generated from the standard

---

[2] Without the monotonicity condition we must add, for each function $f \in \Sigma$, the two following rule schemata : **Leibniz1** $\dfrac{t_i = t_i' \quad f(t_1, \ldots, \gamma, \ldots, t_n) \preccurlyeq t}{f(t_1, \ldots, \gamma', \ldots, t_n) \preccurlyeq t}$

Leibniz2 $\dfrac{t_i = t_i' \quad t \preccurlyeq f(t_1, \ldots, \gamma, \ldots, t_n)}{t \preccurlyeq f(t_1, \ldots, \gamma', \ldots, t_n)}$
with $\gamma \neq \gamma' \in \{t_i, t_i'\}$

rule schemata which denote both that the predicate $=$ is a congruence and that $\wedge$ and $\vee$ are commutative and associative, in addition to the following rules:

Rest $\frac{t=t'}{t\preccurlyeq t'}$ **Antisym** $\frac{t\preccurlyeq t'\ \ t'\preccurlyeq t}{t=t'}$ **Trans** $\frac{t\preccurlyeq t'\ \ t'\preccurlyeq t''}{t\preccurlyeq t''}$ **Lb** $\frac{}{t\wedge t'\preccurlyeq t}$

Ub $\frac{}{t\preccurlyeq t\vee t'}$ **Glb** $\frac{t\preccurlyeq t'\ \ t\preccurlyeq t''}{t\preccurlyeq t'\wedge t''}$ **Lub** $\frac{t\preccurlyeq t''\ \ t'\preccurlyeq t''}{t\vee t'\preccurlyeq t''}$ **Dist** $\frac{t_1\preccurlyeq t_1'\vee t\ \ t_2\wedge t\preccurlyeq t_2'}{t_1\wedge t_2\preccurlyeq t_1'\vee t_2'}$

Subst $\frac{t\ p\ t'}{\sigma(t)\ p\ \sigma(t')}$ with $p\in\{=,\preccurlyeq\}$ and $\sigma:X\to T$

Repl$_\preccurlyeq$ $\frac{t_i\preccurlyeq t_i'}{f(t_1,\ ...,\ t_i,\ ...,\ t_n)\preccurlyeq f(t_1,\ ...,\ t_i',\ ...,\ t_n)}$

We will see that the set P is appropriate for rewriting. As previously stated, for the distributive lattice logic, both sets Up_P and $Down_P$ are defined from the semi-commutation of some inference rules with others. All this is established by the following result. Herein, we only give the statement for the rules describing the behaviour of the binary relation $\preccurlyeq$. The statement for the equality predicate is identical to Proposition 8. We can establish that:

Proposition 13. Subst, Glb, Lub, Dist *and* **Repl**$_\preccurlyeq$ *semi-commute with* **Trans**.

Antisym does not semi-commute with **Trans**. Thus, $Up_\preccurlyeq$ contains all the instances of **Subst, Glb, Lub, Rest, Dist** and **Repl**$_\preccurlyeq$ and $Down_\preccurlyeq$ contains all the instances of **Trans** ; $Up_=$ and $Down_=$ are defined as in the previous example (Section 4), except we add for $Down_=$ all the instances of the rule **Antisym**.
In the distributive lattice theory, given a set of formulae Γ, the relation $\leftrightarrow_\Gamma^\preccurlyeq$ is defined as the closure of Γ under the application of the rules **Subst, Glb, Lub, Rest, Dist** and **Repl**$_\preccurlyeq$ according to Definition 9.
Despite the tedious (but simple) proofs on semi-commutation of proof trees, our method easily allowed to build a bidirectional rewriting relation for the distributive lattice, satisfying the logicality property according to Theorem 12.

6   Conclusion

In this paper, we have given basic properties to characterise logics where rewriting can be used as a deduction mechanism. We have shown that rewriting is equivalent to deriving in the underlying calculus of the logic. This is called *logicality* and is based on a simple property of semi-commutation between inference rule instances. This naturally led to define basic transformation rules which allowed to transform any proof tree into a rewrite tree and then ensure the logicality result. In addition of the two examples developed in this paper, in [5] this generic framework has also been instantiated on the logic of special relations developed in [10]. This logics is interesting because it has been defined as an extension of the usual algebraic logic where it is possible to specify some generalisations of the Leibniz law between any binary relation (as transitivity or typing) and combining them.

An important research issue that we are investigating is to extend this approach in order to define, in a generic manner, the notion of rapid prototyping of property-oriented specifications by using rewriting techniques. In this scope, we are working on an adaptation of Knuth-Bendix completion in a generic way. This mainly requires to axiomatise the usual notions of terminating, Church-Rosser, confluence, and local-confluence, and retrieve standard theorems (mainly Lankford and Newman's theorems).

Acknowledgements. We especially thank Michaël Rusinowitch, Florent Jacquemard, Hélène Kirchner and Claude Kirchner for constructive comments on the idea developed in this paper. We also thank Jean-Louis Giavitto, Pascal Poizat and Sandrine Vial for careful readings of the draft version of the paper.

References

1. M.Aiguier, D.Bahrami and C.Dubois, *On the General Structure of Rewrite Proofs.* Technical report, University of Evry, 2001. ftp://ftp.lami.univ-evry.fr/pub/publications/reports/2001/index.html/lami_58.ps.gz.
2. E.Astesiano and M.Cerioli, *Free Objects and Equational Deduction for Partial Conditional Specifications.* TCS, 152(1):91-138. Amsterdam: Elsevier, 1995.
3. F.Baader and T.Nipkow, *Term Rewriting and All That.* Cambridge University Press, 1998.
4. L.Bachmair and H.Ganzinger, *Rewrite techniques for transitive relations.* 9th IEEE Symposium on Logic in Computer Science, pp. 384-393, 1994.
5. F.Barbier, *Méta-réécriture : application à la logique des relations spéciales.* Master thesis, University of Evry, 2001. Supervised by M. Aiguier and D. Bahrami (In french), avalaible at http://www.lami.univ-evry.fr/~fbarbier/recherche-fr.html
6. S.Kaplan, *Simplifying Conditional Term Rewriting Systems: Unification, Simplification and Confluence.* Journal of Symbolic Computation, 4(3):295-334. Amsterdam: Academic Press, 1987.
7. J.Levy and J.Agusti, *Bi-rewriting systems.* Journal of Symbolic Computation, 22(3):279-314. Amsterdam: Academic Press, 1996.
8. J.Meseguer, *Conditional rewriting logic as a unified model of concurrency.* TCS, 96(1):73-155. Amsterdam: Elsevier, 1992.
9. V.van Oostrom, *Sub-Birkhoff.* Draft, 13 pages, 18 December 2000, available at www.phil.uu.nl/~oostrom/publication/rewriting.html.
10. M.Schorlemmer, *On Specifying and Reasoning with Special Relations.* PhD thesis, Institut d'Investigaciò en Intel.ligència Artificial, University of Catalunya, 1999.
11. G.Struth, *Knuth-Bendix Completion for Non-Symmetric Transitive Relations.* Proceedings of the Second International Workshop on Rule-Based Programming (RULE2001), Electronic Notes in TCS, 59(4). Elsevier 2001.
12. G.Struth, *Canonical Transformations in Algebra, Universal Algebra and Logic.* PhD thesis, Institut für Informatik, University of Saarland, 1998.
13. T.Yamada, J.Avenhaus, C.Loría-Sáenz, and A.Middeldorp, *Logicality of Conditional Rewrite Systems.* TCS, 236(1,2):209-232. Amsterdam: Elsevier, 2000.

Using Symbolic Computation in an Automated Sequent Derivation System for Multi-valued Logic

Elena Smirnova

Ontario Research Center for Computer Algebra
University of Western Ontario
London, Ontario, CANADA N6A 5B7
alena@orcca.on.ca

Abstract. This paper presents a way in which symbolic computation can be used in automated theorem provers and specially in a system for automated sequent derivation in multi-valued logic. As an example of multi-valued logic, an extension of Post's Logic with linear order is considered. The basic ideas and main algorithms used in this system are presented. One of the important parts of the derivation algorithm is a method designed to recognize axioms of a given logic. This algorithm uses a symbolic computation method for establishing the solvability of systems of linear inequalities of special type. It will be shown that the algorithm has polynomial cost.

1 Automated Theorem Provers and Automated Sequent Derivation Systems

The main goal of any automated theorem prover is to derive some statement from existing data or expressions, whose certainty is established because it is an axiom or because it is proved in one of the previous steps. Sometimes one wishes to verify whether a given statement can be derived from existing facts by using specified rules. In this case we can construct a logic containing these rules as rules of its sequent calculus, then present an input statement in the form of a sequent of this logic, and then try to find its derivation by using backward deduction.

The system to be presented here is designed to construct the backward deduction of an input statement, given in the form of a sequent of classical or multi-valued logic.

One of the most popular multi-valued logics is Post's logic [1]. We will consider Post's logic, extended by comparisons of logical values of predicate formulas, that allows us to compare the certainty of several facts or expression. This logic is able to replace fuzzy and continuous logics in practice, since only finite sets of rational numbers from a finite diapason are used. The calculus of these logics may be applied to computer representation of knowledge. There

J. Calmet et al. (Eds.): AISC-Calculemus 2002, LNAI 2385, pp. 64–75, 2002.

may be another approach consisting in using even-valued Post's logics for natural sciences and odd-valued Post's logics (which include a paradoxical value) for humanitarian and applied sciences [2] and [3].

2   Sequent Derivation in Post's Logics Extended by Logical Comparisons

2.1   Logical Values for Data Representation

For graded knowledge representation, data in this logic are represented by integer or rational numbers from a given interval. Fix a natural number k and consider one of two sets, either $L_Z = [-k, k] \cap \mathbb{Z}$ or $L_Q = [-k, k] \cap \mathbb{Q}$ as the set of logical values for our logic.

All logical values that are greater than zero are considered as true. All logical values less then zero, are false. When L contains zero, the logical value that is equal to zero is named *paradoxical* (neither true or false).

Even-valued Post's logics, applied for natural sciences, cannot contain the logical paradoxical value. For this case we take a segment $[-k, k] \setminus 0$ and consider one of sets $L_{Z \setminus \{0\}} = [-k, k] \cap \mathbb{Z} \setminus \{0\}$ or $L_{Q \setminus \{0\}} = [-k, k] \cap \mathbb{Q} \setminus \{0\}$

For the general case, the set of logical values is noted simply as L.

2.2   Predicate Formulas and Sequents for the Representation of Statements

Predicate formulas of this logic are used to represent extended expressions, combining given facts or data by using negation, conjunction, disjunction, four-place connective *if B < A then C else D fi* (called conditional expression), and quantifiers.

The logical value of conjunction and disjunction are defined by the minimum and maximum of logical values of their arguments respectively. The operation of logical negation corresponds to arithmetic negation, i.e., to unary minus. The conditional expression is interpreted as in programming languages. Let the comparison sign $\prec$ be one of the following types: $\leq, <$. A comparison sign is placed between predicate formulas. For example $p(x) \leq q(y)$. Such a formula is called a *logical comparison*, or more briefly comparison. We consider also the relation of equality: $p = q$. Logical values of comparisons are defined in the usual way (i.e. in the classical two-valued logic).

A *sequent* is a chain of predicate formulas (called the members of the sequent), beginning with the sign "$\rightarrow$" and with the members separated by commas. The meaning of a sequent is a disjunction of its members. The formula representing this disjunction is called an *image formula of a sequent*. For example, the image formula of the sequent $\rightarrow F_1, F_2, \ldots F_n$ is $F_1 \vee F_2 \vee \ldots \vee F_n$.

The simplest sequents this logic are sequents containing only *elementary logical comparisons*, i.e. formulas having the form $(a < b)$ or $(a \leq b)$, where a and b are constants, predicate variables, or predicates variables with negation. This type of sequent is called an elementary sequent. An example of an elementary sequent is $\rightarrow (a < b), (p(x) \leq q(x)), (c \leq d), (q(z) \leq 1)$.

2.3 Sequent Derivation, Theorems, and Axioms

A *derivation* in a logic is a consequence of sequents, each of which represents an axiom or is obtained from preceding sequents by applying one of the derivation's rules of this logic.

A sequent is *derivable* in a logic if it is the last sequent of some derivation (i.e. it is possible to construct its derivation in this logic). A sequent derivable in a given logic is also called a *theorem* of this logic.

A formula A, that does not contain any logical comparisons, is derivable in our logic, if it is possible to derive the sequent $\rightarrow (0 \leq A)$.

To recognize whether a given sequent is a theorem of our logic, we have to find its derivation (i.e. after applying all possible rules, we must obtain axioms only).

A sequent is an *axiom* of this logic if the image formula of its maximal elementary subsequent is identical true.

In order to recognize whether a given sequent Γ is an axiom of a considered logic we must first build its elementary subsequent Γ^* by erasing all members containing logical binary connectives, external negations and equality signs, then consider the image formula F_{Γ^*} of an obtained elementary sequent. If a formula F_{Γ^*} is true for any values of its variables, then input sequent Γ is an axiom.

3 Axiom Recognition Algorithm

One of the main problems during automated sequent derivation is axiom recognizing. If in step n of a derivation process the system obtains a sequent, that is an axiom, it must not continue deduction of this sequent. Otherwise if it obtains a sequent, containing only atomic formulas (meaning there is no rule to apply in the next step) and if this sequent is not an axiom then the derivation process can be stopped, because in this case the initial sequent can be derived in a given logic.

This section presents an algorithm for axiom recognition in extended Post's logic.

Let Γ be a sequent of a Post's logic extended with logical comparisons. In the general case that sequent has the form

$$\rightarrow ((a_1^1 < a_1^2), \ldots, (a_k^1 < a_k^2), (a_{k+1}^1 \leq a_{k+1}^2), \ldots, (a_m^1 \leq a_m^2)), A_1, A_2, \ldots A_t$$
$$(1)$$

where $k, m, t \geq 0$, a_i^j for $i \in \{1, 2, \ldots m\}$, $j \in \{1, 2\}$ is an atomic formula or the negation of an atomic formula, and each of A_i is different from elementary comparison. Then the maximal elementary subsequent Γ^* of Γ has the following form

$$\rightarrow ((a_1^1 < a_1^2), \ldots, (a_k^1 < a_k^2), (a_{k+1}^1 \leq a_{k+1}^2), \ldots, (a_m^1 \leq a_m^2)) \qquad (2)$$

The formula-image of this elementary sequent Γ^* is

$$F_{\Gamma^*} = ((a_1^1 < a_1^2) \vee \ldots \vee (a_k^1 < a_k^2) \vee (a_{k+1}^1 \leq a_{k+1}^2) \vee \ldots \vee (a_m^1 \leq a_m^2)) \quad (3)$$

As defined above, to establish that the sequent Γ is an axiom it is necessary to prove that the formula F_{Γ^*} is the identically truth. We suggest considering an equivalent statement: Γ is an axiom if the negation of F_{Γ^*} is identically false. It means that the logical value of $\neg F_{\Gamma^*}$ is false for any values of the variables, contained in this formula.

Let us construct the negation of F_{Γ^*}:

$$\neg F_{\Gamma^*} = \neg((a_1^1 < a_1^2) \vee \ldots \vee (a_k^1 < a_k^2) \vee (a_{k+1}^1 \leq a_{k+1}^2) \vee \ldots \vee (a_m^1 \leq a_m^2)) =$$

$$\neg(a_1^1 < a_1 2)\& \ldots \&\neg(a_k^1 < a_k^2)\&\neg(a_{k+1}^1 \leq a_{k+1}^2)\& \ldots \&(a_m^1 \leq a_m^2) = \quad (4)$$

$$(a_1^2 \leq a_1^1)\& \ldots \&(a_k^2 \leq a_k^1)\&(a_{k+1}^2 < a_{k+1}^1)\& \ldots \&(a_m^2 < a_m^1)$$

Now we must show that this formula cannot be satisfied for any values of the variables $\{a_i^j\}_{1 \leq i \leq m, 1 \leq j \leq 2}$ from the set of logical values L. Or in other words, the system of linear inequalities

$$(5) \quad \begin{cases} a_1^2 \leq a_1^1 \\ \vdots \\ a_k^2 \leq a_k^1 \\ a_{k+1}^2 < a_{k+1}^1 \\ \vdots \\ a_m^2 < a_m^1 \end{cases}$$

has no solution in the set L.

Theorem 1. *The input sequent (1) is an axiom iff the system (5) is unsolvable in the set L.*

The proof of this theorem is evident from the preceding reasoning.

Example 1. The sequent $\rightarrow (x \leq y), (w < z), (y < x)$ is an axiom in extended Post's logic with the set of logical values $L = [-100, 100]$, because the system

$$\begin{cases} y < x \\ z \leq w \\ x \leq y \end{cases}$$

has no solution in this set.

Example 2. The sequent $\rightarrow (-x < y), (x < -x), (y < x)$ is an axiom in extended Post's logic with the set of logical values $L = [-5, 5] \cap \mathbb{Z} \setminus \{0\}$, because the system

$$\begin{cases} y \leq -x \\ -x \leq x \\ x \leq y \end{cases}$$

has no solution in integer numbers from $[-5, 5] \setminus \{0\}$. But this sequent is not an axiom of extended Post's logic with the set of logical values $L = [-5, 5] \cap \mathbb{Z}$, because this system has the solution $x = y = 0$.

Thus the problem of axiom recognition for extended Post's logic reduces to the problem of establishing the insolubility of a system of linear inequalities. To solve this problem we suggest using a symbolic computing method and an algorithm based on this method these are described in the following section.

4   Method for Solvability Testing of Linear Inequalities System of Special Type

Let us consider a system of linear inequalities S_0

$$
\begin{cases}
x_1^1 \prec^1 x_1^2 \\
x_2^1 \prec^2 x_2^2 \\
\vdots \\
x_m^1 \prec^m x_m^2
\end{cases}
$$

where here each x_i^j is either a constant, a variable or the opposite of a variable, and $\prec^i \in \{<, \leq\}$ for $1 \leq i \leq m, 1 \leq j \leq 2$.

The problem to solve is to test the solvability of this system in the given set L.

As we explore solvability of system in the set $L \subset [-k, k]$, in order to respect variables values limitation, for each variable x we add inequalities : $-k \leq x \leq k$ and after that for each inequality $x \prec y$ we add the opposite inequality $-y \prec -x$ (where $\prec$ is the sign of strict or weak inequality). Then we process a new "symmetric" system S, equivalent to input system S_0.

$$
\begin{cases}
x_1^1 \prec^1 x_1^2 \\
-x_1^2 \prec^1 -x_1^1 \\
\vdots \\
x_i^1 \prec^2 x_i^2 \\
-x_i^2 \prec^2 -x_i^1 \\
\vdots \\
x_n^1 \prec^n x_n^2 \\
-x_n^2 \prec^n -x_n^1,
\end{cases}
$$

where here each x_i^j is either a constant, a variable or the opposite of a variable, and $\prec^i \in \{<, \leq\}$ for $1 \leq i \leq n, 1 \leq j \leq 2$.

The (un)solvability of system (7) is establishing according to the following theorem:

Theorem 2. *The system S, having the form (7), is unsolvable iff at least one of the following conditions is fulfilled.*

1. *The system S contains a cycle $x_1 \prec^1 x_2 \prec^2 x_3 \prec^3 \ldots \prec^{m-1} x_m \prec^m x_1$ for some $m \geq 2$ and at least one of inequalities $\prec^i$ is strict.*
2. *The system S contains a cycle $x_1 \leq x_2 \leq x_3 \leq \ldots \leq x_m \leq x_1$ for some $m \geq 2$ and*

     1) this cycle contains two different constants c_1 and c_2

     2) or among the x_i there are a non zero constant and a opposite pair $\{-x, x\}$, where x is a variable

     3) or this cycle does not contain any constant different from zero, but it contains a opposite pair $\{-x, x\}$ or the constant zero and

          a. the set of logical values L does not contain zero

          b. or zero belongs to the set of logical values L and the system S_1 which is obtained from S by replacing all variables from that cycle by zero is insolvable

     4) or this cycle does not contain any opposite pairs, but it contains a unique constant c, different from zero, and the system S_2, obtained from S by replacing all variables from that cycle by this constant c is insolvable

     5) or there no constant in this cycle neither opposite pair and the system S_3 which follows is insolvable; S_3 obtained from S by replacing all variables from that cycle by a new variable x_{new} and all opposite them variables by $-x_{new}$

3. *The system S has no cycles but contains a chain $x_1 \prec x_2 \prec x_3 \prec \ldots \prec x_m$ where x_1 and x_m and only them are constants and $x_m < x_1$*

   The following condition is considered only if $L \subseteq \mathbb{Z}$

4. *The system S has no cycles but contains a chain $x_1 \prec^1 x_2 \prec^2 x_3 \prec^3 \ldots \prec^{m-1} x_m$ where x_i are pairwise distinct, x_1 and x_m and only them are constants and the length of this chain measured as the number of signs '¡' strictly greater than following value*

     1) $x_m - x_1 - 1$ if $0 \notin L$ and $x_m x_1 < 0$

     2) $x_m < x_1$ in other cases.

Notice that the conditions 2.3.b, 2.4 and 2.5 let us reduce the initial system S and iterating the test with the smaller systems S_1, S_2 or S_3 gives a terminating process.

4.1   Algorithm for Solvability Testing

The algorithm for solvability testing is based on verification of the Theorem 2 conditions and lies to encoding of the initial system (7) into a finite oriented graph.

    For verify whether conditions of the Theorem 2 are satisfied , we suggest constructing a graph G, associated with the system (7). The graph G is built as follows. Let V is a set of (signed) variables, appearing in S, and C be the set of constants appearing in the system (7). The set of vertices of graph G is $V \cup C$. Edges of G are labelled by elements in $\{<, \leq\}$. The triple $(x, \prec, y)$ is an edge in G iff $x \prec y$ is an inequality of the system (7).

Example. Let the system (7) is

$$\begin{cases} 0 \le x \\ x < y \\ 2 \le -y \\ x \le 2 \\ -2 \le x \\ y < 2 \end{cases}$$

then corresponding graph G is

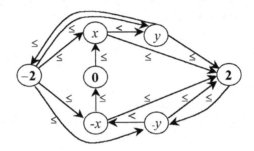

Therefore the conditions of the Theorem 2 may be represented like graph with following patterns

Pattern 1: The graph G contains a cycle with the edge, marked by sign of strict inequality

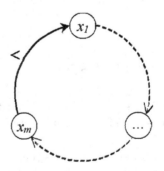

Example: $x \le y \le z < x.$

Pattern 2.1: The graph G contains a cycle with two different constants

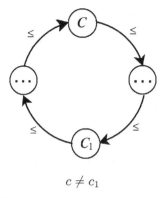

$$c \neq c_1$$

Example: $x \leq 3 \leq y \leq 4 \leq x$

Pattern 2.2: The graph G contains a nonstrict cycle with opposite pair $\{-x, x\}$ and non-zero constant

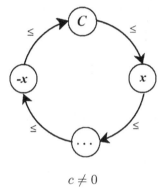

$$c \neq 0$$

Example: $-x \leq 5 \leq x \leq -x$

Pattern 2.3: The graph G contains a nonstrict cycle with opposite pair $\{-x, x\}$ or constant zero

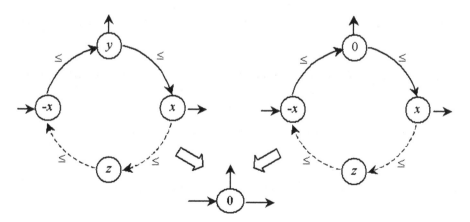

Pattern 2.4: The graph G contains a nonstrict cycle with a unique constant

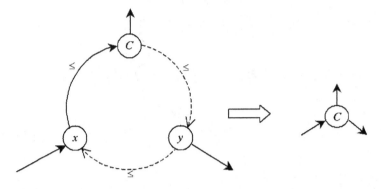

Pattern 2.5: The graph G contains a nonstrict cycle without any opposite pairs or constants

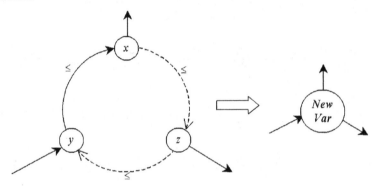

Pattern 3: The graph G contains a path between constants c and c_1 and $c_1 > c$

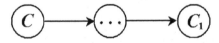

Example: $8 \leq x < y < 3$

Pattern 4 (considered only in case in which $L \subseteq \mathbb{Z}$): the graph G contains a path between constants c and c_1, let r is a number of edges in this path, marked with sign '<' and $c_1 - c < r - 1$, if $c_1 c < 0$ and $0 \notin L$ or $c_1 - c < r$ otherwise

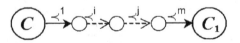

Example: $1 < x < y < 3$

An implementation of solvability test algorithm uses strongly connected components [4] instead of cycles for a better efficiency.

Algorithm works as follow

- First compute the strongly connected components of graph G. If some component satisfies condition 1, 2.1, or 2.2.a of Theorem 2 we are done.
- Otherwise, apply to each component satisfying condition 2.2.b, 2.3 or 2.4 the corresponding modification until we get a system S' which associated graph G' is acyclic.
- Test whether G' satisfies condition 3, or condition 4 if we are in the integer-values case.
- If no condition is satisfied (graph G does not contain any considered pattern) algorithm gives an answer that the input (7) has at least one solution, because the conditions of the Theorem 2 are necessary and sufficient.

4.2 Complexity of Solvability Testing Algorithm

Now we can show that solvability of system (7) can be establish in polynomial time.

Theorem 3 *(Un)solvability of a given system (7) with m inequalities and n variables and constants, is computable in time $O(mn)$*

Proof. Steps 1 and 2 use an algorithm for searching strongly connected component [4] are computed in time $O(m)$. Step 3 uses an algorithm for minimal cost covering tree [5], with is computed in time $O(mn)$.

4.3 Case When the Set of Logical Values Is Not Symmetric Around Zero

Sometimes one wish to deal with a set of logical values, than is not symmetric around zero. For example an interval [0,1] is used in many of multi-valued logics.

In the general case the set of logical values $L = [a, b] \cap \mathbb{Q}(\mathbb{Z})$, where $a, b \in \mathbb{Z}$.

To recognize axioms in such logics we suggest using the same algorithm for system insolvability testing, as for symmetric the case. But the input system S and the set of logical values L need a transformation to make the input conform to the requirement of the previous case.

The following transformation suffices

1. The new set of logical values is $L' = [-k, k] \cap \mathbb{Q}(\mathbb{Z})$, where

$$k = \begin{cases} \dfrac{b-a}{2}, & \text{if } L = [a, b] \cap \mathbb{Q} \\ \left[\dfrac{b-a+1}{2} \right], & \text{if } L = [a, b] \cap \mathbb{Z} \end{cases}$$

2. The input system S is transforming using the following formula :

$$x' = x - \delta(a, b, x)$$

where x is a variable or constant of the initial system S, x' is a member of the new system S' corresponding to x and the shift function is defined as

$$\delta(a,b,x) = \begin{cases} \dfrac{(a+b)}{2}, & \text{if } L = [a,b] \cap \mathbb{Q} \text{ or } L = [a,b] \cap \mathbb{Z} \text{ and } 2|(b-a) \\ \dfrac{a+b}{2} - \dfrac{1}{2} \cdot \operatorname{sgn}\left(x - \dfrac{a+b}{2}\right), & \text{if } L = [a,b] \cap \mathbb{Z} \text{ and } 2 \nmid (b-a) \end{cases}$$

In the case in which logical values are rationals or integers from an interval $[a,b]$, when the difference $(b-a)$ is even, we just make a parallel shift of the system, but in case of integer values when $(b-a)$ is odd, we must perform a non-parallel linear transformation, as shown in Figure 1.

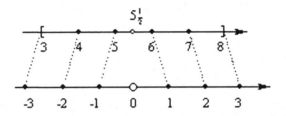

Fig. 1. Transformation of nonzero symmetric set $[3,8] \cap \mathbb{Z}$ to symmetric set $[-3,3] \setminus \{0\} \cap \mathbb{Z}$

5   Applications

The logic considered here, proposed by Kossovski [2], is only one example of a multi-valued first-order logics with a finite number of values. Such logics are included in the database of the automatic theorem prover for automated sequent derivation in classical and non-classical logics.

The goal of this automated system is to give a general idea about Logics and Artificial Intelligence. The prover is intended to teach students how to construct derivation in non-classical logic and also it can be used as an application for research in multi-valued logic. In comparison with other automatic theorem provers, for example Deep Thought [6], the present system does not require the input of huge tables for the valuation function. For instance, to describe a propositional multi-valued logic with five values $\{-2, -1, 0, 1, 2\}$ and only three logical connectives $\{\neg, \&, \vee\}$, a user of Deep Thought must enter 3 tables: one $[5 \times 1]$ for negation and two $[5 \times 5]$ for conjunction and disjunction.

Using the system considered here, a user has to choose the kind of logic to be used and to enter the set of logical values, for example L=[-5,5]. Afterwards the system will use its internal database of inference rules and axiom recognition algorithms. The main ideas used for the design of this prover correspond to the approach described in [7]. Taking advantage of the special native of the system of linear inequalities, used for axiom recognition in the present logic, we were able to create a particular algorithm for solvability testing, rather then having to the use simplex [8] or ellipsoid methods [9]. Demo version of this automatic sequent derivation system is available in the Internet at `www.orcca.on.ca/MathML/Elena/software`.

Acknowledgements. Ideas of considered logics and of data representation in it are due to Dr. N.K. Kossovski. First formulation of Theorem 2 was proposed by A.Tishkov. Proof of the Theorem 2 and 3 and also many ideas of implementation of main algorithm of this system are due to Dr. Danièle Beauquier. Special thanks to Dr. D.J. Jeffrey and Dr. Robert M. Corless for their kindness, support and great help during preparation this paper.

References

[1] A.S. Karpenko, Multivalued logics. Logic and Computer. Issue 4, Moscow, Nauka, 1997.

[2] Kossovski N.K., Tishkov A.V. Logical theory of Post Logic with linear order. TR-98-11, Department of Informatics University Paris-12. 9p.

[3] N.Kossovski, A. Tishkov. Gradable Logical Values For Knowledge Representation. Notes of Scientific Seminars vol. 241 pp 135–149 St.Petersburg University 1996.

[4] R. Tarjan, Depth First Search and Linear Graph Algorithms, SIAM J. Computing 1 (1972), 146–160

[5] R. Tarjan, Data Structures and Network Algorithms, SIAM Philadelphia,(1983) 71–95

[6] Gerberding S. Deep Thought. University of Darmstadt, Dept. of Computer Science, 1996.

[7] Lowerence C.Paulson. Designing a theorem Prover. Oxford, 1995. pp. 416–476.

[8] Michael J.Panik. Linear Programming: Mathematics, Theory and Algorithms. Kluwer Academic Publishers, 1996. pp.125–139.

[9] L.G. Hachiyan, A polinomial algorithm in linear programming, Soviet Mathematics Doclady 20,(1979), pp. 191–194.

[10] Jorge Nocedal, Stephen J. Wright. Numerical Optimization, Springer, 1999.

The Wright ω Function

Robert M. Corless and D.J. Jeffrey

Ontario Research Centre for Computer Algebra
and the Department of Applied Mathematics
University of Western Ontario
London, CANADA
{Rob.Corless,David.Jeffrey}@uwo.ca

Abstract. This paper defines the Wright ω function, and presents some of its properties. As well as being of intrinsic mathematical interest, the function has a specific interest in the context of symbolic computation and automatic reasoning with nonstandard functions. In particular, although Wright ω is a cognate of the Lambert W function, it presents a different model for handling the branches and multiple values that make the properties of W difficult to work with. By choosing a form for the function that has fewer discontinuities (and numerical difficulties), we make reasoning about expressions containing such functions easier. A final point of interest is that some of the techniques used to establish the mathematical properties can themselves potentially be automated, as was discussed in a paper presented at AISC Madrid [3].

1   Notation and Definitions

The Wright ω function is a single-valued function, defined in terms of the Lambert W function. Lambert W satisfies $W(z)\exp(W(z)) = z$, and has an infinite number of branches, denoted $W_k(z)$, for $k \in \mathbb{Z}$. See [4] for a discussion of why the branches were chosen as they are. The Lambert W function is therefore multivalued. The Wright ω function[1] is a single-valued function, defined as follows:

$$\omega(z) = W_{\mathcal{K}(z)}\left(e^z\right) \tag{1}$$

where $\mathcal{K}(z) = \lceil (\operatorname{Im}(z) - \pi)/(2\pi) \rceil$ is the *unwinding number* of z. Note that the sign of this unwinding number is such that $\ln(\exp(z)) = z + 2\pi i\mathcal{K}(z)$, which is opposite to the sign used in [5], because we discovered after that publication that the present sign choice leads to fewer minus signs in formulas.

---

[1] This nomenclature has never, to our knowledge, appeared in print before. We use the letter ω as a cognate of W, and we name this function after Sir Edward M. Wright, for his works [12] establishing the complex branching behaviour of this function as a tool for investigating the roots of $y\exp(y) = z$ (later called the Lambert W function)

J. Calmet et al. (Eds.): AISC-Calculemus 2002, LNAI 2385, pp. 76–89, 2002.
© Springer-Verlag Berlin Heidelberg 2002

2   Graphs and Special Values

A graph of $\omega(z)$ for real z can be produced easily in Maple by the command
`plot([y+ln(y),y,y=0.001..2]);`. A section of the Riemann surface for $\omega(z)$
can be plotted by the following commands:

```
omega := mu + I*nu;
x := evalc(Re(omega+ln(omega)));
y := evalc(Im(omega+ln(omega)));
plot3d( [x,y,mu], mu=-4..2, nu=-4..4,
        colour=black, axes=BOXED,
        style=PATCHNOGRID, labels=["x","y","mu"],
        view=[-2..1, -5..5, -5..3],
        grid=[200,200], style=POINT );
```

See Table 1 for special values.

Table 1. Special values of $\omega(z)$.

z	$\omega(z)$
$-\infty$	0
0	$W_0(1)$
1	1
$2 + \ln 2$	2
$-1/3 + \ln(1/3) + i\pi$	$-1/3 = W_0\left(-\frac{1}{3}e^{-1/3}\right)$
$-1/3 + \ln(1/3) - i\pi$	$W_{-1}\left(-\frac{1}{3}e^{-1/3}\right)$
$-1 + i\pi$	-1
$-1 - i\pi$	-1
$-2 + \ln 2 + i\pi$	$W_0\left(-2e^{-2}\right)$
$-2 + \ln 2 - i\pi$	$-2 = W_{-1}\left(-2e^{-2}\right)$
∞	∞

2.1   Summary of Results of This Note

The main result is a clarification, using this new function, of results due originally
to Wright [12] and independently rediscovered in [11] and [9]. Although $y = \omega(z)$
satisfies the equation (in this paper $\ln(z)$ is the principal branch of the logarithm
of z)

$$y + \ln y = z , \qquad (2)$$

when $z \neq t \pm i\pi$ for $t \leq -1$, there should be a distinction made between the
solutions of the equation, and Wright ω. In other words, (2) is not a satisfactory
definition of ω.

In addition to this basic point, we here present new branch point series (with the correct closure), new asymptotic series (from the equivalent series for the Lambert W function), and new proofs of the analytic properties of $\omega(z)$, using properties of the unwinding number.

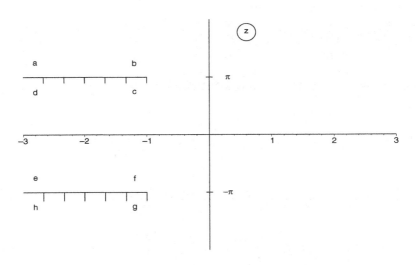

Fig. 1. The z-plane, showing the slit (equivalently, branch cut) we call the "doubling line" (above) and its "reflection", across each of which the Wright ω function is discontinuous. Along both slits, the closure (indicated by short lines extending down from the slits) is taken from below—clockwise around the branch points—to agree with the closure of the unwinding number.

We here summarize some properties of ω, proved in [9]. First, equation (2) has a unique solution, $\omega(z)$, for all $z \in \mathbb{C}$ except on the line L_D defined by $z = t \pm i\pi$ for $t \le -1$. When z is on L_D, the equation has precisely two solutions, these being $\omega(z)$ and $\omega(z - 2\pi i)$; we therefore call L_D the "doubling line". See Figure 1 and Figure 2. On the reflection of the doubling line, namely, the line defined by $z = t - i\pi$, with $t \le -1$, equation (2) has no solution at all[2]. Second, ω is an analytic function of z except on the doubling line *and its reflection*; on these two lines, $\omega(z)$ is discontinuous. This immediately gives the following.

---

[2] Unfortunately, in the paper [6], we got this wrong—we missed the fact that there was no solution on this line. Indeed, at that time, we hadn't realized this function is discontinuous there. Additionally, we were using the opposite sign for the unwinding number, which made the formulas messier.

Theorem: For all $z \in \mathbb{C}$ and integers k,

$$W_k(z) = \omega(\ln_k(z)), \tag{3}$$

where $\ln_k(z) = \ln z + 2\pi i k$. [This logarithmic notation is discussed further in a later section.]

Proof. This holds at least provided z is not in the interval $-\exp(-1) \le z < 0$ and $k = -1$, which is the image in the domain of W of the critical doubling line (and also the image of its reflection). If z is in the interval $-\exp(-1) \le z < 0$, and $k = -1$, then we have instead that $W_0(z) = \omega(\ln|z| + i\pi)$ since $\mathcal{K}(\ln|z| + i\pi) = 0$, and that $W_{-1}(z) = \omega(\ln|z| - i\pi)$ since $\mathcal{K}(\ln|z| - i\pi) = -1$. Phrasing this the other way, we have

$$W_0(z) = \omega(\ln z)$$
$$\text{and}$$
$$W_{-1}(z) = \omega(\ln z - 2\pi i).$$

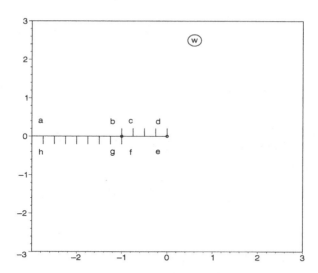

Fig. 2. The ω-plane, showing the images of doubling slit and its reflection. The negative real ω-axis is not, per se, a branch cut (this is the range of the function) but it is a branch cut of $\omega + \ln \omega$, which is why that expression is not exactly the inverse function for ω.

2.2   Properties of ω

We group the properties into analytic properties and algebraic properties.

Analytic properties. Theorems and lemmas:

(i) $\omega(z)$ is single-valued

(ii) $\omega : \mathbb{C} \to \mathbb{C}$ is onto $\mathbb{C} \setminus \{0\}$.

(ii)(a) Except at $z = -1 \pm i\pi$, where $\omega(z) = -1$, $\omega : \mathbb{C} \to \mathbb{C}$ is injective; hence ω^{-1} exists uniquely except at 0 and -1.

(iii)   See Figure 2.

$$\omega^{-1}(y) = \begin{cases} y + \ln(y) - 2\pi i & -\infty < y < -1 \\ -1 \pm i\pi & y = -1 \\ y + \ln(y) & \text{otherwise.} \end{cases}$$

(iv)   (a) ω is continuous (in fact analytic) except at $z = t \pm i\pi$ for $t \leq -1$.

(b) For $z = t \pm i\pi$ and $t \leq -1$, we have

(1) $\omega(t + i\pi^-) = \omega(t + i\pi) = \omega(t - i\pi^-)$

(2) $\omega(t + i\pi^+) = \omega(t - i\pi) = \omega(t - i\pi^+)$

(v)   (a) $\omega + \ln \omega = z \iff \mathcal{K}(\omega + \ln \omega) = \mathcal{K}(z)$.

(b) $\mathcal{K}(\omega + \ln \omega) = \mathcal{K}(z)$ unless $z = t - i\pi$, $t \leq -1$.

(vi)   If $z \neq t - i\pi$ for $t \leq -1$, then $\omega(z) + \ln \omega(z) = z$.

If moreover $z \neq t + i\pi$, $t \leq -1$, then this solution is unique; if $z = t + i\pi$, $t \leq -1$, then $y = \omega(t - i\pi)$ is also satisfies $y + \ln y = z$. There is no y such that $y + \ln y = z$ if $z = t - i\pi$.

2.3   Proofs.

(i)   The functions $\exp z$, $\mathcal{K}(z)$ and $W_k(z)$ (for each fixed k) are single-valued. Hence the composition $W_{\mathcal{K}(z)}(\exp z)$ is single-valued.

(ii) $\mathcal{K}(z)$ covers all of $\mathbb{Z}$ as z covers all of $\mathbb{C}$, and the branches of W partition the plane, except that -1 is hit twice: $W_{-1}(-1/e) = W_0(-1/e) = -1$. Only $W_0(0) = 0$, and 0 is the only point not in the range of e^z; hence there is no finite z such that $\omega(z) = 0$, but no other points in the range are missed.

(iii) (v) $\implies$ (iii) because $\omega e^\omega = e^z$, and hence $\ln(\omega e^\omega) = \ln \omega + \omega - 2\pi i \mathcal{K}(\omega + \ln \omega) = z - 2\pi i \mathcal{K}(z)$ and since $\mathcal{K}(\omega + \ln \omega) = \mathcal{K}(z)$ except on $z = t - i\pi$ for $t \leq -1$, we have $z = \omega + \ln \omega$ for the "otherwise" case; the case $y = -1$ is by computation; and the case where $z = t - i\omega$, $\omega(z) \in (-\infty, -1) \iff \mathcal{K}(z) = -1$ and direct computation from $\omega e^\omega = e^z \in (-1/e, 0)$ gives $z = \omega + \ln(-\omega) - i\pi$ as claimed. Equivalently, $\bar{z} = \omega + \ln(\omega)$.

(iv) (iii) $\implies$ (iv) because ω is the inverse of $y \to y + \ln(y)$ except when $-\infty < y \leq -1$ when ω has a different inverse. Moreover, $y + \ln y$ is continuous except when $y \leq 0$. Its derivative is $1 + 1/y$, which is zero only if $y = -1$. Therefore ω is continuous (analytic) except possibly when $\omega < 0$. This is precisely $z = t \pm i\pi$, $t \leq 1$. Inspection shows that ω really is discontinuous on $z = t \pm i\pi$ for $t < 1$; but $\omega(t + i\pi) = \omega(t + i\pi^-)$ and $\omega(t - i\pi) = $

$\omega(t - i\pi^+)$ are both continuous from below, because $\mathcal{K}(z)$ is. The fact that $\omega(t + i\pi^-) = \omega(t - i\pi^-)$ follows from the analyticity of $W_0(z)$ in $|z| < 1/e$.

(v) (a) $W_{\mathcal{K}(z)}(e^z)e^{W_{\mathcal{K}(z)}(e^z)} = e^z = \omega(z)e^{\omega(z)}$ by definition. Taking logs, $\ln(\omega e^\omega) = \ln e^z$, or $\omega + \ln \omega - 2\pi i \mathcal{K}(\omega + \ln \omega) = z - 2\pi i \mathcal{K}(z)$. Therefore, $\omega + \ln \omega = z \iff \mathcal{K}(\omega + \ln \omega) = \mathcal{K}(z)$.

(v) (b) $\mathcal{K}(W_{\mathcal{K}(z)}(e^z) + \ln W_{\mathcal{K}(z)}(e^z)) = \mathcal{K}(z)$. $\mathcal{K}(a)$ can change only when $a = t + (2k + 1)\pi$ for $k \in \mathbb{Z}$, or when a is itself discontinuous. We distinguish two cases, therefore:

(1) $W_{\mathcal{K}(z)}(e^z) + \ln W_{\mathcal{K}(z)}(e^z)$ can be discontinuous at discontinuities of $\mathcal{K}(z)$, namely $z = t + (2k + 1)\pi$ for $k \in \mathbb{Z}$, or when $W_{\mathcal{K}(z)}(e^z) < 0$. We ignore discontinuities of $\mathcal{K}(z)$ for the moment. $W_{\mathcal{K}(z)}(e^z) < 0$ only when (i) $\mathcal{K}(z) = 0$ and $e^z < 0 \iff z = t + i\pi, t \le -1$, or (ii) $\mathcal{K}(z) = -1$ and $e^z < 0 \iff z = t - i\pi, t \le -1$. Both (i) and (ii) are discontinuities of $\mathcal{K}(z)$ anyway.

(2) $\mathcal{K}(\omega(z) + \ln \omega(z))$ can be discontinuous when $\omega + \ln \omega = t + (2k+1)\pi i \implies \omega e^\omega = -e^t \iff \omega(z) \subseteq$ an image of $\mathbb{R}^-$ under W. Therefore $z \subseteq$ a pre-image of $\mathbb{R}^-$ under e^z.

But this is just $z = t + (2k + 1)\pi i$, which is a place of discontinuity of $\mathcal{K}(z)$. Note that $\mathcal{K}$ is integral-valued. Therefore, if $\omega(z)$ is such that $\mathcal{K}(\omega(z) + \ln(\omega(z))) = \mathcal{K}(z)$ for any z in a strip $(2k - 1)\pi < \operatorname{Im} z \le (2k + 1)\pi$, where $\omega + \ln(\omega)$ is continuous, then we have $\mathcal{K}(\omega(z) + \ln \omega(z)) = \mathcal{K}(z)$ everywhere in that strip. Let us choose $k \in \mathbb{Z}$, and look at the pre-image of $\omega = 2k\pi i$. Then $\omega + \ln \omega = 2k\pi i + \ln(2k\pi) + i\pi/2$ and hence $\mathcal{K}(\omega + \ln \omega) = k$. Since $\omega = W_{\mathcal{K}(z)}(e^z)$ we have $\omega e^\omega = e^z$ and $2k\pi i \cdot e^{2k\pi i} = e^z \iff e^z = 2k\pi i$; moreover $2k\pi i \in$ range $W_{\mathcal{K}(z)}$, and therefore $\mathcal{K}(z) = k$. Therefore

$$z = \ln(2k\pi i) + 2k\pi i$$
$$= \omega + \ln \omega.$$

This establishes that if $\omega(z) = W_{\mathcal{K}(z)}(e^z)$, then $\omega + \ln \omega = z$ except possibly on the edges of the strips $z = t + (2k + 1)\pi i$. Now we have $\mathcal{K}(\omega(z) + \ln(\omega(z))) = \mathcal{K}(z)$ if $(2k - 1)\pi < \operatorname{Im}(z) < (2k + 1)\pi$, and hence $\omega + \ln \omega = z$. Note that $\omega(z) = W_{\mathcal{K}(z)}(e^z)$ is continuous from below as $\operatorname{Im}(z) \to (2k + 1)\pi^-$. Therefore, provided that $\omega(z) \notin \mathbb{R}^-$, $\omega(z) + \ln(\omega(z))$ will be continuous as $\operatorname{Im}(z) \to (2k+1)\pi^-$. Therefore, since $\operatorname{Im}(\omega(z) + \ln \omega(z)) = \operatorname{Im}(z)$ for $(2k - 1)\pi < \operatorname{Im}(z) < (2k + 1)\pi$, we have $\mathcal{K}(\omega + \ln \omega) = \mathcal{K}(z)$ even if $\operatorname{Im}(z) = (2k + 1)\pi$ by continuity:

$$\lim_{\operatorname{Im}(z) \to (2k+1)\pi^-} \mathcal{K}(\omega(z) + \ln \omega(z))$$
$$= \lim_{\operatorname{Im}(z) \to (2k+1)\pi^-} \mathcal{K}(z).$$

Therefore $\mathcal{K}(\omega(z) + \ln \omega(z)) = \mathcal{K}(z)$ unless $\omega(z) < 0$, and $\operatorname{Im}(z) = -i\pi$.

(vi) This now follows immediately.

2.4   Corollary

Define $z(k, \theta) = x + i \cdot (2k + \theta)\pi$. Then $z(k + 1, -1) = z(k, 1)$ since $x + i \cdot (2k + 2 - 1)\pi = x + i \cdot (2k + 1)\pi$, since $\mathcal{K}(x + i \cdot (2k + \theta)\pi) = k$ for $-1 < \theta \leq 1$. Since

$$W_k(e^{x+i(2k+\theta)\pi}) = W_k(e^{x+i\pi\theta}) = W_k(e^x(\cos \pi\theta + i \sin \pi\theta)),$$

we have $W_k(e^{x+i\pi\theta}) \to W_k(-e^x + i \cdot 0^+)$ as $\theta \to 1^-$, and

$$\lim_{\theta \to -1^+} W_{\mathcal{K}(z(k+1,\theta))}(e^{z(k+1,\theta)}) = \lim_{\theta \to -1^+} W_{k+1}(e^{x+i(2k+2+\theta)\pi})$$

since $\mathcal{K}(x+i\cdot(2k+2+\theta)\pi) = k+1$ for $-1 < \theta \leq 1$. Since $W_{k+1}(e^{x+i\cdot(2(k+1)+\theta\pi)}) = W_{k+1}(e^{x+i\pi\theta}) = W_{k+1}(e^x(\cos \pi\theta + i \sin \pi\theta))$ we have

$$W_{k+1}(e^{x+i\pi\theta}) \to W_{k+1}(-e^x + i \cdot 0^-)$$

as $\theta \to -1^+$. By continuity of w, then, unless $k = 0$ or $k = -1$ and $0 > -e^x > 1$

$$W_k(-e^x + i \cdot 0^+) = W_{k+1}(-e^x + i \cdot 0^-).$$

Alternative (direct) proof of (iv) (a).
    Lemma: $W_k(-e^x + i \cdot 0^+) = W_{k+1}(-e^x + i \cdot 0^-)$ unless $-e^{-1} \leq -e^x < 0$ and $k = 0$.
    Proof.
    Images of the lines $y = t$, $x = $ constant < 0, are smooth curves under $w(x + iy)$, by inspection, except if $w(x + iy) < 0$. This is what we have to prove. Can we define e^z as a value on the Riemann Surface for W? Yes, except when $-e^{-1} \leq e^z < 0$, by placing it on the sheet with winding number $\mathcal{K}(z)$. This is a bijection between the Riemann Surface for log and for W, except on $-e^{-1} \leq e^z < 0$. Once this is done, the cut's images on the Riemann Surface can obviously be moved at will. Since $W_k(-e^x + i \cdot 0^-)$ lies on the other side of the cut, we have equality.

Algebraic properties.

– Derivatives and integrals:

$$\frac{dw}{dz} = \frac{w}{1+w}$$

$$\int w^n \, dz = \begin{cases} \frac{w^{n+1}-1}{n+1} + w^n/n \text{ if } n \neq -1 \\ \ln w - 1/w \quad \text{if } n = -1 \end{cases}$$

The derivative formula is valid except on the doubling line and its reflection, when it is valid as a derivative in the real direction only. The integrals can be verified directly by differentiation of both sides. The addition of the constant term $-1/(n+1)$ to the integral of w^n is a trick due, in the case $\int x^n \, dx$, to W. Kahan. Using this trick, the formula for limiting case $n = -1$ is a simple limit of the formula for $n \neq -1$.

– Series about $z = a$, where $a = \omega_a + \ln \omega_a$: the following (computed by Maple) is the beginning of the series for ω which contains second order Eulerian numbers.

$$\omega_a + \frac{\omega_a}{1 + \omega_a}(z - a) + \frac{1}{2!}\frac{\omega_a}{(1 + \omega_a)^3}(z - a)^2$$

$$-\frac{1}{3!}\frac{\omega_a(2\omega_a - 1)}{(1 + \omega_a)^5}(z - a)^3 + \frac{1}{4!}\frac{\omega_a(6\omega_a^2 - 8\omega_a + 1)}{(1 + \omega_a)^7}(z - a)^4$$

$$-\frac{1}{5!}\frac{\omega_a(24\omega_a^3 - 58\omega_a^2 + 22\omega_a - 1)}{(1 + \omega_a)^9}(z - a)^5$$

$$+\frac{1}{6!}\frac{\omega_a(120\omega_a^4 - 444\omega_a^3 + 328\omega_a^2 - 52\omega_a + 1)}{(1 + \omega_a)^{11}}(z - a)^6$$

$$+O((z - a)^7)$$

The general term is [6]:

$$\omega(z) = \sum_{n \geq 0} \frac{q_n(\omega_a)}{(1 + \omega_a)^{2n-1}}\frac{(z - a)^n}{n!} \tag{4}$$

where

$$q_n(w) = \sum_{k=0}^{n-1} \left\langle\!\!\left\langle \begin{array}{c} n - 1 \\ k \end{array} \right\rangle\!\!\right\rangle (-1)^k w^{k+1}. \tag{5}$$

is defined in terms of second order Eulerian numbers.

– Series about ∞: This series was originally due to de Bruijn, and Comtet identified the coefficients as Stirling numbers.

$$\omega \sim z - \ln(z) + \frac{\ln(z)}{z} + \frac{1}{2}\frac{\ln(z)(\ln(z) - 2)}{z^2}$$

$$+\frac{1}{6}\frac{\ln(z)(-9\ln(z) + 6 + 2\ln(z)^2)}{z^3} + \frac{1}{12}$$

$$\frac{\ln(z)(3\ln(z)^3 - 22\ln(z)^2 + 36\ln(z) - 12)}{z^4} +$$

$$\frac{1}{60}\ln(z)(-125\ln(z)^3 + 350\ln(z)^2 + 12\ln(z)^4$$

$$- 300\ln(z) + 60)/z^5 + O(\frac{1}{z^6})$$

The general term is (translating from the Lambert W results of [2,7]) $\omega(z) =$

$$z - \ln z + \sum_{\ell \geq 0}\sum_{m \geq 0} c_{\ell m}\frac{\ln^m z}{z^{\ell + m}} \tag{6}$$

where $c_{\ell m} = (-1)^\ell \left[\begin{array}{c} \ell + m \\ \ell + 1 \end{array}\right]/m!$ is defined in terms of Stirling cycle numbers [8]. This series converges for large enough z, outside the region bounded by the doubling line and its reflection. The proof is unpublished. The series can be rearranged in several ways, following [9] and [6]: $w(z) =$

$$z - \ln z + \sum_{n \geq 1} \frac{(-1)^n}{z^n} \sum_{m=1}^{n} \frac{(-1)^m}{m!} \left[\begin{matrix} n \\ n-m+1 \end{matrix} \right] \ln^m z \,. \tag{7}$$

Using a new variable $\zeta = z/(1+z)$, we get $w(z) =$

$$z - \ln z + \sum_{m \geq 1} \frac{\ln^m z}{m! z^m} \sum_{p=0}^{m-1} (-1)^{p+m-1} \zeta^{p+m} \left\{ \begin{matrix} p+m-1 \\ p \end{matrix} \right\}_{\geq 2} \tag{8}$$

where the numbers in curly braces are 2-associated Stirling numbers. Using $L_\tau = \ln(1 - \tau) = \ln(1 - \ln z / z)$ and $\eta = \sigma/(1 - \tau) = 1/(z(1 - \ln z / z)) = 1/(z - \ln z)$, series (83) and (84) from [6] become

$$w(z) = z - \ln z - L_\tau$$
$$+ \sum_{n \geq 1} (-\eta)^n \sum_{m=1}^{n} (-1)^m \left[\begin{matrix} n \\ n-m+1 \end{matrix} \right] \frac{L_\tau^m}{m!} \tag{9}$$

and

$$w(z) = z - \ln z - L_\tau$$
$$+ \sum_{m \geq 1} \frac{1}{m!} L_\tau^m \eta^m \sum_{p=0}^{m-1} \left\{ \begin{matrix} p+m-1 \\ p \end{matrix} \right\}_{\geq 2} \frac{(-1)^{p+m-1}}{(1+\eta)^{p+m}} \,.$$
$$\tag{10}$$

The series converge for large enough real z, though the detailed regions of convergence are not yet settled. Curiously enough (10) is exact at $z = 1$ and at $z = \infty$, and moreover if we truncate it to N terms it agrees with the N term Taylor series expansion at $z = e$ as well, making one think of 'Hermite' interpolation at 1 and at ∞. Convergence is rapid.

- Series about $-\infty$: from the series $W(z) = \sum_{n \geq 1} (-n)^{n-1} z^n / n!$, for $|\exp(z)| < \exp(-1)$ we have

$$w(z) = \sum_{n \geq 1} \frac{(-n)^{n-1}}{n!} e^{nz} \tag{11}$$

2.5   Branch Point Series for $w(z)$

The Wright w function has branch points at $z = -1 \pm i\pi$. The following series obtain. Near $z = -1 + i\pi$,

$$w(z) = -\sum_{n \geq 0} a_n \left(i \sqrt{2(z+1-i\pi)} \right)^n \tag{1}$$

where the double conjugation gives us the correct closure from below on $t + i\pi$ for $t \leq -1$. Near $z = -1 - i\pi$,

$$w(z) = -\sum_{n \geq 0} a_n \left(-i \overline{\sqrt{2(z + 1 + i\pi)}} \right)^n. \tag{2}$$

In both cases a_n is given by the recurrence relation [10]

$$a_0 = a_1 = 1$$
$$a_k = \frac{1}{(k+1)a_1} \left(a_{k-1} - \sum_{i=2}^{k-1} i a_i a_{k+1-i} \right). \tag{3}$$

The derivation of these series from the results of [10] is straightforward, except for the use of $\sqrt{\overline{z}}$. We here verify that this construction, which is one of a family of transformations modelled on some used by G.K. Batchelor, gives us the correct closure. We know that $\omega(t + i\pi^-) = W_0(-e^t)$ whilst $\omega(t + i\pi^+) = W_1(-e^t)$, and $\omega(t - i\pi^+) = W_0(-e^t)$ whilst $\omega(t - i\pi^-) = W_{-1}(-e^t)$. Putting $z = t + i\pi^+$ in $\overline{\sqrt{2(z + 1 - i\pi)}}$ gives $\overline{\sqrt{2(t + 1 + i \cdot 0^+)}}$, for $t \approx -1$. If $t + 1 \geq 0$ then we have no branch cut to cross—this series will be continuous, therefore, along the line $t+1+i\pi$, $t \geq -1$. If $t+1 < 0$, we are on the branch cut. $\overline{t + 1 + i \cdot 0^+}$ is $t+1+i\cdot0^-$, and $\arg \sqrt{2(t + 1 + i \cdot 0^-)} = -\pi/2$. Therefore $\arg \overline{\sqrt{2(t + 1 + i \cdot 0^-)}} = +\pi/2$, and this means that the series (2) can be written

$$w(z) = -\sum_{n \geq 0} a_n (\rho)^n$$

and by inspection of the signs of the series for $W_{-1}(-e^t)$ and hence $W_{+1}(-e^t)$ just above the branch cut, this is correct. [Here $\rho = \sqrt{-2(t + 1)} > 0$.] Next, consider $z = -1 + i\pi^-$. A similar argument leads to the conclusion

$$w(z) = -\sum_{n \geq 0} a_n (-\rho)^n$$

which is the series for $W_0(-e^t)$ for $t \approx -1$, because its signs alternate. Consideration of $z = t - i\pi^+$ and $t - i\pi^-$ gives, for $t + 1 < 0$,

$$w(z) = -\sum_{n \geq 0} a_n (-\rho)^n \quad z = t - i\pi^+$$
$$= -\sum_{n \geq 0} a_n \rho^n \quad z = t - i\pi^-$$

and continuity if $t + 1 \geq 0$.

Remark. The use of $\overline{\sqrt{(z - a)}}$ to represent a square root function with a closure different from the CCC closure, as explained by Kahan, is a useful tool in a computer algebra setting. However, it relies on the designers to be sophisticated enough to provide symbolic means of representing (and not over-simplifying) these series, and the users to be sophisticated enough to know that $\sqrt{\overline{z}} \neq \overline{\sqrt{z}}$ on the branch cut.

3   Interpolating $W_k(z)$

Finally, we interpret equation (3) as an interpolation scheme for $W_k(z)$. We note that k need not be an integer in that equation; the geometric interpretation is precisely that of a circular cylinder cutting the Riemann surface for W. Note also that $k = 0$ and $k = -1$ are special, and not interpolated by this scheme.

We deduce that $W_k(z)$ is, in some sense, analytic in k, except if $- \exp(-1) \le z < 0$ and $k = 0$ or $k = -1$.

$$\frac{dW_k(z)}{dk} = \frac{d}{dk}\omega(\ln z + 2\pi ik)$$

$$= 2\pi i \frac{\omega(\ln z + 2\pi ik)}{1 + \omega(\ln z + 2\pi ik)} \ .$$

By the analytic properties of ω, this derivative is not continuous on $- \exp(-1) \le z < 0$ at $k = 0$ or $k = -1$. Otherwise, indeed, $W_k(z)$ is analytic in k.

4   Why

Computer algebra is about expressiveness, and simplicity is power. There are an essentially infinite number of applications of the Lambert W function and its cognates.

1. The Lambert W function provides the first example of a function *just* outside the standard body of Risch-like theory: its derivative is rational in x and W, not polynomial. One cannot use the same theorems, but one can hope to use similar methods, to establish its non-elementarity [1].
2. The Lambert W function is the simplest example of a *root of an exponential polynomial*; and exponential polynomials are the next simplest class of functions after polynomials. Computer algebra systems have a real edge over numerical systems (though not everyone knows it) in dealing with polynomials; the next big area will be non-polynomials, starting with exponential polynomials. This is the field of Cylindrical Non-Algebraic Decomposition.
3. The Lambert W function is the first nontrivial example of a multivalued function. The trivial ones (ln and the inverse trigonometric functions) have branching behaviour so simple that it doesn't even need a notation: we can say $\ln(z) + 2\pi ik$ and not have to invent a new notation $\ln_k(z)$ to do so (though in fact we *have* introduced and used this notation—one can't use $\log_k$ because the "log to the base k" interpretation would get in the way—for conciseness and as the thin entering point of the wedge for more complicated functions). The multivalued nature of W "stress tests" naming conventions, numerics on branches, computer-aided analysis, and the results of series computation. Right now, Maple knows the series for $W_0(z)$ about the branch point $z = - \exp(-1)$, but it doesn't know the series for $W_{-1}(z)$

or $W_1(z)$ about the same point, even though these series were all introduced in the same paper [4]. We think that this is because the series are defined *piecewise*: for W_{-1} and W_1, the series about the branch point have to deal with the fact that the range is split by the branch cut, and so the series are (radically) different if $\text{Im}(z) > 0$ or $\text{Im}(z) < 0$; each branch of W has both a Puiseux series and a Taylor series—about the same point! But different series apply above and below the branch cut. This remarkable behaviour puts a significant stress on the ability of `series` to express its answer to the question `series(LambertW(-1,x),x=-exp(-1))` (which it currently refuses to answer).

4.1 Why Invent the Wright ω Function?

It is certain that for some applications, just the ordinary Lambert W function will be superior—this new function cannot supplant the old. Bill Gosper did not succeed in introducing his cognate of W (which he jokingly called "the Dilbert Lambda Function"); Don Knuth has so far been unable to get action on our promise to him to introduce the TreeT function into Maple ($T(x) = -W(-x)$, and this is more convenient for combinatorial applications). So why should we bother with a new one?

In equation (1) we give the definition of the Wright ω function, in terms of W and one new function, the unwinding number $\mathcal{K}(z)$, which we will be needing anyway. So why not just use the right hand side of the definition and not bother with a notation?

1. W is multivalued, but ω is single-valued.
2. Numerically evaluating $\omega(z)$ for large z by way of the definition (1) is like driving from the south of London to the north of London via Waterloo[3]: it's possible (unless there's freezing rain) but unless you have a reason to be in Waterloo, it's probably better to go directly. Less metaphorically, taking $\exp(z)$ for large z gives a significant risk of overflow, and a significant restriction on the numerical range of z that we can do the computation for; but W is like ln, and in some sense just undoes the exponentiation, making it wasted effort in any case. The asymptotics are that $\omega(z) \sim z - \ln z + \cdots$; so we see just how wasted. This is *not a theoretical consideration*: Jon Borwein has had to implement his version of ω precisely to avoid this overflow difficulty in his convex optimization problems.
3. Numerically evaluating `omega( -0.9 + I*Pi )` by way of the formula uncovers a subtle difficulty: because `ceil( (Im(z)-Pi)/(2*Pi) )` will do some symbolic processing, it will compute $\mathcal{K}(z)$ exactly right, and cancel the symbolic `Pi`. But `exp(-0.9+I*Pi)` is left alone, until the user calls `evalf`. Then something awful happens: at 10 Digits, `Pi` rounds to something larger than

---

[3] London, Ontario via Waterloo, Ontario, of course. That sentence reads quite differently if you think of train travel in London, England, for example (thanks to Arthur Norman for pointing that out)

π; this then gives us a *negative imaginary part* on the order of roundoff in the result of the call to exp. This is all explainable in terms of the Maple model of floating-point arithmetic, but it's a disaster nonetheless—one made visible by the next step, the computation of $W_0(x - i \cdot \varepsilon)$, which is *on the wrong side of the branch cut*. The numerical value of $W_0(x - i \cdot \varepsilon)$ is not at all close to the value of $W_0(x + i \cdot \varepsilon)$, and this discontinuity is *spurious*. The ω function is *continuous* at this point. So: we should have a separate routine for the numerical evaluation of ω that guarantees that we get continuity (where ω is continuous), because the definition combines *discontinuous* functions in such a way that their discontinuities (mostly) cancel.

There are other advantages to using the Wright ω function directly.

1. In addition to being single-valued, ω is continuous (indeed analytic) for all z not on the two half-lines $z = t \pm i\pi$ for $t \leq -1$. It is discontinuous across these lines.
2. The Wright ω function has a simpler Taylor series than the Lambert W function does. Indeed, it is the series for the Wright ω function that leads to nearly all the series given in [6].
3. The fabulously simple equation $W_k(z) = \omega(\ln z + 2\pi i k) = \omega(\ln_k z)$ explains the branching behaviour of W perfectly, once we understand the branching behaviour of ω.
4. The solution of the equation $y + \ln y = z$ is given by

$$y = \begin{cases} \omega(z) & z \neq t \pm i\pi, t \leq -1 \\ \omega(z), \omega(z - 2\pi i) & z = t + i\pi, t \leq -1 \\ \text{nonesuch} & z = t - i\pi, t \leq -1 \end{cases} \tag{12}$$

The paper [11] seems to be the first to use this fact.

What are the disadvantages? Well, the principal one is that the counting applications depend on the use of W (or, rather, TreeT) as a generating function. There, the series at the origin is what is important. With this transformation, we have moved this point to $-\infty$. The series are still there—just less convenient. And, that is what introducing this function is all about: convenience. We will need to have all of these functions around—well, certainly TreeT, but probably not Dilbert Lambda. Even Bill Gosper has mostly given up on that one.

5   Concluding Remarks

This paper presents a number of mathematical results describing the properties of the function $\omega(z)$. These results have some intrinsic mathematical interest, and they are written here for the first time, and so in a technical sense the paper contains novel results. However, the results are really interesting only because:

1. Without symbolic computation making the function's definition, simplification rules, and numerical evaluation widely available, the function is merely arcane.

2. Discontinuity (along the branch cuts) is especially visible, and nontrivial, in this function. Therefore it will make a good test case for reasoning about complex-valued expressions.

3. The methods used to prove properties of ω are essentially old-fashioned mathematics, not commonly seen in standard curricula, and may potentially be automated. This is in the spirit of [3] and represents a potentially interesting direction for future research.

Acknowledgements. The code for the numerical evaluation of $\omega(z)$ (which will be discussed in a future paper) was scrutinized by Dave Hare. The colour graph of the Riemann Surface for ω (not shown here, but used in the upcoming Lambert W poster, was produced with a Maple program written by George Labahn (getting the colours to come out continuous, while making Maple show the discontinuity, is not trivial.) Jon Borwein and Bill Gosper provided motivation to look at ω. Our thanks also to Prof. John Wright (Reading & Oxford), and his father Sir Edward Maitland Wright, for permission to use a picture of Sir Edward on the poster of the Lambert W function.

References

[1] BRONSTEIN, M., AND DAVENPORT, J. H. Algebraic properties of the Lambert W function.

[2] COMTET, L. *Advanced Combinatorics*. Reidel, 1974.

[3] CORLESS, R. M., DAVENPORT, J. H., DAVID J. JEFFREY, LITT, G., AND WATT, S. M. Reasoning about the elementary functions of complex analysis. In *Proceedings AISC Madrid* (2000), vol. 1930 of *Lecture Notes in AI*, Springer. Ontario Research Centre for Computer Algebra Technical Report TR-00-18, at http://www.orcca.on.ca/TechReports.

[4] CORLESS, R. M., GONNET, G. H., HARE, D. E. G., JEFFREY, D. J., AND KNUTH, D. E. On the Lambert W function. *Advances in Computational Mathematics 5* (1996), 329–359.

[5] CORLESS, R. M., AND JEFFREY, D. J. The unwinding number. SIGSAM *Bulletin 30*, 2 (June 1996), 28–35.

[6] CORLESS, R. M., JEFFREY, D. J., AND KNUTH, D. E. A sequence of series for the Lambert W function. In *Proceedings of the ACM ISSAC, Maui* (1997), pp. 195–203.

[7] DE BRUIJN, N. G. *Asymptotic Methods in Analysis*. North-Holland, 1961.

[8] GRAHAM, R. L., KNUTH, D. E., AND PATASHNIK, O. *Concrete Mathematics*. Addison-Wesley, 1994.

[9] JEFFREY, D. J., HARE, D. E. G., AND CORLESS, R. M. "Unwinding the branches of the Lambert W function". *Mathematical Scientist 21* (1996), 1–7.

[10] MARSAGLIA, G., AND MARSAGLIA, J. C. "A new derivation of Stirling's approximation to $n!$". *American Mathematical Monthly 97* (1990), 826–829.

[11] SIEWERT, C. E., AND BURNISTON, E. E. "Exact analytical solutions of $ze^z = a$". *Journal of Mathematical Analysis and Applications 43* (1973), 626–632.

[12] WRIGHT, E. M. "Solution of the equation $ze^z = a$". *Bull. Amer. Math Soc. 65* (1959), 89–93.

Multicontext Logic for Semigroups of Contexts

Rolf Nossum[1] and Luciano Serafini[2]

[1] Agder University College, Department of Mathematics,
Gimlemoen, N-4604 Kristiansand, Norway
Rolf.Nossum@hia.no
[2] ITC-IRST, Centro per la Ricerca Scientifica e Tecnologica,
Via Sommarive 18, I-38050 Povo, Italy
Serafini@itc.it

Abstract. A multicontext logic with algebraic structure is proposed, where contexts are either primitive or composed from other contexts. Composition of two contexts can support various intuitions: sequence concatenation, set union, multiset union, etc.
A local models semantics for algebraic context composition is defined, with a corresponding deductive calculus containing multilanguage bridge rules. Soundness and completeness results are proved for the case of semigroups of contexts, i.e. where context composition is an associative operation. Other properties of context composition, besides associativity, are defined by additional algebraic equations.

Keywords: Integration of Logical Reasoning and Computer Algebra, Logic and Symbolic Computing, Reasoning

1 Introduction

The Multi-Context Systems (MCS) formalism for contextual reasoning was proposed in [9] (under the name of Multi Language systems), and its semantics and proof theory have been developed in a series of papers [10,8,18]. MCSs are rich in vocabulary, having languages differentiated by context. They support the integration between reasoning *in* context and annotation *about* context. Most MCS defined in the literature presume some sort of structure of the context within which reasoning takes place, but here the richness varies a lot. In MCS there is a relatively simple indexing scheme which identifies context, and this has been extended to a partial order in some cases. Previous works do not provide a theoretical foundation for sets of contexts with a more complex structure.

However, in many applications, where contexts are used to represent pieces of knowledge that are dynamically modified, combined, copied, etc, it is useful to store (in the labels) the information about structural relations between context, as for instance, the fact that a context is a copy of another, or that a context is the union of other two different contexts. For this purpose we have to refine the labeling mechanisms so that we allow, for instance, the label $c \oplus d$ to denote a context obtained by combining the two contexts c and d.

J. Calmet et al. (Eds.): AISC-Calculemus 2002, LNAI 2385, pp. 90–101, 2002.

In this paper, we are proposing MCSs with an enriched context structure, allowing contexts to combine into new composite contexts according to equationally specified patterns.

This paper is structured as follows: We start by giving two motivational examples, and then go on to introduce the algebraic notation and logical language we are using. Then, we define a semantics for interpreting formulas labeled by contexts, and a corresponding deductive calculus with inter-contextual deduction rules. Then a class of algebras called AFG algebras is introduced, and a completeness proof for that class is carried out. Finally, a comparison is made with some related work on formal models of contextual reasoning.

2 Partiality Motivates Algebraic Contexts

Predicates about the world are partial, both in the absolute sense that no known theory covers everything in the universe, and also in the relative sense that usually, many facts which could have been expressed in the theory at hand are left unexpressed.

The term 'context' is used in a variety of senses in the literature, and we point out two competing ones:

In the philosophy of language, 'context' has been used to denote the total state of affairs which applies to an utterance being interpreted, while a projection onto selected facets of reality has been called an 'index' [13,16].

This is contrast to emergent terminology in AI, where 'context' is almost invariably used to denote a partial state of affairs, and where the idea of a 'most general context', i.e. the context representing the total state of affairs, is sometimes denounced [14].

Without taking a metaphysical stand, we shall proceed to deal with representations of partial states of affairs, and refer to such representations as contexts. For the purposes of this paper, a context is either a primitive entity c (compare the 'micro-contexts' of [12,11]), or a composite entity $x \oplus y$ obtained by combining other contexts x and y.

Whenever chunks of context combine into a larger joint context, issues of partiality, granularity, and specificity arise:

- what is the language of the joint context?
- what is true in the joint context?
- is one way of forming a joint context equivalent to another way?
- how do we move from one composite context to another?

In [2] it is suggested that contextual reasoning patterns come in three broad categories: localised reasoning, shifting, and push/pop. Localised reasoning is confined to one specific context, and shifting exchanges one context for another according to intercontextual rules. The third category, push/pop, includes patterns of reasoning where progression is from formulas asserted in one context to formulas asserted in an incrementally augmented/depleted context.

We propose to take an algebraic look at the operation of incrementally adding information to a context, or combining an existing context with another. Let us illustrate:

Example 1 *In this example, a context is defined by weather reports gathered from meteorological offices at airports. One important parameter for air traffic is horizontal visibility on the ground.*

In order for a scheduled flight to depart on time, it is necessary that certain weather minima are fulfilled, such as sufficient horizontal visibility for take-off and landing at the departure and destination airports.

A pilot planning a flight from Moscow via Paris to Washington must take into account three weather reports. Let us denote the contexts of local weather reports from each town by Mo, Pa, Wa, respectively. Beginning in context Mo, the pilot learns about Pa, and forms an accumulated context that we denote by

$$Mo \oplus Pa$$

Looking up the weather in Washington, this context is augmented further to

$$(Mo \oplus Pa) \oplus Wa$$

In this example, the order in which weather reports are looked up does not matter:
$$(Mo \oplus Pa) \oplus Wa = (Mo \oplus Wa) \oplus Pa$$

and retrieving the same weather report twice is equal to having it once:

$$Mo \oplus Mo = Mo$$

In fact, it seems adequate to think about a series of such amendments in terms of the set of increments that are made overall, so the $\oplus$ operation in this example is a kind of set union. Set union is associative, commutative and idempotent, and the algebra of sets has equations corresponding to these three properties:

$$(u \oplus v) \oplus w = u \oplus (v \oplus w) \tag{1}$$

$$u \oplus v = v \oplus u \tag{2}$$

$$u \oplus u = u \tag{3}$$

In other examples (see below), there will be other algebraic equations defining other properties of context augmentation. Far from all context systems combine in the same way as sets do.

Before we leave this example, let us observe another trait having to do with translation between contexts: Visibility reports are routinely gathered and exchanged between airports worldwide. We can represent them as triples $\langle v, p, t \rangle$, where v is a measure of visibility, p is a geographical location, and t is a time-point. It is natural to think of these triples as members of a relation in a relational database, with attributes for visibility, location and time. When different *contexts are combined, triples from each component relation are put together in a joint relation. But, in a Russian database the visibility attribute would be called 'vidinost', and would be measured in meters, while in an American database the corresponding 'visibility' attribute would be measured in miles. The semantics of the ⊕ operation should take this into account by providing a correspondence in the form of translations between entities of different local context languages.*

This motivates a mapping which translates between local context languages, and later on this will be formalized. For more details of the formalization of federated databases in a multilanguage environment, see [8,17].

Example 2 *Let us take an example with a different flavour; the context defined by a person's beliefs. Here, the intuitive interpretation of the ⊕ operation is the following. Given two contexts u and v, the context u⊕v is the context u extended with a description of v, (from u's perspective). For instance if u is the context of John's beliefs, and suppose that John does not know about the existence of Mary. Let v be the context of Mary's beliefs. Suppose that John meets Mary, and then he has to do two things:*

1. *define a new context (denoted $u \oplus v$) representing John's view of Mary's beliefs;*
2. *relate this new context with his beliefs, i.e. with u.*

This might involve things not expressed (or even expressible) in the original context u, for instance:

- *extending the language. u does not contain any predicate for Mary's beliefs (this is the case if John does not know Mary), John has to add a new predicate (or modality) to express Mary's beliefs.*
- *extending the set of true formulae. John has to insert some new axioms expressing what he actually believes about Mary's beliefs.*

With this intuitive interpretation we can stipulate epistemic variants of belief by admitting equations on ⊕ terms. Admitting or rejecting equations like the following amounts to placement within an epistemic taxonomy.

For instance, we have that the idempotence equation

$$u \oplus u = u$$

states perfect introspection, while the associativity equation

$$(u \oplus v) \oplus w = u \oplus (v \oplus w)$$

means that for all u, v, w, from u's point of view, v is a credible witness of w's beliefs, i.e. mutual perfect introspection.

3 Algebras of Contexts

Our concern in this paper is the context combination operator $\oplus$ and its interaction with other features of MC formalism. We shall give a semantical structure which is complete for certain intercontext deduction rules, when the properties of context combination are given as algebraic equations of a certain class. As shown by the previous examples, it is natural to think about algebraic concepts when two or more entities combine to form a new entity. If a countable set C of atomic contexts is given a priori, we can take it as the carrier of an algebra with one binary operator $\oplus$ and equations

$$y_i = z_i, \quad 1 \leq i \leq N \tag{4}$$

for some $N > 0$, where y_i and z_i are terms on $\oplus$.

As an example consider bags (multisets) over C, generated by the $\oplus$ operation. In a bag, as opposed to a sequence, the order of elements is immaterial. The relevant equations express associativity and commutativity of $\oplus$:

$$(u \oplus v) \oplus w = u \oplus (v \oplus w) \tag{5}$$
$$u \oplus v = v \oplus u \tag{6}$$

In a proper set over C, where repeated context entries don't count, the $\oplus$ operation is also idempotent:

$$u \oplus u = u \tag{7}$$

We'll use E to denote the set of algebraic equations governing the $\oplus$ symbol. The set of all $\oplus$-terms over C can be denoted $C^\oplus$, and the set of equivalence classes imposed by the algebraic equations E is then denoted $C_E^\oplus$.

Intuitively, C is the set of primitive labels for contexts. Each label $c \in C$ is associated with a context. Notice that different labels can be associated to the same context. The set $C^\oplus$ is the set of *terms* for denoting contexts. Each $x \in C^\oplus$ is associated with a context, and, as for primitive labels, different terms can be associated to the same context. Finally, $C_E^\oplus$ can be viewed as the set of "canonical names" for contexts. Each element $x \in C_E^\oplus$ is also associated with a context ctx, and x can be thought as a canonical name for ctx, or equivalently, the equivalence class (under the set of equation E) of the terms in $C^\oplus$ associated with ctx. Notice that, for any $x, y \in C_E^\oplus$, if x is different from y then the context associated to x is different from the context associated to y. Keeping this in mind, in the rest of the paper, when it is not ambiguous, we use the term "context" as a shortcut of the description "label for context".

We proceed to define a semantics and a proof system for a large class of algebras of contexts. For simplicity of presentation we restrict ourselves to the case in which the languages associated to each context is propositional, and leave the first order case out for now.

3.1   Context Languages

As a typographical convention, we use a, b, c, d, sometimes subscripted, to denote primitive contexts from C, while t, u, v, w, x, y, z and their subscripted variants are used liberally to denote primitive contexts from C, composite contexts from $C^{\oplus}$, or equivalence classes of contexts from $C_E^{\oplus}$.

For each context $u \in C_E^{\oplus}$ we have a propositional language L_u, that is used to express facts in this context.

Definition 1 (Well formed formulae) *Well formed formulae are defined as follows (for all $u \in C_E^{\oplus}$). If ϕ is a propositional formula in L_u, then for all $y \in C^{\oplus}$ $y : \phi$ is a well formed formula, and ϕ is called a y-formula.*

Definition 2 (Language mapping) *For all $c \in C$, there is a partial recursive function l_c that maps $u \oplus c$-formulae into u-formulae for arbitrary $u \in C_E^{\oplus}$.*

Intuitively, a language mapping from $u \oplus c$ to u states which part and how the content of the context $u \oplus c$ is represented in the context u. Considering the belief example (Example 2), the usual language mapping is the reification function, i.e. the total function that associates to each $u \oplus c$-formula ϕ the u-formula $bel(c, \phi)$.

3.2   Local Model Semantics

Every equivalence class of contexts, i.e. each $u \in C_E^{\oplus}$, has its own formula language L_u. The semantical structure we are about to define, takes as its basic building blocks the local interpretations of each language L_u. We can identify interpretations with subsets of L_u, i.e. the true formulas in each interpretation.

The semantical structure for the entire system of languages reflects the way in which contexts are augmented by adding ground contexts by the $\oplus$ operation. We start by defining ground extensions of context terms:

Definition 3 (x-continuation) *Given $x \in C_E^{\oplus}$, an x-continuation is a context*

$$(\ldots (x \oplus c_1) \oplus c_2 \ldots) \oplus c_h$$

where $0 \leq h$ and $c_i \in C$ for $1 \leq i \leq h$. When $h = 0$, this is just x. Note that unless $\oplus$ is associative, the parentheses are not redundant.

Definition 4 (x-chain) *For $x \in C_E^{\oplus}$, an x-chain m is a function which maps every x-continuation y to a set m_y of interpretations of L_y (the local models of context y), such that for some x-continuation y, m_y is not empty, and for all x-continuations y, and ground contexts $c \in C$:*

1        $m \models y : l_c(\phi)$ *if and only if* $m \models y \oplus c : \phi$
2   *the cardinality of m_y is at most 1.*

Definition 5 (Satisfiability) *An x-chain m satisfies a formula $y : \phi$ where y is an x-continuation, in symbols $m \models y : \phi$, if for any $s \in m_y$, $s \models \phi$ according to the definition of satisfiability for propositional formulae.*

Definition 6 (Logical consequence) *A formula $x : \phi$ is a logical consequence of a set of formulae Γ, in symbols $\Gamma \models_E x : \phi$ if, for any z-chain m, such that x is an z-continuation, if $m \models \{y : \gamma \in \Gamma | x \neq_E y \text{ and } y \text{ is a } z\text{-continuation}\}$, then for all $s \in m_x$, $s \models \{\psi | z : \psi \in \Gamma, x =_E z\}$, implies that $s \models \phi$.*

3.3   Reasoning between Contexts

The notion of x-continuations induces a partial order among contexts, each context preceding its continuations. Let us see how one moves between composite contexts which are related in a partial order. We rely on a natural deduction calculus extended with indices as described in [18], extended with the following bridge rules for any set of algebraic equations E.

$$\frac{u : l_c(\phi)}{u \oplus c : \phi} \; Rup_c \qquad \frac{u \oplus c : \phi}{u : l_c(\phi)} \; Rdw_c$$

$$\frac{u \oplus c : \phi \leftrightarrow \psi}{u : l_c(\phi) \leftrightarrow l_c(\psi)} \; RRI \qquad \frac{v : \phi}{u : \phi} \; u =_E v$$

Derivability from a set Γ of formulae can be defined from a set E of equations $\Gamma \vdash_E u : \phi$, if there is a deduction of $u : \phi$ from Γ, that uses u-rules (as defined in [18]) and the above bridge rules.

4   Semigroups with Ground Equations

Whenever associativity of the $\oplus$ operation is given or implied, the resulting algebra is called a semigroup. In the present paper, we are restricting ourselves to certain semigroups of contexts called AFG algebras.

As shown in [15], AFG algebras have sets and bags (multisets) of contexts as special cases, as well as the 'flat contexts' of [4]. Technically, AFG algebras simplify the proof of completeness, and in the proof it is pointed out where associativity is used.

Definition 7 (AFG algebras) *An associative finite ground algebra, abbreviated AFG algebra, is one that satisfies these criteria:*

- *it is associative, i.e. contains the equation*

$$(u \oplus v) \oplus w = u \oplus (v \oplus w)$$

- *every equation apart from associativity is restricted so that the variables in it can only be instantiated by constants, not by terms containing $\oplus$.*
- *the number of equations is finite.*

In semigroups, parentheses are redundant, so we may write for instance $(a \oplus b) \oplus (c \oplus d)$ as $a \oplus b \oplus c \oplus d$, or for that matter as $abcd$. The latter form is common in pure semigroups, i.e. where terms represent strings of atomic symbols and $\oplus$ represents concatenation of strings.

There are many other AFG algebras besides strings, however, and we prefer to retain the generic $\oplus$ symbol when writing terms from $C^{\oplus}$.

Observe that in sets, the axiom of commutativity can be restricted to context constants:

$$c \oplus d = d \oplus c \tag{8}$$

because with associativity we can get any permutation of

$$c_1 \ldots c_m$$

by a series of applications of (8) on adjacent elements.

Idempotence can then be adequately taken as an AFG equation too:

$$c \oplus c = c \tag{9}$$

because with associativity and commutativity it is possible to collect equal elements of

$$c_1 \ldots c_m$$

so that they are adjacent, and then repeatedly deleting an element where adjacent ones are equal by applying (9).

It is sometimes convenient to include a special context ϵ, such that

$$\epsilon \oplus u = u = u \oplus \epsilon \tag{10}$$

For example, in applications where there is an outermost supercontext, enclosing all other contexts, that could be ϵ.

5   Soundness and Completeness

We can prove the following soundness and completeness result for AFG algebras:

Theorem 1 (Soundness and Completeness) *For any set of AFG equations* E, $\Gamma \models_E u : \phi$ *if and only if* $\Gamma \vdash_E u : \phi$.

5.1   Soundness

Soundness of Rdw_c and Rup_c are direct from item 1 of the chain conditions, and RRI is sound by virtue of item 2 of the chain conditions. Soundness of the remaining bridge rule

$$\frac{v : \phi}{u : \phi} \quad u =_E v$$

follows because when $u =_E v$, any u-chain is also a v-chain.

5.2   Completeness

The completeness proof relies on canonical models which respect the bridge rules of our Multi Context system. The basic building blocks will be maximal consistent sets of well-formed formulae, adapted to the multilanguage environment. Let us state the versions of consistency and maximality that we need.

Definition 8 (x-consistency) *A finite set Δ of well-formed formulae is said to be x-consistent iff $\Delta \not\vdash_E x : \bot$, and an infinite set is x-consistent iff every finite subset is x-consistent.*

Definition 9 (x-maximality) *A set Δ of well-formed formulae is said to be x-maximal iff Δ is x-consistent and for all well-formed labelled formulae $y : \delta$ such that $\Delta \cup \{y : \delta\}$ is x-consistent, $y : \delta \in \Delta$.*

Theorem 2 (Lindenbaum) *Any x-consistent set of wffs can be extended to an x-maximal set.*

Proof: Start with an x-consistent set Δ_0 and an enumeration of all well-formed formulae $\langle x_i : \delta_i \rangle, i \geq 1$, and define inductively

$$\Delta_i = \Delta_{i-1} \cup \{x_i : \delta_i\} \text{ if } \Delta_{i-1} \cup \{x_i : \delta_i\} \text{ is } x - \text{ consistent}$$
$$\Delta_{i-1} \text{ otherwise} \tag{11}$$

Now

$$\Delta = \bigcup_{i=0}^{\infty} \Delta_i$$

is x-consistent, because otherwise by definition there would be a finite subset Δ^f such that $\Delta^f \vdash_E x : \bot$, and an index n such that $\Delta^f \subseteq \Delta_n$, contradicting x-consistency of Δ_n. To prove x-maximality, suppose $\Delta \cup \{x : \delta\}$ is x-consistent for some δ. Then $\Delta_{i-1} \cup \{x : \delta\}$ is also x-consistent, where i is the index of $x : \delta$ in the enumeration of wffs, therefore $x : \delta \in \Delta_i \subseteq \Delta$. $\square$

Canonical model. Now let us choose an arbitrary x-consistent wff $x : \delta$ and construct an x-chain for it. To begin with, we expand $\{x : \delta\}$ to an x-maximal set Δ by the construction in the previous lemma.

Definition 10 (Canonical model) *For all x-continuations $y =_E x \oplus c_1 \ldots \oplus c_h$*

  - *let $\Delta_y = \{\lambda \mid x : l_{c_1}(\ldots l_{c_h}(\lambda)) \in \Delta\}$*
  - *let S_y be the set of interpretations of the language L_y*
  - *let $T_y = \{s \in S_y \mid s \models \Delta_y\}$ be the subset of interpretations that validate Δ_y*
  - *and let m be the function that maps y to $\emptyset$ if $T_y = \emptyset$ and to $\{t\}$ otherwise, where t is some arbitrary member of T_y.*

  Our canonical model is the x-chain m.

Δ_y is well-defined, so m is really an x-chain. To see this, we prove that

$$x : l_{c_1}(\ldots l_{c_h}(\lambda)) \in \Delta \quad \text{iff} \quad x : l_{d_1}(\ldots l_{d_k}(\lambda)) \in \Delta$$

whenever

$$x \oplus c_1 \ldots \oplus c_h =_E x \oplus d_1 \ldots \oplus d_k. \tag{12}$$

In fact,

$$x : l_{c_1}(\ldots l_{c_h}(\lambda)) \in \Delta$$

iff, by h applications of Rup,

$$((x \oplus c_1) \ldots \oplus c_h) : \lambda \in \Delta$$

iff, by associativity,[1]

$$x \oplus c_1 \oplus \ldots \oplus c_h : \lambda \in \Delta$$

iff, by (12),

$$x \oplus d_1 \oplus \ldots \oplus d_k : \lambda \in \Delta$$

iff, by associativity,

$$((x \oplus d_1) \ldots \oplus d_k) : \lambda \in \Delta$$

iff, by k applications of Rdw,

$$x : l_{d_1}(\ldots l_{d_k}(\lambda)) \in \Delta.$$

As regards the model conditions, the m we have defined here trivially fulfills condition 2, and condition 1 is fulfilled because for $c \in C$ and an x-continuation $y = x \oplus c_1 \oplus \ldots \oplus c_h$, we have

$$y : l_c(\lambda) \in \Delta \quad \text{iff} \quad y \oplus c : \lambda \in \Delta$$

by Rup, Rdw, and x-maximality of Δ.

The x-chain m satisfies the wff $x : \delta$ (take $h = 0$), so we have completeness.

6   Related Work

Algebras of contexts were proposed in [15] for single-language axiomatic context systems. In that work the modal logic with the *ist* modality defined in [12,3,7] is augmented with an algebra on the domain of contexts (the first argument of the *ist* modality). The results extend to arbitrary equational varieties. The *ist* formalism lacks much of the expressivity of a multilanguage formalism for context reasoning. The present work contains the first formalization of multicontext logics with algebraically structured contexts.

---

[1] The construction does not depend essentially on associativity, so the results of this paper will be generalizable to arbitrary equational varieties by adjusting the definition of language mappings and using the initiality of term algebras. This is a work in progress.

A first proposal for an algebra of contexts is described in [6]. This work is more focused on the specific nature of the operation (the union) rather than a generic operation characterized via a set of equations, as it is here. On the other hand it seems that properties such as associativity, commutativity, idempotence, are relevant when one has to define operations between contexts.

A further theoretical framework, similar to a theory of contexts combination is proposed in [5], where contexts are formalized via a modal operator. The fact that ϕ is true in the context c is represented by the modal formula $\Box_c \phi$. Under this intuitive interpretation of modal formulas, combining contexts becomes combining modalities. An example of formula with a combined modality is $\Box_{c_1 \cup c_2} \phi$. This formula intuitively means that ϕ holds in the context $c_1 \cup c_2$, i.e., in the context obtained by combining c_1 and c_2 via a union operator. The main differences between this work and our approach are two. First, we have a general calculus for any operation definable via a set of equations, while [5] considers specific operations such as union, inverse, etc. Second, we deal with contexts with different languages, while in [5] all contexts share a common global language.

Algebras for contexts are relevant for the semantic web. The main assumption of the semantic web is that it is populated by a set of well-designed modular ontologies, each of which partially describes, from a specific perspective, a piece of knowledge. Constructing a new ontology is often a matter of assembling existing ones. Instead of building ontologies from scratch, one wants to reuse existing ontologies. Operations for combining ontologies are: ontology inclusion, ontology restriction, and polymorphic refinement. In [1], an argument is made in favour of encapsulating ontology in autonomous but partially coordinated contexts. Clearly to combine ontologies we need a way to combine contexts, which is what the present work is about.

7   Conclusion

In this paper we have described a general algebra for context combination in which the relation between source contexts and combined contexts is stored in the structure of the labels. This information was exploited semantically and proof-theoretically in order to infer relations between the content of different contexts. We have provided a semantics based on Local Model Semantics for the general case, and a proof method, based on Multi Context System for a limited, but significant, case, namely the set of contexts that are built via an operation axiomatisable by an AFG algebra.

The results of this paper can be lifted to other groupoids, and equational varieties in general, at the expense of slightly more complicated definitions and proofs. Furthermore, the results will generalize to local languages other than the propositional ones considered here, if only a minimal notion of satisfiability is defined. Work in these directions is in progress.

References

1. M. Benerecetti, P. Bouquet, and M. Bonifacio. Distributed context-aware applications. *Human Computer Interaction*, 16:213–228, 2001. Special issue on Context-aware computing.
2. M. Benerecetti, P. Bouquet, and C. Ghidini. Contextual reasoning distilled. *Journal of Theoretical and Experimental Artificial Intelligence*, 2000.
3. Saša Buvač, Vanja Buvač, and Ian A. Mason. The semantics of propositional contexts. In *Proceedings of the Eight International Symposium on Methodologies for Intelligent Systems*, volume 869 of *Lecture Notes in Artificial Intelligence*. Springer Verlag, 1994.
4. Saša Buvač, Vanja Buvač, and Ian A. Mason. Metamathematics of context. *Fundamenta Informaticae*, 23(3), 1995.
5. T. Costello and A. Patterson. Quantifiers over contexts. In *KR & R*, 1998.
6. C. Dichev. Theory relations and context dependencies. In *Proceedings of the 6th Australian Joint Conference on Artificial Intelligence (AI-93)*, pages 266–272, Melbourne, 1993.
7. Dov Gabbay and Rolf Nossum. Structured contexts with fibred semantics. In P. Bonzon, M. Cavalcanti, and R. Nossum, editors, *Formal Aspects of Context*. Applied Logic Series, Kluwer, 2000.
8. Chiara Ghidini. *A Semantics for Contextual Reasoning: Theory and Two Relevant Applications*. Ph.d. dissertation, Dipartimento di Informatica e Studî Aziendali, Università degli Studî di Roma "La Sapienza", Roma, 1998.
9. Fausto Giunchiglia. Contextual reasoning. *Epistemologia*, special issue on I Linguaggi e le Macchine(XVI):345–364, 1993. IRST Technical Report 9211-20, IRST, Trento, Italy.
10. Fausto Giunchiglia and Luciano Serafini. Multilanguage hierarchical logics (or: how we can do without modal logics). *Artificial Intelligence*, 65:29–70, 1994.
11. Ramanathan Guha and Douglas Lenat. Language, representation and contexts. *Journal of Information Processing*, 15(3):340–349, 1992.
12. Ramanathan V. Guha. *Contexts: A Formalization and Some Applications*. PhD thesis, Stanford University, 1991.
13. David Kaplan. Demonstratives: an essay on the semantics, logic, metaphysics, and epistemology of demonstratives and other indexicals. In Joseph Almog, John Perry, and Howard Wettstein, editors, *Themes from Kaplan*, pages 481–563. Oxford University Press, Oxford, 1989.
14. J. McCarthy and S. Buvač. Formalising context, (expanded notes). In A. Aliseda, R. van Glabbeek, and D. Westerståhl, editors, *Computing Natural Language*, volume 81 of *CSLI Lecture Notes*, pages 13–50. Stanford University, Center for the Study of Language and Information, 1998.
15. Rolf Nossum. A uniform quantificational logic for algebraic notions of context. Skriftserien 82, Agder University College, N-4604 Kristiansand, Dec 2001. ISBN 82-7117-448-7.
16. John Perry. Indexicals, contexts, and unarticulated constituents. In Atocha Aliseda, Rob van Glabbeek, and Dag Westerståhl, editors, *Computing Natural Language*, pages 1–11. CSLI Publications, Stanford, California, 1998.
17. Luciano Serafini and Chiara Ghidini. Context-based semantics for information integration. In Pierre Bonzon, Marcos Cavalcanti, and Rolf Nossum, editors, *Formal Aspects of Context*, pages 175–192. Kluwer Academic Publishers, Dordrecht, 2000.
18. Luciano Serafini and Fausto Giunchiglia. ML systems: A proof theory for contexts. *Journal of Logic, Language and Information*, July 2001.

Indefinite Integration as a Testbed for Developments in Multi-agent Systems

J.A. Campbell

Department of Computer Science
University College London
Gower Street, London WC1E 6BT, UK
jac@cs.ucl.ac.uk

Abstract. Coordination of multiple autonomous agents to solve problems that require each of them to contribute their limited expertise in the construction of a solution is often ensured by the use of numerical methods such as vote-counting, payoff functions, game theory and economic criteria. In areas where there are no obvious numerical methods for agents to use in assessing other agents' contributions, many questions still remain open for research. The paper reports a study of one such area: heuristic indefinite integration in terms of agents with different single heuristic abilities which must cooperate in finding indefinite integrals. It examines the reasons for successes and lack of success in performance, and draws some general conclusions about the usefulness of indefinite integration as a field for realistic tests of methods for multi-agent systems where the usefulness of "economic" criteria is limited. In this connection, the role of numerical taxonomy is emphasised.

1   Introduction

Apart from a few obviously algorithmic topics (evaluation and simplification of algebraic expressions, differentiation), symbolic mathematical computing was in its early days a well-recognised sub-area of artificial intelligence because it was an excellent source of tests and illustrations of heuristics. Because symbolic computation has progressed by moving away from heuristics towards the use and improvement of algorithms for mathematical operations, the degree of recognition within AI has decreased: AI is basically not about algorithms. This is understandable, because the primary interest of a subject is usually the technical content of the subject itself. The fact that the state of its art happens to make it a useful area of application for AI (or not) is secondary. When this state of the art ceases to be heuristic, the history of AI says that the heuristics themselves have typically ceased to be of interest for AI research.

But "typically" does not mean "always". There are situations where heuristics left behind by the progress of a topic towards an algorithmic state of the art are still useful - especially if they can help to throw light on current *general* problems in AI. One such general problem-area is the behaviour of autonomous agents in reaching consensus in a multi-agent system, where some coordination

J. Calmet et al. (Eds.): AISC-Calculemus 2002, LNAI 2385, pp. 102–116, 2002.

of varieties of expertise is needed for the solution of problems but where an individual agent has no more than one particular expertise.

This paper suggests that a good testbed for questions of consensus and coordination in multi-agent systems is indefinite integration, if the examples of particular expertise that are used are single heuristics of the kind taught to students of calculus. That is, indefinite integration is of value for studies in multi-agent systems if we set aside the progress in algorithmic integration in finite terms that started with the work of Moses (on the SIN package) [8] and Risch [13], and revive the heuristic understanding of the subject that was current in about 1965 and which is still taught at the level of first-year undergraduate calculus courses.

Section 3 below surveys the relevant topics from agency and multi-agent systems. Section 4 puts a version of heuristic indefinite integration into this framework, and section 5 summarises what has happened when the version was tested on a representative sample of exercises in integration. Section 6 discusses the experience gained in the work, mentions topics that deserve further research, and argues for further attention to be paid to integration as a testbed for ideas that may enhance future multi-agent systems.

2 Related Work

Although the standard methods for indefinite integration are now fully algorithmic [2,5], and commercially-available packages like Mathematica make use of them, there is still some activity on and around integration by heuristic means. This is typically either to improve the flexibility of services available to users, with the help of large repositories of information (as in the TILU project [17] of R. Fateman, T.H. Einwohner and collaborators), or for reasons connected with human problem-solving and mathematical education [4,16]: it is desirable to understand better the use of heuristics, and even to automate some of the tutoring of students in such uses. Indefinite integration is an obvious target-area. The emphasis is on embodiment of heuristics that students do or should adopt.

The exercise reported in the present paper has the different motivation of studying the interaction of simple (but not necessarily student- friendly) heuristics or heuristic fragments in a multi-agent framework. But there is certainly scope for a future combination of the two motivations, e.g. in considering how software agents could communicate with and help human agents.

On that last point, there is already relevant work [10] which uses a multiple-agent scheme for the modelling of (social) interactions in mathematical creativity, and various software projects (e.g. [1]) for computational algebra, mainly in connection with development of proofs. These are of interest even if they do not treat integration as a problem-area; many of the questions concerning agents are not specific to a particular topic.

The design of autonomous agents and systems of such agents is a large field, as various monographs (e.g. [18]) show. The most relevant results are mentioned in Section 4. Two topics that have not received much attention, despite their likely importance in the future, are the determination of preferences between alternatives when economic or other simple numerical criteria are not evident, and the use of existing examples or cases to permit agents to decide on current actions. Heuristic integration was chosen as a test field precisely because both of those considerations occur routinely within it.

3   Consensus and Coordination among Multiple Agents

The outline of the usual cycle in an approach of a multi-agent system to a problem is:

(i)   it is decided what the next relevant task or subproblem P is;
(ii) agents receive information on (i), assess its relevance to themselves, and state their opinion about participation in any attempt to solve P;
(iii) (optionally) statements from (ii) may involve questions about possible sharing with other agents the effort towards solving P, and there may be several iterations of (ii) and (iii) until a final pass through (ii) occurs, leading to some conclusive output (offer, promise, refusal to contribute to the work on P) from each agent;
(iv) P is allocated to one agent, or some part of the work is allocated to each member of a (sub)set of the agents;
(v)   the work allocated in (iv) is performed or attempted, and the results are made available.
(vi) (optionally) some assessment of the results of (v) is carried out, and its conclusions are fed back into (i) in the next iteration of the cycle if the original goal of the computation has not yet been reached.

Although straighforward in outline, this description hides questions of technical detail which are in some respects still subject to research and improvement. In particular:

1. how is the decision in (i) made?
2. how do agents determine relevance and the opinions that follow rom this, in (ii)?
3. what forms can the iterations in (iii) take, and what are the conditions for the iterations to terminate?
4. how is the allocation in (iv) decided?
5. how is the quality of the results from (v) assessed in (vi)?

There are some standard answers to those questions under particular conditions. If the conditions are fulfilled for a particular application, multi-agent systems have functioned quite well for it in practice - hence the rapid recent expansion of interest in such systems. Nevertheless, it is still desirable to extend the range of conditions under which they can function well. This is a motivation for the present paper.

Indefinite integration is a good test example for a number of reasons, as indicated in the experience reported in Section 5. In particular, unlike some problems for which the underlying knowledge compels only one realistic way of answering questions (1) and (2) above, it allows many plausible alternative approaches to be evaluated and compared. Conclusions from any such comparison are relevant for multi-agent systems in general. (The present project has not yet taken much advantage of this flexibility). Because skill in heuristic integration relies significantly on ability to recognise similarities of an exercise with past completed or failed exercises, it is a good source of particular or case-like knowledge for research on how agents can exploit cases in addition to their usual stocks of generalisations via rules, inheritance networks etc.

The summary in the rest of Section 3 unites scientific contributions from various authors. Further background, and references, can be found in the book edited by Weiss [18] or any of the (few, so far) other good general monographs on multi-agent systems.

3.1   Control of a Multi-agent System

The main distinction with respect to (1) is between dictatorial and democratic methods. At one extreme, a super-agent with an understanding of an overall problem-area uses this understanding to decide what should be done next, which agent(s) should do it, and how well the various agents have been dealing with their assigned jobs. At the other extreme, each agent has a vote of equal weight concerning the next step for the system, and the voting majority always wins. (In principle this implies that the agents have at least some of the understanding that a super-agent would possess - but even when this is not so, as in politics, the outcome is usually somewhat better than totally random confusion). There are various intermediate possibilities, e.g. elected or selected semi-super-agents with final responsibilities over parts of the overall expertise.

The typical present approach is more dictatorial than democratic. The original dictator was the "blackboard manager" in the blackboard- architecture scheme [9] pioneered in HEARSAY (often regarded as the first successful example of distributed AI - for real-time speech recognition - and the immediate ancestor of multi-agent systems). This manager had enough knowledge of the application to be able to manage the blackboard (a central combined repository of results and agenda) dynamically and rearrange the agenda creatively.

The HEARSAY application was near-linear; it consisted of several tasks that (for a short passage of speech) could be performed in sequence, with limited overlap and feedback between stages. The manager's main unifying knowledge was in scheduling, and in responding to demands for feedback (e.g. of the form "this phrase of text makes no sense; can you compose another one out of the phonemes that have just been used?") from the software processing one of the tasks. Where an application does not have this near-linear character, it is usually not possible to build a convincing dictatorial super-agent.

It is more common to find administrative super-agents that behave dictatorially in enforcing the results of a tendering process, but without resistance because agents are engineered to respect the conventions of the process of calling for bids to a tender specification and having the tender for a task awarded to the bid that is best according to a fully-known prior criterion. This describes the "contract net" protocol [14], due originally to R. Smith. For a contract net, a super-agent is just a convenience that reduces the overheads of running a multi-agent system. (Given full transparency of the bids, all the agents could access the information and agree on what the award of the tender should be).

The contract net is semi-democratic, because a bid amounts to a weighted vote. This is a simple quantitative criterion. Quite often in knowledge-intensive subjects, however, experts are reluctant to reduce their opinions to numbers: they see their exchanges of opinions with other experts (or the teaching of apprentices) in terms of statements in the language of their expertise. Of course, the opinions must ultimately be expressible as single numbers, because when an opinion or result X is finally preferred to another one Y, it must then be possible to associate some numbers x and y with them such that $x ¿ y$. But experts in many areas would prefer that to happen only at the last step, after all other considerations, negotiations etc. have been exhausted.

For the study reported in this paper, the intention was to choose a test topic for which this was substantially true and where no obvious real-number measure of progress towards solving problems existed. At the same time, for the results to have some general interest it was desirable to avoid a topic where the knowledge was highly specialised or was complicated to express. This is part of the explanation for the choice of indefinite integration as a topic.

3.2   Agents' Reactions to Statements of Tasks

A non-autonomous agent in AI can be regarded as an object, In the sense = of object-oriented computing. If it receives a message containing a problem that it recognises as being appropriate to its internal method(s), it answers with a message containing the result that it computes. If an object is part of a multi-agent system, it may be rebuilt slightly so that it can distinguish between messages of the types "Does this problem match your method(s)?" (to which the expected response is Boolean) and "Solve this problem". For neither type does it display autonomy; its designer would always be able to predict whether its Boolean "bid" in stage (ii) under a contract-net protocol would be 1 or 0.

An autonomous agent is more than an object: it has some internal characteristics that influence its responses to external messages. A common scheme for expressing the characteristics is the BDI model [12], in which the information essential for autonomy is represented inside the agent in terms of beliefs (about the current problem, other agents, and itself), desires (goals for itself) and intentions (relatively short-term plans or means for trying to achieve its goals). This scheme is both general and flexible; it gives considerable freedom to experiment with the details of how to represent and use all three components. It is therefore respected here.

In a standard contract-net scheme, an agent's first response to the statement of a task would be a numerical bid (sometimes 0, or "I do not wish to be involved"). This implies that the agent can carry out a numerical computation about the value (in terms of its own view of the world) of being awarded the tender for the task, and therefore that its relevant BDI information is numerical or can yield numbers for the computation.

The most thoroughly-developed formalism for measuring advantage numerically is economic, with reference to payoff functions, penalties and profits, properties of markets and games, bargaining based on price, etc. This has already led to an accepted set of techniques, from economics and game theory, for multi-agent systems. It is therefore more interesting now to examine alternatives where these techniques do not capture the primary sense of the knowledge, including BDI knowledge, in an application. Integral calculus is one such application.

3.3 How Agents with Different Views Arrive at a Consensus

The simplest way of obtaining consensus has been described above: the contract-net protocol. It relies on the quantitative considerations in Section 3.2. In a situation where the stage (iii) of the process summarised at the beginning of Section 3 is absent, the contract net can bring about a consensus in one step.

If stage (iii) is needed, then the iterations of stages (ii) and (iii) amount to an exercise of bargaining, where trade-offs are involved. For example, an agent with a desire to get the payoff that might follow from being allocated all of a task T with parts T1 and T2 where it is rather more (cost-)effective at doing T1 than T2, but having a limited capital on which to bid, might appreciate from what other agents had said in a first round of (ii) that it could not hope to compete with other bids for the entire task, and might therefore try to interest high bidders B for T in a deal where it subcontracts to do T1 on terms that might be advantageous to B. One could even find an agent introducing T1 and T2 into the debate because of its knowledge of the structure of T despite the fact that a tender specification had mentioned only T and not the parts.

Stage (iii) is probably, for AI, the most interesting stage of the behaviour of a multi-agent system. Most of the interest is in the properties of the iteration. For various models of total or bounded economic rationality of agents, termination is assured (or arbitrary termination can be imposed otherwise, along the same lines as in Section 3.5).

Negotiation among agents, which is a large subject in itself [7], has almost always referred to models of that kind. The situation is different, and deserves more research, if we do not have that tight and quantitative economic framework at our disposal. There are obviously applications where such a framework is not evident; the issue is mentioned near the end of Section 3.1. This is a further motivation for experimenting with integral calculus.

3.4 Task Allocation

For a system using the contract net, the allocation itself is simple. After the agents have devoted effort to formulating bids, coalitions and/or subcontracting arrangements, the bids can be compared automatically and a criterion for the best bid applied to select the winner(s). This criterion may be that the lowest bid is the best, or some modification (e.g. use of the second-lowest bid in deciding how/ where a contract will be awarded) selected to minimise adverse effects of bidding that may be rigged against the entity that makes the specification for the tender.

Other ways of allocating tasks have also been used. If some part of a system is given dictatorial powers, that part may assign tasks to agents unilaterally, with reference to (say) known competences of the agents and the competences required for those tasks. This has the drawback of requiring the dictator to understand the whole computation well enough to make good assignments consistently and effectively. In a position intermediate between dictatorship and a pure contract net, assignment depends on some combination of bidding and evidence of agents' ability to perform tasks of the kind currently under consideration. A very simple form of evidence is the labelling of any "known competences" that may be attached to an agent when it is created. Richer forms may involve records of previous histories of good and bad performances by the individual agents during solution of past problems, or indications that agents can produce convincing solutions to trial exercises which are cut-down or abstracted versions of the latest task that is to be assigned.

The study reported here was intended in part to explore the usefulness of such intermediate approaches, which are at present less well understood than dictatorships or contract-net schemes. It turns out that indefinite integration has some features that make it natural to adopt an intermediate approach.

3.5   Assessing the Quality of Intermediate and Final Results

In some applications, a payoff function exists that rates the value of any result. Progress can then be ensured by demanding that the results of successive steps increase the payoff, or at least that they do not decrease it significantly and systematically. Alternatively - and equivalent to it - the attempt to solve a problem can be formulated as a search in a state space, where the nature (e.g. the set of constraints that only a goal can satisfy) of the goal state is known and where an estimate of the distance from the latest state to the goal is available. The search is then directed towards new states that show the biggest reduction in the estimated distance to the goal.

Termination occurs when the goal is reached, or it can be enforced when the percentage of improvement in the payoff between successive steps falls below a given threshold, or by "satisficing": acceptance of the first result found whose payoff value is above some preset threshold for acceptability.

These considerations rely on numerical estimates. When there are no obvious means of numerical estimation, as in the situations described near the end of Section 3.1, some knowledge-intensive alternative must be used. The subject of multi-agent systems gives no general guidelines for these situations. It is therefore desirable to experiment with realistic examples of the situations, to define possible alternatives and to draw conclusions about how they work in practice. This is one further justification for studying indefinite integration, where natural payoff functions and clear formulations of appropriate state spaces are not evident.

4   Indefinite Integration as a Multi-agent Application

Students learn various techniques for taking the next step in a multiple-step process of finding an indefinite integral of a given expression. Good students learn also how to select the best one for each stage of the process. Less effective students may learn one technique well and try to apply it to every step where it has any effect. No student of this kind is likely to get good marks in tests, but a collective made up of one-technique students with different techniques may be able to pool their efforts and produce acceptable results. This is the view under examination in the present work. The equivalent of a one-technique student is an agent that has one approach for any problem of indefinite integration.

Examples of one-technique approaches embodied in agents include: access to definitions that can be regarded as basic (integrals with respect to x for constants, polynomials in x, trigonometric functions, and instances such as $1/x$ and $1/(1 + x^2)$); integration by parts; guess and differentiate; half-angle substitution ($x \rightarrow$ trigonometric function of some half-angle); trigonometric product substitution; algebraic transformation/matching and access to a memory of solved examples).

A scheme based on this model of behaviour is described in detail in Section 4.1. The scheme's consequent implications for agents and for knowledge held in the multi-agent system are outlined in Section 4.2.

4.1   The Operation of a Multi-agent Computation for Integration

The pooling process starts with a contribution from each agent to the attempted solution of a problem. This contribution is either null (an indication that the problem contains no features that the agent can relate to its own expertise) or a one-step development of the problem. For example, an agent specialised for a trigonometric substitution offers a restatement of the problem in which the substitution has been made.

Following this first round of contributions, each agent receives the opportunity to comment on the contributions of all the other agents. Agents comment on their own contributions in the same way as they comment on others.

To maintain the essential character of a system of agents with limited expertise, agents do not have any advanced knowledge of integration that can be used to generate comments. Examples of individual agents' simple observations applicable at this point are: interest in a contribution because its form allows an agent to use its own technique easily on that contribution; approval because a contribution has reduced the length or apparent difficulty of the problem (or disapproval if that quantity has been increased); detection of similarity between a contribution and some stored information about the solution of a past problem.

At present, these observations are implemented simply; essentially by inspection of syntactic details such as lengths of expressions E, presence or absence of functions or sub-expressions that an agent can integrate and/or that occur in its own case knowledge, and changes in these details between E and what it becomes after an agent applies its own integration method once to E. We would not regard a student with just this limited knowledge of integration as a good student; nevertheless, in computations, it is quite effective (while not being particularly efficient). There is certainly room for research on adding more advanced or realistic knowledge, especially if we wish to build a system that performs at the level of a group of good students. But the point of the present project is different: investigation of whether and how a group of rather unsophisticated agents can produce overall behaviour that does not seem unsophisticated.

In effect, each agent always has access to the standard differentiation and algebraic simplification procedures, and can therefore check by differentiation whether its latest contribution is the answer to the overall problem. The first contribution that is announced as the answer after this checking then receives no further comment: the system returns it as such, and the computation terminates.

After a round of comments, the comments themselves are evaluated further. Contributions that receive no approvals or indications of interest (in the senses described above) are eliminated. An agent expressing interest in a contribution is given the opportunity to operate once on that contribution, and the outcome O is noted. Detection of similarity with something in the records of past solutions is treated in the same way, because it is equivalent to an expression of interest: the detecting agent indicates an O which is an outcome of the detection.

The evaluations can be regarded as leaf nodes in a tree whose root is the problem in its current state. The nodes representing contributions that have received only approvals are at depth 1; those that have just received further attention are at depth 2. The depth-2 nodes are again evaluated further.

In principle this kind of look ahead can be pursued recursively to depth n until all depth-n leaf nodes have approval ratings and no agent indicates that it wants to evaluate any such node further. However, in practice there is no guarantee that the recursion will terminate for arbitrary problems of integration, and no problem tested so far has

shown any apparent benefit from being followed beyond a depth of 2. This is a provisional conclusion, based on experience obtained until now.

When the look ahead is completed, the contributions surviving from the first round possess expressions of interest and approval, or merely approval. Approvals are just votes or numbers; expressions of interest can be counted also, but in addition, because of the look ahead beyond depth 1, they carry non-numerical information. How to rate them by comparison with approvals is a question of the kind raised near the end of Section 3.1; this is discussed further in section 6. But it can be assumed that an expression of interest is not worth less than an approval. This permits approval-only contributions whose numbers of approvals are less than the maximum number of approvals plus expressions of interest found among the contributions to be eliminated.

If more than one contribution remains, there is no obvious way to select the winner (which is then used as the starting-point for the next cycle). This is an issue for further research, which is considered in Section 6. A simple way is to rate a contribution with A approvals and E expressions of interest as worth A

$+ nE$ for some n (e.g slightly greater than 1), and to break any tie by a random choice.

The overall process has the obvious weakness that agents may not be sufficiently expert about problems outside their competence to be able to state reliable approvals or expressions of interest. This reliability could be improved if agents were given more and wider knowledge - but that would defeat the first purpose of the present study, which is to investigate whether a system of specialised agents can produce anything better than simple specialised behaviour. This is as interesting as, say, investigating the behaviour of a programme committee for a conference covering a wide and new field, with respect to selecting a set of papers that does a good job of representing the state of the art of the whole field.

4.2   Choices for Representation of Knowledge

If the multi-agent system contains only techniques and no memory for results, the performance is poor. In tests, it has then solved only simple problems, and not even all the simple problems tested have been solved. This is not a surprise. In the most convincing applications of agent technology in AI, a stock of relevant knowledge is held somewhere: in a communal store, or spread over the individual agents, or both.

Holding generalised knowledge (e.g. in rules) improves performance in many fields. But when a field does not have many such rules, and if they would permit good performance merely by being built into a single and fairly simple program, there would be no advantage in dividing them up to make some more complicated (e.g. multi-agent) software for the same job. This is true of indefinite integration.

This leaves particular knowledge (examples of previous solutions), which is a significant part of human problem-solving knowledge for integration. It is well suited for use in a multi-agent system, and is employed here.

The actual representation used is a repository held independently of the agents. An item of knowledge should be at least a statement of an integrand and its integral, but typical entries also contain sequences of steps leading from the former to the latter. Some entries have been developed by hand; others have been extracted from traces of the computation of the system. Negative knowledge has been useful unexpectedly often: if an integrand is associated with a transformed version V in a sequence that is recorded as leading nowhere, and an agent detects a similarity between V and a contribution to a current problem-solving exercise, that agent can (and does) express disapproval of the contribution.

The present format of the repository allows for annotations of the entries. It is intended as a representation for cases, in the sense of case-based reasoning [6]. Some annotations have been added to negative entries, to explain why they have led to unsatisfactory behaviour, though with limited use so far. The main concept used in treating the entries as cases is the determination of similarity (between parts of the entries and contributions to a current problem of integration). Similarity is estimated by a distance metric set up on the characteristics (e.g. function types) that figure in these items, by methods of numerical taxonomy [15]. This is a subjective estimation because the construction of distance metrics is subjective, but that is in the basic nature of numerical taxonomy. The justification here, as in the other areas where it has been applied, is that its results are convincing.

Metrics express distances between entities A and B, e.g. the distance between physical locations in terms of their N coordinates in an N- dimensional space. One adds together the pth. powers of the absolute values of the differences between the coordinates of A and B in the individual dimensions, and takes the pth. root of the sum. The most common metrics are the Euclidean ($p = 2$) and Manhattan ($p = 1$) metrics. $p = 1$ is used here. The greater/smaller the distance between A and B, the smaller/greater is their similarity.The "art" of numerical taxonomy consists of arriving at a good scaling between dimensions for properties that are (unlike map coordinates) not of the same type, mapping non-numerical information onto a numerical scale, and making good choices for the representative properties themselves. For example, if one of the properties is at first sight Boolean - say, explicitly trigonometric character of an integrand, so that its contribution to the distance between two integrands is either 1 or 0 - should "explicitly exponential character" be a separate property (as a school student might believe) or not (as a specialist in integration might believe)? And how should specialist knowledge that the Boolean picture is oversimplified be exploited so that distances between 0 and 1 become possible? As stated in the previous paragraph, the answers contained in the present system are subjective, involving trial and error. They can certainly be improved (e.g. the answer to the last question above is at present the answer of the student and not the specialist), but they have not led to unsatisfactory behaviour.

The fact that case-like and "past history" knowledge is held outside the agents is a detail of implementation. The knowledge acts like the agents' bank. Each agent has an account, consisting of all the items of knowledge that it can access. An item is labelled with identifiers for the agent or agents that can see it during normal operation. In addition, each agent is given access to a certain number of items, selected randomly, which are not in its own account, if that agent has searched through that account without finding anything relevant to its reason for initiating the search. Some improvement of problem- solving effectiveness has been noticed (via agents' comments on contributions that other agents have made during the basic problem- solving cycle) when that number is raised from 2 to 3, but it is too early to draw general conclusions from that behaviour.

5   Experience of the Multi-agent System in Use

Examples have been taken from B. Peirce's compilation [11] of standard indefinite integrals, to cover roughly the range of types and difficulty of problem that are seen in first-year undergraduate courses in calculus (including polynomials, trigonometric and inverse trigonometric functions, exponentials, logarithms, fractional powers, products, and rational expressions whose integrals are not particularly complicated).

An example of a set of contributions from specialised agents is presented below, to give an indication of the method at work. Consider the first step for

$$\int x \cos 2x \, dx.$$

Some of the contributions are
(Integration by parts)

$$(1/2)x \sin 2x - (1/2) \int \sin 2x \, dx \tag{1}$$

(half-angle substitution)

$$(1/2) \int \tan(t/2) \ \sec^2(t/2) \ \cos(2 \tan t/2) \, dt \tag{2}$$

(adding 0 and rewriting)

$$\int (x + a) \cos 2(x + a - a) \, dx - a \int \cos 2(x + a - - - a) \, dx \tag{3}$$

(guess and differentiate)

$$(1/2)x \sin 2x \tag{4}$$

(access to past history)

$$\int x \cos x \, dx = x sinx - \int \sin x \, dx = x sinx + cosx \tag{5}$$

(2) and (3) each attract no expressions of interest, and (2) additionally gets a disapproving vote for excessive length. The other three are investigated over one more step of looking ahead. (1) receives the substitution $2x \to z$, which produces the new subproblem of integrating $\sin z$ with respect to z; (4) is differentiated, leading to the subproblem of finding an integral for $\sin 2x$, and (5) receives the substitution $x \to 2z$, which leads to a calculus-free problem of algebraic simplification and (re)substitution. Algebraic operations are part of the underlying environment, not part of the agents' expertise, and are applied at each step before agents see the results of the previous step. On the next step, therefore, it is observed that (5) has already led to a solution. (The process is robust, here and in general: if some knowledge, such as that supporting (5), had been absent but the remaining knowledge could support solution, a solution could be found - as would happen from (1) and (4) if they passed through one more step of looking ahead).

The case-like stock of knowledge mentioned in Section 4.2 has 22 entries at present, taken with no particular selection criteria from Peirce's examples.

On a sample of 45 problems, each being either an example quoted by Peirce or derived from one by algebraic transformations and substitution, the system reached the correct answer 36 times with cutoffs (e.g. the look-ahead depth) set conservatively, as at present. In the remaining 9 examples, 5 were victims of this conservative attitude, but (because the scheme does not always follow a sequence of steps to a solution that a mathematician would regard as efficient) would have required generous cutoffs in order to reach their targets. Of the others, 3 had not found an appropriate search path because the knowledge base (range of single-agent expertises or relevant information in the case-like stock of knowledge) was not yet rich enough for the problems, 1 was in effect in a loop for the same reason, and 1 had traversed a large loop from a trigonometric problem P to the eventual discovery that this was equivalent to a problem $0 + P$. (A cure for that particular instance of agents' failure to cooperate effectively is easy enough to produce, but it has not been examined thoroughly for side-effects that might appear elsewhere).

There was no apparent correlation between difficulty of a problem, from the viewpoint of integral calculus, and its presence in the set of 9 exercises for which the system did not compute an integral.

The computer-algebra part of the software is a development of an old set of programs [3] originally written to deal with integration of functions of a complex variable. It would be preferable to employ a good general computer algebra system, if it were easy, or even possible, to control (e.g. for single steps of integration) or access (e.g. arbitrary syntactic patterns of expressions and their parts) that system externally from an AI-based program.

6   Discussion

By the standards of symbolic mathematical computation, the results in Section 5 are not impressive. But that is (fortunately) not the point. The paper reports a successful use of an old approach to a traditional symbolic computing application, as a demonstration that that application is a good testbed for experimentation with and assessment of alternative ideas about how to build systems of autonomous agents with different expertises that contribute to the solution of exercises in a non-trivial area - particularly where there is no natural or immediate numerical way (involving notions of payoffs and similar economic criteria) to prefer one agent's contribution over another. How to deal with that latter situation is a live research issue in AI, which can only benefit from detailed and controlled further experimental work. Indefinite integration has all the right properties to be a test field.

A good example of an associated question for research is in how to treat and rate non-quantitative comments about the qualities of different items when it is necessary to choose the one item that is "best" or most popular. In the present work (and probably in general), a qualitative comment contains or implies ratings by several different criteria, including criteria applied by an entity other than the source of the comment (e.g. criteria of distance from or relevance to a goal, demands for resources in following up the commented suggestion, perceived quality of related past comments from the same source). In the work reported here, it has been found that identifying and distinguishing the various possible criteria gives significantly better results than turning each comment immediately into a slightly enhanced numerical vote (see the end of Section 4.1). It has also been found that evaluating the criteria with respect to each other and to numerical data like votes of approval can be reduced (at least for the material studied so far) to the measurement of distances between qualitative items (or "operational taxonomic units", in the language of [15]), after first establishing a convincing scaling between the different criteria or dimensions.

There are further topics in which research of general interest - not just interest for indefinite integration - is possible, For example, there is scope for a more substantial investigation of the effects of the volume, kind, representation (case-based or otherwise) and accessibility (for all agents, versus accessibility only for individual agents or groups of agents) of knowledge in a multi-agent system on the performance of that system. Case-based, historical, episodic etc. knowledge is clearly relevant in knowledge-based computations where the emphasis is on particular rather than generalised knowledge - as is usually true in multi-agent computations - but multi-agent systems research has paid rather little attention to this knowledge and its possible exploitation, to date.

Acknowledgement. It is a pleasure to acknowledge the hospitality of Prof. J. W. Perram and the Maersk Mc-Kinney Moller Institute for Production Technology, University of Southern Denmark, where this work was begun.

References

1. C. Benzemüller, and V. Sorge. Critical Agents Supporting Interactive Theorem Proving. In *Progress in Artificial Intelligence*, eds. P. Barahona and J. J. Alferes, 208-221 (Springer-Verlag, Berlin, 1999)
2. M. Bronstein. *Symbolic Integration I: Transcendental Functions* (Springer-Verlag, Berlin, 1997)
3. J. A. Campbell, J. G. Kent, and R. J. Moore. *B.I.T.*, **16**, 241-259 (1976)
4. A. Cardon, and C. Moulin. *Adaptation of a Learning System to the Student's Behaviour.* Paper No. 185, PSI-LIRINSA, INSA de Rouen, Mont Saint-Aignan, France (1996)
5. J. H. Davenport, Y. Siret, and E. Tournier. *Computer Algebra Systems and Algorithms for Algebraic Computation.* (Academic Press, London, 1993)
6. J. Kolodner. *Case-Based Reasoning* (Morgan Kaufmann, Los Altos, CA, 1993)
7. S. Kraus. *Strategic Negotiation in Multiagent Environments* (MIT Press, Cambridge, MA, 2001)
8. J. Moses. *Comm. A.C.M.*, **14**, 548-560 (1971)
9. H. P. Nii. Blackboard Systems: The Blackboard Model, Problem Solving and the Evolution of Blackboard Architecture. *AI Magazine*, 38-53 (August 1986)
10. A. Pease, S. Colton, A. Smaill, and J. Lee. = A Multi-Agent Approach to Modelling Interaction in Human Mathematical Reasoning. Report EDI- INF-RR-0056, Division of Informatics, University of Edinburgh, Scotland (2001)

11. B. Peirce. *A Short Table of Integrals* (Ginn & Co., Boston, MA, 1957)
12. A. S. Rao, and M. P. Georgeff. Modeling Rational Agents with a BDI- Architecture. In *Principles of Knowledge Representation and Reasonin g*, eds. J. Allen, R. Fikes, and E. Sandewall, 473-484 (Morgan Kaufmann, San Mateo, CA, 1991)
13. R. Risch. *Trans. Amer. Math. Soc.*, **139**, 167-189 (1969)
14. R. Smith. *IEEE Transactions on Computers*, **29**, 1104-1113 (1980)
15. P. A. Sneath, and R. R. Sokal. *Numerical Taxonomy* (Freeman, San Francisco, 1976)
16. V. Tarasov. Applying Intelligent Agent Technology to Create Instruction Programs. *Proc. FIPW'297-98: Developments in Distributed Systems and Data Communications*, 190-204 (Petrozavodsk, Russia, 1998)
17. A TILU Web site is at `torte.cs.berkeley.edu:8010/tilu`
18. G. Weiss, editor. *Multiagent Systems* (MIT Press, Cambridge, MA, 1999)

Expression Inference – Genetic Symbolic Classification Integrated with Non-linear Coefficient Optimisation

Andrew Hunter

Department of Computer Science,
University of Durham
Science Labs, Durham, UK
andrew1.hunter@durham.ac.uk
http://www.durham.ac.uk/andrew1.hunter/index.html

Abstract. Expression Inference is a parsimonious, comprehensible alternative to semi-parametric and non-parametric classification techniques such as neural networks, which generates compact symbolic mathematical expressions for classification or regression. This paper introduces a general framework for inferring symbolic classifiers, using the Genetic Programming paradigm with non-linear optimisation of embedded coefficients. An error propagation algorithm is introduced to support the optimisation. A multiobjective variant of Genetic Programming provides a range of models trading off parsimony and classification performance, the latter measured by ROC curve analysis. The technique is shown to develop extremely concise and effective models on a sample real-world problem domain.

Keywords. Symbolic Regression; Classification; Genetic Programming; ROC Curves; Multiobjective Optimisation.
Topic. Symbolic Computations for Expert Systems and Machine Learning.

1 Introduction

A large number of techniques are currently used for non-linear classification, including neural networks, decision trees and parametric statistical models. The general task, given a data set, $\mathcal{D}$, which includes a number of cases, each holding a target value for a dependent or output variable, t, together with a vector of values for potential independent or input variables, $\mathbf{x} = \{x_i\}$, is to infer a model $y = f(\mathbf{x})$ that best predicts the target variable.

The model inference task is often split into two separate parts: selection of the model architecture; and optimisation of the coefficients. For example, the Back Propagation algorithm provides an efficient method to calculate the gradient of the error function of a neural network with respect to the weights (the coefficients of the model); this gradient may be "plugged into" a non-linear

J. Calmet et al. (Eds.): AISC-Calculemus 2002, LNAI 2385, pp. 117–127, 2002.

optimisation algorithm such as Quasi-Newton BFGS [13] to determine the co-efficients. However, the algorithm requires a fixed network structure, which is often selected "by hand." Algorithmic approaches to determine the structure include: constructive algorithms, that add units [6]; pruning approaches, that build an over-specified model and then remove parts [15]; and feature selection algorithms that identify and remove irrelevant variables [5] [10].

A useful general framework views non-linear models as the functional composition of input variables and adjustable coefficients. Models such as neural networks and decision trees use particular restricted sets of functions, structural conventions governing permitted compositions, and define algorithms to infer the coefficients and/or model form.

An important issue in many application domains is *comprehensibility* of the inferred model. For example, in medical applications neural networks may not be acceptable, even if performance is good, as the clinician cannot comprehend how the model comes to a decision. We may then prefer to generate a *symbolic classifier* – a mathematical expression in the input variables that is relatively easy to comprehend. In a recent paper [9] we introduced one such restricted model suitable for classification – a logistic function applied to a polynomial. The requirement for comprehensibility also implies that there are two broad inferential objectives: to maximize the "comprehensibility" of the inferred model, and to maximize its "performance" – and that we may need to trade-off performance in these two objectives.

Genetic Programming [12] is a general framework for model inference, where a model is represented by an expression-tree, and a set of genetic operators specialized for tree structures is used in searching the model space. Koza [12] originally suggested symbolic regression as an application of GP, but the technique is seldom used, primarily because the inclusion of adjustable coefficients in GP is problematic. Rodríguez-Vázquez *et. al.* [14] suggested using Genetic Programming with the operator set limited to $\{\Sigma, \Pi\}$ to generate polynomial models *sans* coefficients for a regression application. The GP is translated into the equivalent polynomial, and then the coefficients are added and optimised by least squares. Recently, we extended this approach to classification by applying a logistic activation function and optimising coefficients in the maximum likelihood framework using Quasi-Newton methods [9].

This paper introduces a more general framework for handling coefficients in symbolic regression. We embed coefficients at the nodes of the expression tree, and introduce an error propagation algorithm that efficiently calculates the gradient of the error function with respect to the coefficients. Genetic Programming is used to generate expression trees, each of which is optimised by Quasi-Newton techniques using error propagation. The technique neatly divides the model inference process into architecture design (for which GP is well-suited) and coefficient optimisation, for which Quasi-Newton is preferable. We use a Multi-objective version of the Genetic Programming algorithm to control the trade-off between model performance (measured by ROC curve analysis) and parsimony, the latter being closely related to comprehensibility. The result is an exploratory inferential algorithm that generates a large number of alternative symbolic classifiers, and retains a set representing a range of trade-offs between performance and parsimony.

2 GP Expression Representation

Genetic Programming represents a model using an expression-tree – internal nodes are drawn from a set of permissible functions (operators), and terminals from the set of variables, $\{x_i\}$. The standard technique to handle coefficients is to introduce a special terminal symbol, the *ephemeral constant*[12], the value of which may be altered by evolution. This is an ineffectual search method, however. In this paper, we instead embed coefficients at nodes in the tree (as multiplicative coefficients at all terminals, and as special coefficients of some specific operators; for example, the decision threshold of the IFT$_S$ operator)[1]. For example, the GP shown in figure 1, which (discarding coefficients) is expressed in Polish notation as $(*x_1(+(+x_7x_8)x_3))$, translates to the quadratic polynomial:

$$y = c_2c_5x_1x_7 + c_2c_6x_1x_8 + c_2c_7x_1x_3 \tag{1}$$

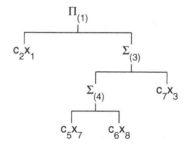

Fig. 1. A simple Genetic Program

If the only operators are $\{\Sigma, \Pi\}$ it is appropriate to omit the embedded coefficients, translate the expression tree into the corresponding polynomial, and then to add coefficients, which avoids the redundancies evident in equation 1. Similar translate-and-embed procedures may be available for other choices of operator sets. However, we consider models with the operators drawn from the set $\mathcal{O} = \{\Sigma, \Pi, \Phi, \text{IFT}_S\}$, which cannot always be reduced to a simpler form; the definition of these operators is given in table 1.

Genetic Programming has a number of desirable properties for model inference. First, it is constructive, in the sense that it starts with simple trees, and then attempts to combine successful models together. Optimisation of the coefficients makes this process viable – crossover is not destructive, as in most GP

---

[1] A valid alternative is to embed coefficients on edges, as occurs in neural networks

applications, since the coefficients are "tweaked" to form the best possible combination of two crossed-over models. The inherent tendency of GP to "bloat" (increase tree size arbitrarily) is a useful property in constructive search, and can be controlled and exploited by the techniques discussed later in this paper.

Table 1. Definition and Gradient of Operators. The x_i are inputs, c_i coefficients, and a the output (activation). Coefficients, if any, are shown after the semi-colon.

Operator	Definition	Gradient $(\nabla(\mathbf{x}; \mathbf{c}))$
$\Sigma(x_1, x_2)$	$x_1 + x_2$	$[1, 1]$
$\Pi(x_1, x_2)$	$x_1 x_2$	$[x_2, x_1]$
$\Phi(x_1; c_1)$	$\frac{1}{1+e^{-(x_1+c_1)}}$	$[a(1-a); a(1-a)]$
$\mathrm{IFT_S}(x_1, x_2, x_3; c_1)$	$\Phi(x_1, -c_1)x_2 + (1 - \Phi(x_1, -c_1))x_3$	$[-a(1-a)(x_2 - x_3),$ $\Phi(x_1, -c_1),$ $(1 - \Phi(x_1, -c_1));$ $-a(1-a)(x_2 - x_3)]$

3   Coefficient Error Propagation

To optimise the coefficients of the expression tree, we isolate them into a vector, $\mathbf{c}$, design an error function, $E(\mathbf{c})$ together with its gradient with respect to the coefficients, $\nabla\mathbf{c}$, and then use the Quasi-Newton non-linear optimisation procedure. For classification, it is desirable that the output be in the range $[0, 1]$, so that it can be interpreted as a probability, which is easily achieved in the two-class case by adding an implicit logistic operator, Φ, at the root of the tree. We choose the cross-entropy [2] error function, given in equation 2 (t_n and y_n are the target and predicted output for the nth case respectively), so that the optimisation corresponds to maximum likelihood estimation of the model coefficients.

$$E = -\sum_n t_n \log y_n + (1 - t_n) \log(1 - y_n) \tag{2}$$

We write the error function in terms of a functional composition of the variables and coefficients, and differentiate with respect to the latter. Applying the chain rule, the partial derivative w.r.t. a given coefficient can be written as the product of terms starting at the root and involving all ancestors of the node where the coefficient is embedded. Evaluation can be performed very efficiently, in a similar fashion to back propagation in neural networks, by first executing the expression tree, storing activation levels at each node, and then propagating errors downwards. At each node, the partial derivative of the node operator with respect to each child is multiplied by the error propagated from its parent, then passed to the child. In some cases the operator may have a coefficient, and the partial derivative of this is also calculated. The partial derivatives of the operators used in this paper are given in table 1.

Example. The differential of the GP shown in figure 1 with respect to coefficient c_7 is given by the equation below, where a_i is the activation of node i:

$$\frac{\partial E}{\partial c_7} = \frac{\partial E}{\partial a_1}\frac{\partial a_1}{\partial a_3}\frac{\partial a_3}{\partial a_7}\frac{\partial a_7}{\partial c_7} = \frac{\partial E}{\partial a_1}.a_2.1.x_3 = \frac{\partial E}{\partial a_1}.c_2 x_1 x_3 \tag{3}$$

Error propagation begins with the delta value $\frac{\partial E}{\partial a_1}$, which for the cross-entropy error function is given by:

$$\frac{\partial E}{\partial a_1} = \frac{y_n - t_n}{y_n(1 - y_n)} \tag{4}$$

With the addition of an implicit logistic operator at the root, this simplifies to:

$$\frac{\partial E}{\partial a_1} = \left(\frac{y_n - t_n}{y_n(1 - y_n)}\right).(y_n(1 - y_n)) = y_n - t_n. \tag{5}$$

The error propagation algorithm has $\mathcal{O}(TN)$ execution time, where N is the number of cases, and T the number of nodes in the tree.

4   Operators

Expression Inference is a very general framework, and can handle any differentiable operators. However, we may validly ask which operators are "most useful?" Specific problem domains might have a requirement for specialised operators such as trigonometric functions (e.g. signal processing) or logarithms and exponentials (for suitably distributed variables). Nominal variables should be treated using indicator variables and logical functions such as $\{\wedge, \vee, \neg\}$. However, we limit our attention real continuous input variables.

First, we include the "polynomial operators", Σ and Π, since polynomials are reasonably comprehensible models with a sound theoretical basis (e.g. the decision boundaries in classification problems with normal distributions are quadratics; higher order polynomials are arbitrarily flexible). We add the logistic sigmoid, Φ, due to its helpful role in "squashing" values.

The hard threshold-if operator, IFT_H (see equation 6), is extremely useful in "splitting" the model into parts that correspond to different partitions of input space, thus allowing the inference of a "mixture of experts" [11]. Such decomposition can greatly aid comprehensibility. Unfortunately, IFT_H is not differentiable with respect to the conditional operand or its threshold coefficient. Instead, we use the soft threshold-if operator, IFT_S, whose equation is shown in table 1, which is differentiable. This is somewhat less comprehensible that IFT_H, but retains an appealing "fuzzy-decision rule" interpretation.

$$\text{IFT}_\text{H}(x_1, x_2, x_3; \theta) = \begin{cases} x_2, \ x_1 \geq \theta \\ x_3, \ x_1 < \theta \end{cases}, \tag{6}$$

5   Multiobjective Genetic Programming

Multiobjective optimisation algorithms find models that maximize a number of objective function measures, $M = \{m_1, m_2, ..., m_k\}$, *without* combining them into a single objective. They establish a number of solutions representing different trade-offs between objectives. The decision about acceptable trade-offs between objectives may then be made *a posteriori*, after inference, rather than *a priori*. The key concept in Multiobjective optimization is the *Pareto optimal set* [1]. Given two models, A and B, we say that $A \succ B$ (*A dominates B*) if A has at least one objective value strictly greater than B, and all objective values at least equal to B:

$$A \succ B \iff \forall i,\ m_i(A) >= m_i(B);\ \exists i,\ m_i(A) > m_i(B) \qquad (7)$$

The *Pareto-optimal set* consists of all non-dominated solutions (i.e. those that are not dominated by any other solution). Multiobjective algorithms search along a *Pareto front* — the non-dominated subset of the models found during the search.

Multiobjective Genetic Programming implements this concept in the context of GP. If one objective is tree size, Multiobjective GP can control the inherent tendency of GP to bloat [3] [4], and produces a range of models of different parsimony versus performance trade-offs. Parsimony is closely related to model comprehensibility. Rodríguez-Vázquez [14] used the MOGA designed by Fonseca and Fleming [7] to optimise a number of performance and complexity measures in GP, and also demonstrated the division of effort between GP structure design and coefficient optimisation by least squares in the context of polynomial modelling. We use a variant of the niched Pareto GA developed by Horn [8], modified for GP evolution by Hunter [9] based on the variant for feature selection developed by Emmanouilidis [5].

The niched Pareto GA [8] maintains a population, and a separate *non-dominated set*, which contains copies of all solutions found on the Pareto front (if this set reaches a maximum size, it is pruned by niching). Selection is implemented by generating a separate *mating pool*; once generated, the mating set is subjected to crossover, mutation or cloning to generate the new population, and the non-dominated set is updated. A specialised form of tournament selection is used. The tournament members are selected from the union of the population and the non-dominated set; the inclusion of the latter introduces an elitist element to the algorithm, as high-performance individuals are preserved unchanged and used in subsequent breeding. Rather than competing with each other, as in a standard tournament algorithm, the tournament members are compared with a larger *dominance set* randomly selected from the current population (Horn's analysis indicates that this approach performs better than the naive tournament). If there exists a single tournament member non-dominated by all members of the dominance set, it wins the tournament and is copied to the mating pool. If two or more tournaments members are non-dominated, they are each candidate winners; if all tournament members are dominated, they are all candidate winners. In such cases, a *niching* strategy is used to decide which candidate should be placed in the mating pool. Each candidate is compared against existing members of the mating pool, and the one with the best *niche score*, $\mathcal{N}_i$, which reflects its dissimilarity from existing members of the mating pool, is chosen. The niche score in general may be phenotypic or genotypic.

To adapt the niched Pareto GA to GP selection, we define a niching similarity measure based on counting the number of nodes that the two trees have in common (see equation 8; $\mathcal{S}_G(k)$ is the symbol – operator or terminal – at node k, # is the number of matches). We then locate the K nearest neighbors in the mating pool (the neighborhood, $\mathcal{K}$), using the similarity measure, and set the niche score to the mean distance; see equation 9. The value K is selected using equation 10, where M is the current size of the mating pool. Strictly, to handle sub-tree size variability, we should use a more sophisticated tree similarity metric; however, the simple measure deployed here has the benefit of easy and rapid evaluation.

$$\mathcal{D}(G_i, G_j) = \#\{k : \mathcal{S}_{G_i}(k) = \mathcal{S}_{G_j}(k)\} \tag{8}$$

$$\mathcal{N}_i = \frac{\sum_{\mathcal{K}} \mathcal{D}(G_i, G_k)}{K} \tag{9}$$

$$K = \begin{cases} M/5, & M < 25 \\ 5, & M \geq 25 \end{cases}, \tag{10}$$

We assess performance by using the concept of *ROC dominance*. Receiver Operating Characteristic [16] curves are a standard technique, widely used in medical applications, which characterise the performance of a classifier as the decision threshold is altered, by displaying the sensitivity versus 1−specificity. A perfect classifier hugs the left-hand side and top of the graph, as there is no need to trade sensitivity and specificity – a threshold can be selected which classifies all cases correctly. In a recent paper [9] we described how to exploit multiobjective optimisation in designing classifiers with differing ROC performance. Rather than aiming for best performance in a single measure, such as the cross-entropy error rate, we compare ROC curves in the domination tournament. One model is said to *ROC dominate* another if its ROC curve has no points below the other and at least one point strictly above it. Using this concept can yield alternative models that perform well at different sensitivity levels. If two models of equal complexity have overlapping ROC curves, then they each perform best at different parts of the sensitivity range; such models are mutually non ROC dominant, and so both will be preserved by the algorithm.

6   Empirical Results

We conducted ten experiments using the new algorithm. Control parameters for the runs are given in table 6. The data used was the standard Ionosphere data set (available from the UCI Machine Learning Repository); only the first ten variables were used, and the cases were divided into 200 for training, and 151

Table 2. GA Settings

Population	100
Non-dominated set size	25
Tournament size	2
Dominance group size	10
Generations	200
Crossover rate	0.3
Mutation rate	0.3

for model selection and test. Average execution time was about two hours per run on an 800MHz PC.

The algorithm generated solutions with numbers of nodes in the range 1-135; non-dominated fronts on each run of the algorithm contained trees ranging up to 35 nodes. The algorithm purged larger trees, indicating that it is very good at controlling bloat. The joint non-dominated front across all models searched in all experimental runs contained solutions with 1–14 nodes – all extremely parsimonious. It also contained a significant number of repeated models, and also of models with the same complexity and marginal overlap of a few points between the ROC curves, which prevents domination. We therefore define an *equivalent non-dominated set* as a minimal sized subset which contains models equivalent to all members of the full non-dominated set, where *equivalent* implies an equal number of nodes, and ROC curves which are either strictly equal, or overlap at less than 5% of the points on the curve. Using this definition, we construct an overall equivalent non-dominated set, which is the basis of the results reported below. The overall equivalent non-dominated set contains thirty-two models with between 1 and 14 nodes, with models of 3 and 6–8 nodes most heavily represented.

6.1 Predictive Performance

To assess the efficacy of the algorithm, we compared the results with two benchmarks – a logistic regression, and a thirty hidden-unit Radial Basis Function neural network, both using all ten input variables. These benchmarks were chosen as logistic regression is a standard statistical technique in widespread use in medical applications, and the RBF network is a powerful (if poorly comprehensible) technique that provides an effective "gold standard" for comparison of performance. All predictions were performed on a hold-out set of 151 cases. Figure 2b shows the ROC curve for a model with 7 nodes – the logistic quadratic $\Phi(-0.88x_1(-1.2x_7 - 1.6x_8 - 1.1x_3))$ (the root Φ is implicit in all our models); the corresponding tree is shown in figure 1. Performance is somewhat inferior to the RBF network, but the model has many fewer free parameters (four, as compared with 311 for the RBF), and uses only four of the available ten variables.

Figure 2a is the three node model $\Phi(0.87x_6 + 1.19x_7)$ – a logistic regression with only 2 variables. Both these models have good performance and relatively high comprehensibility, compared with the logistic regression and neural network alternatives. The fact that the two models have only one variable in common is typical of the problems encountered in searching for models in domains with so many interrelated variables, and highlights the difficulty of the task – we would not find model b via a forward selection algorithm which visited model a, for example. The more complex models had better performance than the first model shown above on parts of the ROC curve, and included examples using the Φ and $\mathrm{IFT_S}$ operators; figure 2c shows the ROC curve for the most complex, 14-node model. An astute choice of these models would give broadly comparable results to the neural network at any point on the ROC curve.

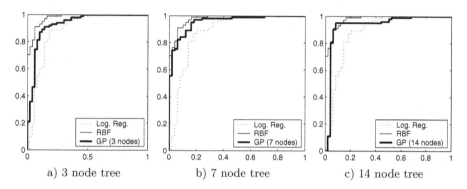

a) 3 node tree   b) 7 node tree   c) 14 node tree

Fig. 2. ROC Comparison against benchmark models.

6.2   Consistency

It is difficult to assess with certainty whether such algorithms search the Pareto front effectively, as the size of the search space makes it impossible to exhaustively sample to establish the true front as a standard of comparison (in a ten variable problem there are 2^{61} quadratic expressions alone). A useful indicator is to measure the consistency of the algorithm – the proportion of experimental runs that find each point on the front. Figure 3a shows the proportion of the united front found by each run. This is quite low (average 18%), indicating poor consistency – the algorithm is not exploring the admittedly extremely large search space very effectively (it is worth stressing that the differing runs all find good performance models, but they are often of higher complexity than the non-dominated front member with comparable performance). It may be that a more constrained heuristic search technique, rather than the "generalist" GA, would improve performance.

Figure 3b shows the proportion of the equivalent non-dominated front found on each generation. It is clear that the algorithm has not converged even after

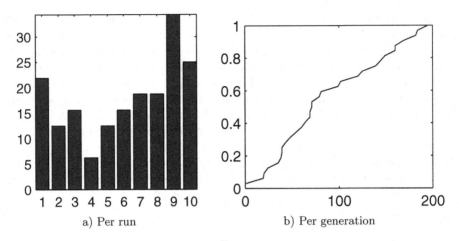

a) Per run   b) Per generation

Fig. 3. Proportion of equivalent non-dominated front covered

two hundred generations, and we could reasonably expect to find more improved solutions if we ran for longer.

7   Conclusion

We have described a general technique to perform inference of mathematical expressions. Expression trees are evolved using the Genetic Programming paradigm. Coefficients are embedded at each node and separately optimised using an efficient error propagation algorithm. The use of multiobjective Genetic Programming allows the generation of a range of alternative solutions, which trade off model parsimony versus ROC curve performance. We have demonstrated that the method can achieve performance levels approaching those of a neural network on a highly non-linear classification problem with significant interdependencies between variables, but with a much more parsimonious and comprehensible model, and can also produce a range of diverse complexity models with reasonably good performance.

Future work will include the integration of binary and ordinal variables, the integration of constraints on operator placement to improve inference, and the analysis of alternative heuristic structure search mechanisms, including the integration of top-down "purity" based methods derived from decision tree theory and bottom-up model merging algorithms.

References

1. A. Ben-Tal. Characterization of pareto and lexicographic optimal solutions. In Fandel and Gal, editors, *Multiple Criteria Decision Making Theory and Application*, pages 1–11. Springer-Verlag, 1979.
2. Christopher M. Bishop. *Neural Networks for Pattern Recognition*. Clarendon Press, Oxford, 1995.

3. Stefan Bleuler, Martin Brack, Lothar Thiele, and Eckart Zitzler. Multiobjective genetic programming: Reducing bloat by using spea2. In L. Spector, E. Goodman, A. Wu, W.B. Langdon, H.-M. Voigt, M. Gen, S. Sen, M. Dorigo, S. Pezeshk, M. Garzon, and E. Burke, editors, *Proceedings of the Congress on Evolutionary Computation, CEC-2001*, pages 536–543, IEEE Press, 2001. Piscataway, NJ.

4. Edwin D. De Jong, Richard A. Watson, and Jordan B. Pollack. Reducing bloat and promoting diversity using multi-objective methods. In L. Spector, E. Goodman, A. Wu, W.B. Langdon, H.-M. Voigt, M. Gen, S. Sen, M. Dorigo, S. Pezeshk, M. Garzon, and E. Burke, editors, *Proceedings of the Genetic and Evolutionary Computation Conference, GECCO-2001*, pages 11–18, San Francisco, CA, 2001. Morgan Kaufmann.

5. Christos Emmanouilidis, Andrew Hunter, and John MacIntyre. A multiobjective evolutionary setting for feature selection and a commonality-based crossover operator. In *Proc. of the 2000 Congress on Evolutionary Computation*, pages 309–316, Piscataway, NJ, 2000. IEEE Service Center.

6. S.E. Fahlman and C. Lebiere. The cascade-correlation learning architecture. In D.S.Touretzky, editor, *Advances in Neural Information Processing Systems*, volume 2, pages 524–532, San Mateo, CA, 1990. Morgan Kaufmann.

7. C. Fonseca and P. Fleming. An overview of evolutionary algorithms in multiobjective optimization. *Evolutionary Computation*, 3(1):1–16, 1995.

8. Jeffrey Horn and Nicholas Nafpliotis. Multiobjective optimization using the niched pareto genetic algorithm. Technical Report IllIGAL 93005, University of Illinois, Urbana, IL, 1993.

9. Andrew Hunter. Using multiobjective genetic programming to infer logistic polynomial regression models. In *Proceedings of the European Conference on Artificial Intelligence, ECAI 2002*. IOS Press, 2002.

10. A. Hunter, L. Kennedy, J. Henry, and R.I. Ferguson. Application of neural networks and sensitivity analysis to improved prediction of trauma survival. *Computer Methods and Algorithms in Biomedicine*, 62:11–19, 2000.

11. R.A. Jacobs, M.I. Jordan, S.J. Nowlan, and G.E. Hinton. Adaptive mixtures of local experts. *Neural Computation*, 3:79–87, 1991.

12. J.R. Koza. *Genetic Programming*. MIT Press, Cambridge, MA., 1992.

13. William H. Press, Brian P. Flannery, Saul A. Teukolsky, and William T. Vetterling. *Numerical Recipes: The Art of Scientific Computing*. Cambridge University Press, Cambridge, UK, 1986.

14. Katya Rodríguez-Vázquez, Carlos M. Fonseca, and Peter J. Fleming. Multiobjective Genetic Programming : A Nonlinear System Identification Application. In John R. Koza, editor, *Late Breaking Papers at the Genetic Programming 1997 Conference*, pages 207–212, Stanford University, California, 1997. Stanford Bookstore.

15. A.S. Weigend, D.E. Rumelhart, and B.A. Huberman. Generalization by weight-elimination with application to forecasting. In R.P. Lippmann, J.E. Moody, and D.S. Touretzky, editors, *Advances in Neural Information Processing Systems*, volume 3, pages 875–882, San Mateo, CA, 1990. Morgan Kaufmann.

16. M. Zweig and G. Cambell. ROC plots: A fundamental evaluation tool in clinical medicine. *Clinical Chemistry*, 39(4):551–577, 1993.

A Novel Face Recognition Method

Li Bai and Yihui Liu

School of Computer Science & IT
University of Nottingham
Nottingham NG8 1BB, UK
bai@cs.nott.ac.uk

Abstract. This paper introduces a new face recognition method that treats 2D face images as 1D signals to take full advantages of wavelet multi-resolution analysis. Though there have been many applications of wavelet multi-resolution analysis to recognition tasks, the effectiveness of the approach on 2D images of varying lighting conditions, poses, and facial expressions remains to be resolved. We present a new face recognition method and the results of extensive experiments of the new method on the ORL face database, using a neural network classifier trained by randomly selected faces. We demonstrate that the method is computationally efficient and robust in dealing with variations in face images. The performance of the method also decreases gracefully with the reduction of the number of training faces.

1   Introduction

In the past decade, considerable amount of research has been done on developing face recognition methods. Recognizing faces in images and videos is an important task with many applications, including access control, suspect identification, and content-based retrieval. It is a difficult task because the appearance of a face varies dramatically because of illumination, facial expression, head pose, and image quality. To further compound the problem, we often have only a few images of a person from which to learn the distinguishing features and then have to recognize the person in all possible situations.

Face recognition methods may be categorised into three main approaches: facial feature-based approach whereby biometric details of a face concerning the eyes, nose, and mouth are extracted from the test face and matched against those of the stored model faces; structural approach whereby face images are represented as graphs and recognition relies on an elastic graph matching process; holistic approach in which the whole face image (or its transformation) is passed on to the classifier.

Facial feature-based approach is the most intuitive approach to face recognition. However, not only are normal facial features difficult to extract but also unreliable especially when there are significant variations present in images. High level features such as those based on texture, shape, or wavelet coefficients have been reported to perform better. Liu[4] proposes a method that uses both shape

J. Calmet et al. (Eds.): AISC-Calculemus 2002, LNAI 2385, pp. 128–135, 2002.

and texture information to assist face recognition. A shape vector is produced for each face image though a manual process and a texture vector is produced by warping each image to the average image of the whole training set. The shape and the texture vectors produced are then combined to make the final feature vector for the image. Because of the high dimensionality of the face vectors produced the eigenface method is employed to reduce the dimension.

The most well known structural approach to face recognition is the dynamic link structure, first proposed by Von Der Malsburg [9]. It belongs to the family of elastic graph matching methods. Image points are sampled and those on a rectangular grid are selected. Gabor wavelet responses at these selected image points are taken as the features of the image. Initially image points of a test face are randomly connected to those of a stored model and the connections are then updated during the matching process. The goal of the matching process is to maximise the similarity of Gabor wavelet responses, as well as the similarity of topological relationships, at the image points of the test face and the corresponding image points of the model face.

There have since been many variants of the dynamic link architecture as the approach associates face images on the basis of their structural relationships. The approach is believed to be robust to variations. However, Zhang et al's [8] experiments using elastic matching reports a 80% classification accuracy on the ORL face database using 5 faces per person for training and 5 for testing.

We mention the eigenface method [1][6] as an example of holistic approach to face recognition. The purpose of this approach is to find a set of basis vectors that capture image variance more optimally than the Euclidean coordinate axes, in order to reduce pixel correlation redundancy) therefore image dimension. These basis vectors are called eigenfaces, and each face image is then represented by its projection coefficients onto the eigenfaces. As the number of eigenfaces is much smaller than the number of pixels of an image, the dimension of face vectors are thus reduced. The dimension can be reduced further by discarding those eigenfaces corresponding to small eigen values. Experiments show that the eigenface method is not robust in dealing with variations in lighting conditions. Various attempts to improve the eigenface method have been made [3].

In summary, there is still the need to develop robust face recognition methods. We present a robust face recognition method based on wavelets.

2   The Wavelet1D Method

Almost all the published face recognition methods based on wavelets use 2D wavelet decomposition. This decomposes the approximation coefficients at level j into four components: the approximation at level $j + 1$ and the details in three orientations (horizontal, vertical, and diagonal). More precisely, the rows of the jth approximation coefficient matrix are convolved with both a low-pass filter and a high-pass filter and the results are column downsampled (only even indexed columns are kept); The columns of each downsampled results are then convolved with both a low-pass and a high-pass filter and the results are row downsampled (only even indexed rows are kept). The resulting four matrix are the level $j + 1$ approximation and the details in three orientations. The size of each coefficient, either approximation or detail, at the 1st decomposition level is $1/4^1$ that of the original image, at the 2nd level $1/4^2$, and at the 3rd level is $1/4^3$ of that of the original, and so on. Intuitively the decomposition coefficients obtained could be used to represent the original image, however, this would result in an even higher dimensional face vector than the original face vector. This is because that the combined size of 1st level coefficients alone (both approximation and details coefficients) is equal to that of the original image, and to represent the original image sufficiently one would have to use the decomposition coefficients at several levels. Some researchers abandon coefficients and resort to extracting facial features from approximation and detail images instead [2], which we believe suffers the same problem as extracting features from the original image.

We propose a more effective method, namely the wavelet1D method. We begin by producing 3 approximation images $\tilde{\zeta}_{i1}, \tilde{\zeta}_{i2}, \tilde{\zeta}_{i3}$ for each $m \times n$ face image ζ_i and replace ζ_i with the average of the approximation images: $\zeta_i = 1/3 \sum \tilde{\zeta}_{ij}$ We then turn ζ_i into a $1D$ vector by coalescing its row vectors $\zeta_i = (\zeta_{kj})$, where $k = 1 \rightarrow m, j = 1 \rightarrow n$. Finally we decompose ζ_i to the 5th level and take the approximation coefficients at that level as the feature vector for the image. The feature vector is shown in Fig.1 below.

a. A face image, its 3 approximations and the average

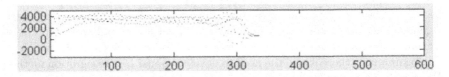

b. The plot of the wavelet coeffients

Fig. 1. Facial feature extraction

3   Neural Network Classifier

We use a radial basis function network to classify the images. The network has two layers of neurons. Each neuron in the hidden layer is created for an input vector and implements the radial basis transfer function. The weight vectors connecting the input to the hidden layer neurons are the transpose of the input matrix. Input to a hidden layer neuron is therefore the distance between the input vector and the weight vector connecting to the hidden layer neuron. If the distance is small the output of the hidden layer neuron is 1 otherwise 0. The second layer neurons implement the linear transfer function and the weight vectors connecting to the second layer neurons are calculated using the input (so radial basis function outputs) and the target matrices.

A neural network normally associates an input object with its class, not necessarily with any concrete object of the class in the training set. As there are 10 varying faces per subject in the ORL face database it would be useful to know the nearest object to the input. To make this possible we use a composite network with several networks linked into a hierarchical structure. The subnetwork at the first level of the hierarchy is the normal radial basis function network that associates a test input with one of the 40 classes. The 40 subnetworks at the second level of the hierarchy are created for the 40 classes to associate the input with the nearest object of its class. Each of the second level subnetworks is trained separately with the faces of its own class. Its first layer implements the normal radial basis transfer function and the second layer implements the competitive transfer function and its second-layer weights are set to the matrix made of target vectors.

4   Experiments

We present our experimental results of the wavelet1D method. We use the ORL face database for the experiments. The ORL database is created at the Olivetti & Oracle Research Laboratory in Cambridge. It contains 400 face images altogether with 10 face images for each of the 40 individuals. It is by far the most challenging database for face recognition, covering variations in lighting condition, facial expression, head pose ($\pm 20^0$), and eye ware. These variations occur even across the images of the same individual. The experiments are extensive because several parameters have to be optimised for both wavelet feature extraction and the neural network classifier.

With the wavelet1D method we need to decide on the level to which an image should be decomposed to produce meaningful and robust features for classification. There is no better way of doing this other than decomposing each image to the 4th, 5th, and the 6th level and observe classification results using the features extracted at each level.

For the neural network classifier, firstly, we have to decide which faces to use to train the neural network classifier. Some researchers manually select training faces of each class in order to cover most of the variations of the class. This would mean that invariance is the result of crafty network training rather than the inherent property of the method. To demonstrate that our method is invariant to variations we select training faces purely by random and use whatever left afterwards to test the network. Secondly, the number of training faces and the radial basis function network spread parameter need to be set. Most of the published works on the ORL face database use 5 faces per class for training and 5 for testing. To demonstrate that our method works well with fewer training faces, we do three sets of tests using the training/testing ratios 5/5, 4/6, and 3/7 respectively. In the 5/5 test there are 200 training faces, 200 test faces, in the 4/6 test there are 160 training faces, 240 test faces, and in the 3/7 experiment there are 120 training faces and 280 test faces. We run each of these tests several times with different network spread values to decide on the best value of network spread.

As stated above we need to conduct three sets of tests corresponding to the three training/testing ratios 5/5, 4/6, and 3/7, and for each of these three tests we need to find the best wavelet decomposition level by decomposing face images to the 4th, 5th and 6th level respectively and observing the performance at each level. The dimensions of face vectors at these three decomposition levels are 658, 336 and 175 respectively. Thereafter for each of the test and level combination we also try a few network spread values to find the best. The best performance is achieved at level 5 for each of the three tests.

We now make the test results more transparent by showing in Figure 2,3,4 the classification errors of the three tests, using the optimal decomposition level 5. For the 5/5 test we show the four wrongly classified faces. For the 4/6 and 3/7 tests, rather than showing the images involved, we use network activation graphs and misclassification graphs to illustrate the errors.

The activation graph plots the output values of the radial basis function. The output values describe the classes of the test faces. The network gives 40 output values, representing network certainty values of the test face being of each of the 40 classes. Ideally only one output should be 1 and all others should be 0, but this extreme case will only happen if the test face is one of the training face. Typically nearly all the neurons respond to a test face so we pick the neuron that gives the maximum response to the test face and the test face is associated with the class represented by that neuron. The '*'s in the graph represents the maximum output value when the test face is correctly classified and the 'o's in the graph represents the maximum output value when the test face is wrongly classified. For example in the third column there is one 'o'. It means that one of 6 test faces of class 3 is wrongly classified, the maximum output value is around 0.3. The activation graph is divided into 40 columns. For the 4/6 test the activations of 6 test faces are distributed in a column, for the 3/7 test 7 faces in a column. The misclassification graph shows the test face of which class is wrongly classified as which other class.

5   Discussion

Most of the published face recognition results on the ORL face database use the 5/5 ratio for classifier training and testing, there is no mentioning how the methods perform when the number of training faces is reduced. We have given performance statistics of our method for varying training/testing ratios using randomly selected training faces. The results demonstrate that our method is

Fig. 2. Mis-classified faces of the 5/5 test (4 errors, 98% accuracy)

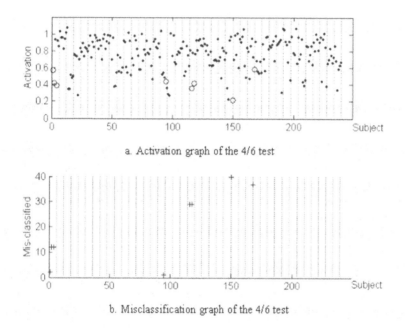

a. Activation graph of the 4/6 test

b. Misclassification graph of the 4/6 test

Fig. 3. Mis-classified faces of the 4/6 test (8 errors, 96.67% accuracy)

accurate and performs well with a small number of training faces. To put this in perspective we mention two of the methods that produce comparable accuracy to our method. One is Tolba's [5] that has achieved a 99.5% recognition rate on the ORL face database using 5 training faces per class, by combining multi-classifiers, the Learning Vector Quantization and Radial Basis Function network, as well as by multi-training of these classifiers. The classifiers are trained with the original training data as a whole first, then again with two groups of data

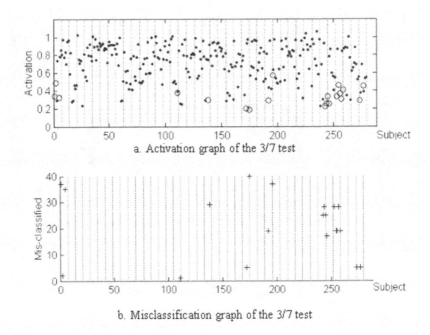

a. Activation graph of the 3/7 test

b. Misclassification graph of the 3/7 test

Fig. 4. Mis-classified faces of the 3/7 test (20 errors, 92.86% accuracy)

by splitting the original training set. One group contains previously correctly identified data and one previously wrongly identified. Thereafter, a front-end classifier has to be trained on the outputs of the two classifiers to produce the final classification. Both their two classifiers assign some test faces to classes that bear little resemblance to the test faces. The other is O.de Vel's line-based method [7] that extracts image features by sampling the image (assuming the boundary of the face is known) with a set of line segments and take some pixel values on the line segments. The number of sampling line segments and the positions of them on the face make a dramatic difference to the performance, so both the training and recognition process have to be repeated on an image many times with different line populations before a classification decision can be made. The classifier is trained by manually selected 5 training faces from each class, and the selected training faces cover variations in head positions and facial expressions.

References

1. P. Belhumeur et al: Eigen Faces vs. Fisherfaces. Recognition Using Class Specific Linear Projection. IEEE Transactions on Pattern Analysis and Machine Intelligence, Vol. 19, July 1997
2. C.Garcia, G.Zikos, G.Tziritas: Wavelet packet analysis for face recognition. Image and Vision Computing, 18(2000) 289-297

3. P.McGuire and M.T.D'Eluteriom: Eigenpaxels and a neural network approach to image classification. IEEE Tansactions on Neural Networks, VOL.12, NO.3, May 2001
4. Chengjun Liu, Harry Wechsler: A Shape and Texture Based Enhanced Fisher Classifier for Face Recognition. IEEE Transactions on Image Processing, VOL.10, NO.4, April 2001
5. A. S. Tolba and A.N.Anu-Rezq: Combined Classifiers for Invariant Face Recognition. 0-7695-0446-9/99, 1999
6. M.Turk, and A.Pentland: Eigenfaces for Recognition. Journal of Cognitive Neuroscience, 3 (1), 1991
7. Olivier de Vel and Stefan Aeberhard, Line-based Face Recognition under Varying Pose. IEEE Transactions on Pattern Analysis and Machine Intelligence, VOL.21, NO.10, October 1999
8. J.Zhang, Y.Yan, and M. Lades. Face Recognition: Eigenface, Elastic Matching, and Neural Nets. Proc. IEEE, VOL.85, 1997
9. Von der Masburg C, Pattern recognition by Labeled Graph Matching. Neural Networks, vol. 1, 141-148, 1988

Non-commutative Logic for Hand-Written Character Modeling

Jacqueline Castaing

Galilée University, LIPN-UMR 7030
93430 Villetaneuse, FRANCE,
jc@lipn.univ-paris13.fr

Abstract. We have proposed a structural approach for on-line handwritten character recognition. Models of characters are writer-dependent. They are codified with the help of graphic primitives and represented in a data base. The originality of our approach comes from the ability for the system, to explain and justify its own choice, and to deal with all different writing systems, such as the Latin alphabet, or the Chinese or Japanese scrip for example, providing that an appropriate data base has been built up. For this reason, our recognizer can be very helpful for learners of "exotic" scripts. In this paper, we propose to analyse the recognition process in an appropriate logical framework, given by non-commutative Logic. We first point out the class of sequents which allows us to describe accurately the recognition process in terms of proofs, then, we will give some results about the complexity of the recognition problem depending on the expressive power of the representation language.

Keywords: Linear Logic, Character Recognition, Distance Computing, Proofs
Topics: Foundations and Complexity of Symbolic Computation, Logic and Symbolic Computing

1 Introduction

We have proposed in [4] a structural approach for on-line handwritten character recognition. Models of characters are *writer-dependent*, They are codified with the help of graphic primitives. As our recognizer is dedicated to learners of "exotic" scripts, the Chinese or Japanese one see [3] for example, the primitives are chosen in order to point out the specific features of the Master's models. So, learners have just to reproduce the good models. The originality of our approach comes from the ability of our recognizer to explain and justify its own choice, and to deal with all different writing systems, providing that an appropriate data base has been built up. The classification (or recognition) of an input character I is then carried out by computing distances between I and the models represented in the data base. As the number of models introduced in the data base may be very high, we use a symbolic learning algorithm see [2], to extract from the knowledge base a smaller number of more general models.

J. Calmet et al. (Eds.): AISC-Calculemus 2002, LNAI 2385, pp. 136–153, 2002.

We first attempted to formalize our approach by means of classical logic. But standard logic, with contraction and weakening rules, doesn't allow us to analyse accurately the classification procedure. In this paper, we show how we use Ruet's non-commutative Logic see [18], to build a logical system in order to achieve our goal. The logical rules of the system are selected mainly to facilitate the design of the recognition procedure. Input characters have now a straightforward representation by means of formulas of the form $P_1 \bullet ... \bullet P_n$, where the strokes $P_1,...,$ and P_n, separated by the non-commutative conjunctive operation $\bullet$ neither can disappear nor can be duplicated. Proofs can then be carried out, step-by-step, related to the rules involved.

In section 2 of this paper we present a survey of OCR (Optical Character Recognition); so, the reader can evaluate the originality of our work. In section 3 we indicate how from the numerical coding we get the symbolic representation of handwritten characters. We only deal with Latin letters. In section 4, we introduce the character recognition procedure and we will give some experimental results. Section 5 presents the non-commutative linear system. Section 6 analyses recognition process by means of proofs Finally, section 7 examines complexity results and makes clear what can be learnt from our theoretical approach. The symbolic learning algorithm used is out of scope of this paper. For a good understanding of sections 5 and 6, we suppose that the reader is familiar with (linear) sequent calculus.

2 A Quick Survey of OCR (Optical Character Recognition)

In OCR, different approaches depending on the type of data, *handwritten* or *printed characters*, and on the application considered have been investigated see [20]. In the case of printed characters, the recognition rate is about 98% for monofont texts, and varies between 80% and 98% for multifont texts. In the case of handwritten characters, off-line and on-line recognition methods are distinguished. For off-line recognition the reader can refer to [21]. In our framework, we are only interested in on-line recognition. A handwriting recognition system can further be broken down into the categories of *writer-dependant*, or *writer-independant*. A writer-independant system is trained to recognize handwriting in a great variety of writing styles, while a writer-dependant system is trained to recognize handwriting of a single individual. The infinite variations of shapes limit the recognition rate to 98% for word recognition. For isolated character recognition (neither grammar nor dictionary is used), the recognition rate can get low to 88%. Some particular applications such that the interface for Personal Digital assistant (PDA) may reach the score of 99%, but on unnatural character set (e.g., graffiti). The technologies usually employed are based on stochastic methods, such as hidden Markov model (HMM) first developped for speech recognition [15], on connectionist methods [11], and on k nearest neighbors procedure. Some approaches permit to take into account the structural information. Frey & Slate method [7], can be compared to our approach as they use graphic primitives and rules of the form "IF ... THEN". However, their method only deals with printed characters.

We give below some recent results taken from [6]. Table 1 shows some isolated character recognition accuracies. Table 2 gives recent outcomes for handwritten word recognition. The lack of common datasets makes it very difficult to

Table 1. Some recent results on isolated character recognition

Author	Method	Accuracy	Comments
Scattolin & Krzyzak [19]	Nearest Neighbor	88.67%	33 writers used
Chan & Yeung[5]	Manually Designed Structural Models	97.4%	150 writers used
Prevost & Milgram [14]	Combined Online and Offline Nearest Neighbor Classifiers	96.4%-98.7%	data from Unipen data used

Table 2. Some recent results on handwritten word recognition

Author	Method	Accuracy	Comments
Nathan & al. [12]	Hidden Markov Model	81.1% on unconstrained word examples	25 writers Writer-independent. 21K word vocab.
Rigoll & al.[16]	Hidden Markov Model	95% on 200 examples	Writer-dependant 3 writers 30K vocab.
Hu, Brown and Turin [10]	Stroke-level HMM	94.5 on % 3,823 unscontrained word examples	Writer-independant. 18 writers 32 word lexicon

compare these results. (At the time of these results, the database coming from Unipen project was not publicly available for word examples).

In handwritten word recognition, many approaches consist in chaining character models together to form words. For these methods, a necessary step is the process of segmenting the input in such a way that isolated characters can be presented to the recognition system. So, the style of writing (in particular cursive handwriting) makes the problem of segmentation very difficult.

Our motivation is two-fold. First, we are obviously concerned with how to recognize handwritten data correctly. Second, and more fundamentally, we are concerned with how to explain the recognition step in order to help learners to master different writing scripts. To this end, we provide a recognition system which is writer-dependant, in fact Master-dependant. The data base only contains character models. Our approach is based on distance computing, the isolated character recognition rate is 94,3%. We also run our experimentation on small test sets containing 120 words from a total word lexicon of 400 words. We obtained word recognition rates up to 90% by restricting the segmentation hypotheses.

We now describe our approach.

3   Knowledge Representation

An electromagnetic data tablet is used as an input device. The tablet samples pen movement in terms of x,y-coordinate values, and det.

3.1   Numerical Representation

Each character written down is numerically represented by sets of coordinate values separated by pen-up and pen-down marks.

Example 1. The character "a" (see Fig.1) has the following numerical representation:
Pen-Up
// "a"
Pen-Down
(38 114) (34 113) (33 113) (32 114) ..... (43 125) (44 125)
Pen-Up

Fig. 1. The character "a"

For our experimental evaluation, we use a numerical data base built up by the author of this paper. Each latin character is written down two hundred times.

3.2   Symbolic Representation

The choice of a symbolic codification can be driven by a variety of considerations. As we aim at helping learners, we choose primitives in order to point out the writing order of the strokes in a character, and to describe the cursive forms of characters. In our experimentation involving the latin letters, we select the following primitives:

1. To capture the direction of the strokes drawn by the graphic pen, we borrow from Freeman's alphabet the graphic primitives. The Freeman's alphabet used(see Fig. 2) has eight directions denoted numerically by the number 0, 1, 2, 3, 4, 5, 6, 7, and symbolically by the primitives h, f, t, e, k, g, d and v. As our method is based upon some distance between graphic primitives, we give the following definition:

Definition 1. *The distance between two graphic primitives p_1 and p_2, denoted by $|p_1, p_2|$, is the minimal number of steps to take for reaching p_2 from p_1, depending on the way taken to reach p_2 from p_1 (clockwise or counterclockwise in the figure Fig.2).*

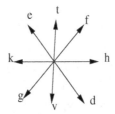

Fig. 2. Freeman's Alphabet

2. To this basic alphabet, we add "cultural" elements, called *descriptive primitives* which are the following one: the primitives s, o, u, x mark respectively the absence of hole, the presence of one hole, the presence of two holes, the presence of three holes or more; the primitives q and r are related to the size of the holes; the primitives i, j, w give the position of the last stroke (highest, lowest, medium position); the two last primitives z and y corresponds to pen-up signal (presence or absence). So, characters in the data base are codified with the help of primitives definite over an alphabet of 19 letters P = { h, f, t, e, k, g, v, d, s, o, u, x, q, r, i, j, w, y, z}. Let us give an example.

Example 2.
The character "a" (see Fig.3) is described by the string "gdfvh". Its particular cursive form can be fully captured by the sequence "oqqrrrjy". The symbolic code of "a" is then given by the string "gdfvhoqqrrrjy".

Definition 2. *The distance between two descriptive primitives, denoted by $\lfloor p_1, p_2 \rfloor$, is given by:*
if $p_1 = p_2$ then $\lfloor p_1, p_2 \rfloor = 0$ otherwise $\lfloor p_1, p_2 \rfloor = 1$.

Fig. 3. Symbolic Representation of the character "a"

3. The size of the strings is normalized to 22: if the length of a string is less than 22, we duplicate the Freeman's values corresponding to the strokes of greatest length untill to reach the fixed value 22.

Example 3. So, the complete representation of our character "a" is now given by the string: "gggdddffffvvvhoqqrrrjy".

3.3   Logical Representation

If we interpret normalized strings with the help of standard connectives, conjunction($\&$) and implication ($\rightarrow$), we may infer g $\&$ g $\rightarrow$ g, and f $\&$ g $\rightarrow$ g $\&$ f from the definition of the connective $\&$. If standard conjunction is roughly applied to graphic primitives, strokes may disappear in a string, and/or the writing order may no longer be respected. To avoid this "bad" effect, we propose to transform the representation of strings. In section 6, we will introduce the appropriate logical framework based upon non-commutaive Logic.

From the initial alphabet of primitives P, we define a set of new predicates, $p_{i,j}$, (i $=$ 1, 22 et j $=$ 0, 18), to mean that the primitive P_j occurs at the ith position of a string. Each normalized string is interpreted by the conjunctive formula W $= \bigwedge_{k=1}^{22} p_{k,i_k}$, where $i_1, ..., i_{22} \in \{0, 18\}$, and the conjunction $\wedge$ is the standard one. From now in our paper, a string refers to such a normalized conjunctive form.

To mean that a string W is used to codify a latin letter $\mathcal{L}$ definite over the set of latin letters $\mathcal{LL}$, $\mathcal{LL} = \{\mathcal{A}, ..., \mathcal{Z}\}$, we invoke Horn clauses, i.e, clauses of the form L $\rightarrow$ R, where the premise L is W, the conclusion R is $\mathcal{L}$, and the implication $\rightarrow$ is the standard one.

Example 4. The character "a" given in the previous example has the following meaning in our knowledge data base:

$p_{1,5} \wedge p_{2,5} \wedge p_{3,5} \wedge p_{4,7} \wedge p_{5,7} \wedge p_{6,7} \wedge p_{7,1} \wedge p_{8,1} \wedge p_{9,1} \wedge p_{10,1} \wedge p_{11,6} \wedge p_{12,6} \wedge p13, 6 \wedge$
$p_{14,0} \wedge p_{15,9} \wedge p_{16,12} \wedge p_{17,12} \wedge p_{18,13} \wedge p_{19,13} \wedge p_{20,13} \wedge p_{21,15} \wedge p_{22,17} \rightarrow \mathcal{A}.$

Distances between primitives are captured by means of formulas involving implication. We will make clear this idea in section 7 by means of proof-trees.

- If the distance between two primitives p_1 and p_2 is equal to 1, we set $p_1 \rightarrow p_2$.
- In the general case, the transitivity of the connective $\rightarrow$ is invoked to compute the distance between two graphic primitives p_1 and p_2.
- Let E be the set of implications introduced. In our application, the cardinality of E is et to the value 49, $\mathcal{C}(E) = 49$ depending on the number of primitives involved.

We show now, how we proceed to classify input characters by means of these formulas.

4   Recognition Process

Let us begin with a definition. Let I and J be two strings. I= $\bigwedge_{k=1}^{22} p_{k,i_k}$ and J $= \bigwedge_{k=1}^{22} p_{k,j_k}$

Definition 3. *We extend the notion of* **distance** *definite over the set of primitives to the strings I and J. The distance between I an J denoted by* $\| \ I, J \ \|$ *is given by:*
$\| \ I, J \ \| = \| \ I, J \ \|_{1,14} + \| \ I, J \ \|_{15,22}$*, where* $\| \ I, J \ \|_{1,14} = \sum_{k=1,14} \lceil p_{k,i_k} p_{k,j_k} \rceil$*, and* $\| \ I, J \ \|_{15,22} = \sum_{k=15,22} \lfloor p_{k,i_k} p_{k,j_k} \rfloor$

Example 5. Let $a_0 = p_{1,5} \wedge p_{2,5} \wedge p_{3,5} \wedge p_{4,7} \wedge p_{5,7} \wedge p_{6,7} \wedge p_{7,7} \wedge p_{8,1} \wedge p_{9,1} \wedge p_{10,1} \wedge p_{11,1} \wedge p_{12,6} \wedge p13,6 \wedge p_{14,6} \wedge p_{15,9} \wedge p_{16,12} \wedge p_{17,12} \wedge p_{18,13} \wedge p_{19,13} \wedge p_{20,13} \wedge p_{21,15} \wedge p_{22,17}$, and

$a_1 = p_{1,4} \wedge p_{2,4} \wedge p_{3,5} \wedge p_{4,6} \wedge p_{5,6} \wedge p_{6,7} \wedge p_{7,7} \wedge p_{8,1} \wedge p_{9,1} \wedge p_{10,1} \wedge p_{11,12} \wedge p_{12,6} \wedge p13,6 \wedge p_{14,7} \wedge p_{15,9} \wedge p_{16,12} \wedge p_{17,12} \wedge p_{18,13} \wedge p_{19,13} \wedge p_{20,13} \wedge p_{21,15} \wedge p_{22,17}$, be two variants of the same letter a. The distance between a_0 and a_1, $\| \ a_0, a_1 \ \| = |p_{1,5}p_{1,4}| + |p_{2,5}p_{2,4}| + |p_{4,7}p_{4,6}| + |p_{5,7}p_{5,6}| = 4$.

We first sketch the recognition process informally, then we will detail further.

4.1   Basic Idea

Let I be the input character to be classified, $I = \bigwedge_{k=1}^{22} p_{k,i_k}$ Let $\mathcal{L} \in \mathcal{LL}$ be a letter codified in the data base by means of the family of clauses $L_i \rightarrow \mathcal{L}$, i =1,200. We proceed as follows:

1. For each letter $\mathcal{L} \in \mathcal{LL}$, we compute the mean distance between I and $\mathcal{L}$ given by the formula: $\mathcal{D}_{\mathcal{L}} = \frac{1}{200}\sum_{k=1}^{200} \| \ I, L_i \ \|$.
2. Let $\mathcal{L}_0 \in \mathcal{LL}$ be such that : $\mathcal{D}_{\mathcal{L}_0} = \min (\mathcal{D}_{\mathcal{L}})$, $\mathcal{L} \in \mathcal{LL}$ (If $\mathcal{L}_0$ is not unique, we consider the first letter in the alphabetical order). The input character I is identified with the letter $\mathcal{L}_0$.

4.2   What We Have Implemented

To achieve the recognition process in some bounded time, we have reduced the number of rules in the data base, not only by removing redundant rules but also by generalizing them with the help of our learning algorithm.

Example 6. Let us consider again the clause corresponding to the character "a" given in the previous examples, $p_{1,5} \wedge p_{2,5} \wedge p_{3,5} \wedge p_{4,7} \wedge p_{5,7} \wedge p_{6,7} \wedge p_{7,7} \wedge p_{8,1} \wedge p_{9,1} \wedge p_{10,1} \wedge p_{11,1} \wedge p_{12,6} \wedge p13,6 \wedge p_{14,6} \wedge p_{15,9} \wedge p_{16,12} \wedge p_{17,12} \wedge p_{18,13} \wedge p_{19,13} \wedge p_{20,13} \wedge p_{21,15} \wedge p_{22,17} \rightarrow \mathcal{A}$.
Our learning algorithm generalizes this rule into a more general one of the following form:
$p_{1,5} \wedge p_{5,7} \wedge p_{20,13} \rightarrow \mathcal{A}$.

So, in our experimentation, from the set of rules of the form $L_i \rightarrow \mathcal{L}$, we point out a smaller subset of rules, (253 rules), of more general form $L'_i \rightarrow \mathcal{L}$, where the premises L_i and L'_i are linked by the formula $L_i \rightarrow L'_i$. Those of the predicates of L_i which remain in the premise L'_i are called **relevant** predicates. The recognition process is carried out in the same way as indicated in section 3.1 by focusing only on relevant predicates. It means that when we compute the distance between strings, we set to 0 the distances between non relevant predicates. Let $n_\mathcal{L}$ be the respective number of new rules pointed out by our learning algorithm for each class of latin letter $\mathcal{L} \in \mathcal{LL}$. Our solution can be sketched as follows:

1. For each letter $\mathcal{L} \in \mathcal{LL}$, we compute the mean distance between the input character I and $\mathcal{L}$ given by the formula: $\mathcal{D}_\mathcal{L} = \frac{1}{n_\mathcal{L}} \sum_{i=1}^{n_\mathcal{L}} \| I, L_i \|$.
2. Let $\mathcal{L}_0 \in \mathcal{LL}$ be such that : $\mathcal{D}_{\mathcal{L}_0} = \min (\mathcal{D}_\mathcal{L})$, $\mathcal{L} \in \mathcal{LL}$ (If $\mathcal{L}_0$ is not unique, we consider the first letter in the alphabetical order). The input character I is identified with the letter $\mathcal{L}_0$).

4.3   Experimental Results

We give the results obtained by our recognizer compared to those obtained with the k nearest neighbors procedure (kNN) using the same data bases. The experimental evaluation is based on a standard "cross-validation". We have a set of n instances (n = 3240). We have iterated 40 times the following procedure: i) n*0.9 instances are randomly selected; the other instances are used as test instances, ii) learning, iii) testing. Thus, we report hereunder averages on 40 trials. Our results are summerized in arrays of 4 lines. The first line gives the recognition rates of our recognizer and the kNN procedure. The second line gives the number of rules learned, third line the number of premises of these latter, and the last line the learning time.

Table 3. Comparison with KNN

Method	Accuracy	Nb of rules	Nb of Premises	Learning time
OURS	0.943	253	14	1516 seconds
kNN	0.955	3240	22	0

4.4   First Conclusion

We can formally describe the recognition process by means of proofs involving Horn clauses, for instance by applying the Robinson's method [17] called *resolution*. But, this approach doesn't allow us to :

- capture fully the recognition process without "trick" (transformulation of formulas)
- analyse accurately the recognition problem in order to answer to the following questions:
  - what happens if we don't normalize strings,
  - and how to analyse the complexity of the recognition problem depending on the expressive power of the representation language.

5   Non-commutative Logic for OCR: $\mathcal{NCL}_{ocr}$

In this section, we introduce the logical framework $\mathcal{NCL}_{ocr}$ in which proofs are carried out. The representation of written characters is now straightforward by using the non-commutative conjunction • and the linear implication —∘.

5.1   Preliminary

In Gentzen calculus, the structural rules - exchange, contraction and weakening-are explicitly applied. In Girard's logic [9], the contraction and weakening rules are forbidden: an immediate repercussion is the distinction of two connectives for conjunction ⊗ (times), & (with), and two connectives ℘ (par), ⊕ (plus) for disjunction. But, unrestricted exchange rules force the commutativity of these connectives. In some applications in computer science, such as in our application, we actually desire non-commutative connectives. Some works on non-commutativity can be found in [1]. In Ruet[18], two equivalent versions of non-commutative linear calculus have been introduced. The first version is defined on presentations, i.e, partial orders; the second version works with order varieties of occurrences of formulas. In practice, we work with a fragment of the first version because the structural rules used, seesaw and entropy presented below (section 4.4), allow us to explicitly change the presentation, so we can focus on any formula to apply a rule, and to apply cyclic exchange.

5.2   Language

The formulas of $\mathcal{NCL}_{ocr}$ are built from the latin letters $\mathcal{A}$,..., and $\mathcal{Z}$, the set of primitives P, the constant ⊤, the negation ⊥, and the following linear connectives: (we just give those we need)

- non-commutative (but associative) conjunction • (next) and disjunction ∇(sequential),
- commutative multiplicative conjunction ⊗(times) and disjunction ∂ (par), (obviously associative)
- exponentials !(of course) and ? (why not).

De Morgan rules hold, so we have the following results:

- $(p)^{\perp} = p^{\perp}$, $(p^{\perp})^{\perp} = p$, where p is an atom,

- $(A \bullet B)^{\perp} = B^{\perp} \nabla A^{\perp}$ $(A \nabla B)^{\perp} = B^{\perp} \bullet A^{\perp}$
- $(A \otimes B)^{\perp} = B^{\perp} \partial A^{\perp}$ $(A \partial B)^{\perp} = B^{\perp} \otimes A^{\perp}$
- $(!A)^{\perp} = ?A^{\perp}$ $(?A)^{\perp} = !A^{\perp}$.

For any formula A, we have $A = A^{\perp\perp}$. We also use the linear implication $\multimap$ under the form $A \multimap B = A^{\perp} \partial B$.

With these connectives, $\mathcal{NCL}_{ocr}$ is a multiplicative fragment of non-commutative logic.

5.3   A Formulation in $\mathcal{NCL}_{ocr}$

Let us show now how the recognition problem can be formulated in $\mathcal{NCL}_{ocr}$.
Normalized input characters I are then fully captured with the help of nested formulas of the form: $(i_1 \bullet ... \bullet (i_{21} \bullet i_{22})...)$, where the primitives i_k for k=1,22 belong to the set P. As the connective $\bullet$ is associative, we can set I $= (i_1 \bullet ... \bullet i_{22})$. In such a formula, the graphic primitives involved follow each other **according to the rewriting order introduced by the pen movement**; and in absence of the contraction and weakening rules, these primitives can neither disappear nor be duplicated.

Distances between connected primitives are captured by means of *rewriting rules*:

- $E_1 = \{$h $\multimap$ f, f $\multimap$ h ... d $\multimap$ h, h $\multimap$ d $\}$, for graphic primitives.
- $E_2 = \{$s $\multimap$ o, o $\multimap$ s, ..., y $\multimap$ z, z $\multimap$ y $\}$, for descriptive primitives.

The set E $= E_1 \cup E_2$ of rewriting rules gives the *context* in which proofs ared carried out.

Clauses in the data base have the following meaning :

$(l_1 \bullet ... \bullet l_{22}) \multimap \mathcal{L}$. where only relevant primitives appear in the premise, the non relevant ones being set to the constant value $\top$ (which holds for any predicate).
Let us give an example of representation in $\mathcal{NCL}_{ocr}$.

Example 7. The character "a" represented in standard logic by the clause (see Example-4) :

$(p_{1,5} \wedge p_{2,5} \wedge p_{3,5} \wedge p_{4,7} \wedge p_{5,7} \wedge p_{6,7} \wedge p_{7,1} \wedge p_{8,1} \wedge p_{9,1} \wedge p_{10,1} \wedge p_{11,6} \wedge p_{12,6} \wedge p_{13,6} \wedge p_{14,0} \wedge p_{15,9} \wedge p_{16,12} \wedge p_{17,12} \wedge p_{18,13} \wedge p_{19,13} \wedge p_{20,13} \wedge p_{21,15} \wedge p_{22,17}) \rightarrow \mathcal{A}$.
has the following more appropriate meaning:

$(g \bullet g \bullet g \bullet d \bullet d \bullet d \bullet d \bullet f \bullet f \bullet f \bullet f \bullet v \bullet v \bullet v \bullet o \bullet q \bullet q \bullet r \bullet r \bullet r \bullet j \bullet y) \multimap \mathcal{A}$,
A generalized form of the character "a" may then be represented as follows:
$(g \bullet \top \bullet \top \bullet \top \bullet d \bullet \top ... \bullet \top \bullet r \bullet \top \bullet \top) \multimap \mathcal{A}$, where the constant $\top$ holds for any predicate.

Let I $= ((i_1 \bullet ... \bullet (i_{m-1} \bullet i_m)...)$ be the linear representation of an input character; let L be the premise of a clause: L$= ((l_1 \bullet ... \bullet (l_{n-1} \bullet l_n)...)$
Instead of computing the distance between I and L, $\| I, L \|$ (for m $=$n $= 22$), we propose to carry out the proofs of the sequents !E, I $\vdash$ L, where !E means that **we can apply the rewriting rules of the context E the number of times we need in the proofs**. In the next section, we will show that the main step of our recognition process, can be in fact achieved formally by means of

proofs of this family of sequents. As we only deal with right-hand side sequents, we transform the formulas with the help of De Morgan rules. The dual of the formula I is

$$I^\perp = ((i_1 \bullet ... \bullet (i_{m-1} \bullet i_m)...)^\perp = (...(i_m^\perp \nabla i_{m-1}^\perp)...) \nabla i_1^\perp).$$

Each !ed formula of the set E has a dual in the set Δ which then contains the following formulas: $\{(f^\perp \otimes h), (h^\perp \otimes f), ..., (z^\perp \otimes y), (y^\perp \otimes z)\}$. So, we are left with sequents of the form $\vdash ?\Delta, I^\perp, L$ to be proved in the calculus we introduce now.

5.4   Rules

Sequents are of the form $\vdash \Gamma$, where Γ is a partially ordered set of occurrences of formulas.

- **Identity-Cut**

$$\vdash A^\perp, A \qquad \frac{\vdash \Gamma, A \qquad \vdash A^\perp, \Delta}{\vdash \Gamma, \Delta} cut$$

- **Structural Rules**

$$\frac{\vdash \Gamma, \Delta}{\vdash \Gamma; \Delta} seesaw \qquad \frac{\vdash \Gamma[\Delta; \Sigma]}{\vdash \Gamma[\Delta, \Sigma]} entropy \qquad \frac{\vdash \Gamma[?\Delta, \Sigma]}{\vdash \Gamma[?\Delta; \Sigma]} center1 \qquad \frac{\vdash \Gamma[\Delta, ?\Sigma]}{\vdash \Gamma[\Delta; ?\Sigma]} center2$$

- **Non-commutative Rules**

$$\frac{\vdash \Gamma; A \qquad \vdash B; \Delta}{\vdash \Gamma; A \bullet B; \Delta} \bullet \qquad \frac{\vdash \Gamma; A; B}{\vdash \Gamma; A\nabla B} \nabla$$

- **Commutative Rules**

$$\frac{\vdash \Gamma, A \qquad \vdash B, \Delta}{\vdash \Gamma, A \otimes B, \Delta} \otimes \qquad \frac{\vdash \Gamma, A, B}{\vdash \Gamma, A\partial B} \partial$$

- **Exponentials**

$$\frac{\vdash \Gamma, A}{\vdash \Gamma, ?A} d \qquad \frac{\vdash ?\Gamma, A}{\vdash ?\Gamma, !A} ! \qquad \frac{\vdash \Gamma, ?A, ?A}{\vdash \Gamma, ?A} c \qquad \frac{\vdash \Gamma}{\vdash \Gamma, ?A} w$$

- **Constant**

$$\overline{\vdash \Gamma, \top}$$

Remark 1. **About seesaw**: It is the key of the system. It is reversible and its inverse, called *co-seesaw*, is a particular case of entropy:

$$\frac{\vdash \Gamma; \Delta}{\vdash \Gamma, \Delta} co\text{-}seesaw$$

Together, they imply cyclic exchange in the usual sense. Intuitively the *seesaw* rule means that proofs with at most 2 conclusions can freely pivotate, thus for such subproofs of a larger proof, commutative and non-commutative compositions should be indistinguishable.

For proofs with more than two conclusions, only cyclic exchange is permitted, for instance the sequent $\vdash (\Gamma; \Delta); \Sigma)$ can be mapped to $\vdash \Sigma; (\Gamma; \Delta))$ but not to $\vdash (\Delta; \Gamma); \Sigma)$. In this general case, commutative and non-commutative compositions are to be distinguished.

Remark 2. **About the Constant rule**: we restrict the use of the Constant rule to the case where the set of formulas Γ contains only one negative atomic formula.

The usual general results about Cut elimination and Gentzen's subformula property hold obviously.

Let us give now some examples of proofs in $\mathcal{NCL}_{ocr}$.

Example 8. The sequent $((g \bullet d) \bullet f) \vdash (g \bullet (\top \bullet f))$ mapped to the sequent $\vdash (g \bullet (\top \bullet f), (f^{\perp} \nabla (d^{\perp} \nabla g^{\perp}))$ is provable in $\mathcal{NCL}_{ocr}$ without applying rewriting rule.

We proceed as follows:

- We first apply the *entropy* rule, then we apply twice the ∇ rule follow by the *seesaw* rule :

$$
\cfrac{
\cfrac{
\cfrac{
\cfrac{
\vdash (g \bullet (\top \bullet f)); f^{\perp}; d^{\perp}, g^{\perp}
}{
\vdash (g \bullet (\top \bullet f)); f^{\perp}; d^{\perp}; g^{\perp}
}\, seesaw
}{
\vdash (g \bullet (\top \bullet f)); f^{\perp}; (d^{\perp} \nabla g^{\perp})
}\, \nabla
}{
\vdash (g \bullet (\top \bullet f)); (f^{\perp} \nabla (d^{\perp} \nabla g^{\perp}))
}\, \nabla
}{
\vdash (g \bullet (\top \bullet f)), (f^{\perp} \nabla (d^{\perp} \nabla g^{\perp}))
}\, entropy
$$

- we exchange the formulas (it is a cyclic exchange) in $\vdash (g \bullet (\top \bullet f)); f^{\perp}; d^{\perp}, g^{\perp}$ to obtain the sequent:$\vdash g^{\perp}, (g \bullet (\top \bullet f)); f^{\perp}; d^{\perp}$ to be proved.

$$
\cfrac{
\cfrac{
\vdash g^{\perp}; g \qquad \vdash (\top \bullet f); f^{\perp}; d^{\perp}
}{
\vdash g^{\perp}; (g \bullet (\top \bullet f)); f^{\perp}; d^{\perp}
}\, \bullet
}{
\vdash g^{\perp}, (g \bullet (\top \bullet f)); f^{\perp}; d^{\perp}
}\, entropy
$$

the sequent $\vdash g^{\perp}; g$ is an identity (after applying the *seesaw*) rule. We are left with the sequent $\vdash (\top \bullet f); f^{\perp}; d^{\perp}$ to be proved. In order to do a cyclic exchange, we again apply the *seesaw* rule to obtain the sequent $\vdash (\top \bullet f); f^{\perp}, d^{\perp}$ and, after exchanging the formulas, we are left with the sequent $\vdash d^{\perp}, (\top \bullet f); f^{\perp}$ to be proved.

$$
\cfrac{
\cfrac{
\vdash d^{\perp}; \top \qquad \vdash f; f^{\perp}
}{
\vdash d^{\perp}; (\top \bullet f); f^{\perp}
}\, \bullet
}{
\vdash d^{\perp}, (\top \bullet f); f^{\perp}
}\, e
$$

- it remains to apply the *seesaw* rule to obtain the sequents $\vdash d^{\perp}, \top$ and $\vdash f, f^{\perp}$, which proofs can be achieved easily by applying respectively the Constant rule and the Identity rule.

Example 9. The sequent $!E, g \vdash d$ mapped to $\vdash ?\Delta, g^{\perp}, d$ is provable in $\mathcal{NCL}_{ocr}$. Only commutative rules are invoked.

- We apply recursively the weakening rule on ?ed formulas of Δ untill we obtain the sequent $\vdash ?(v^{\perp} \otimes g), ?(d^{\perp} \otimes v), g^{\perp}, d$ to be proved.
- Then, by applying the exponential rule (d) on the formula $?(v^{\perp} \otimes g)$, we are left with the sequent $\vdash (v^{\perp} \otimes g), ?(d^{\perp} \otimes v), g^{\perp}, d$ to be proved. As the formulas freely commutate, we map this sequent to the sequent $\vdash d, ?(d^{\perp} \otimes v), (v^{\perp} \otimes g), g^{\perp}$.
- We apply the $\otimes$ rule on the formula $(v^{\perp} \otimes g)$ for sharing context, so we obtain the two sequents $\vdash d, ?(d^{\perp} \otimes v), v^{\perp}$ and $\vdash g, g^{\perp}$ to be proved. The latter is an identity. Let us give the proof-tree of the sequent $\vdash d, ?(d^{\perp} \otimes v), v^{\perp}$.

$$\frac{\dfrac{\vdash d, d^{\perp} \qquad \vdash v, v^{\perp}}{\vdash d, (d^{\perp} \otimes v), v^{\perp}} \otimes}{\vdash d, ?(d^{\perp} \otimes v), v^{\perp}} d$$

We can conclude by applying the Identity rule twice.

6   Recognition in $\mathcal{NCL}_{ocr}$

In this section, we characterize the proofs of the sequents $\vdash ?\Delta, I^{\perp}, L$; then, in the next section we will list what can be gained from our theoretical approach in terms of proofs.

6.1   Proofs in $\mathcal{NCL}_{ocr}$

Lemma 1. *There exists a cut-free proof of the sequent $\vdash ?\Delta, I^{\perp}\partial L$ in $\mathcal{NCL}_{ocr}$ iff:*

1. *$m = n$*
2. *for every $k : 1 \le k \le m$, the sequents $\vdash ?\Delta, i_k^{\perp}, l_k$ are provable.*

Proof. For proving the lemma-1, we need the following result:

Lemma 2. *The sequent $\vdash ?\Delta, I^{\perp}, L$ is provable in $\mathcal{NCL}_{ocr}$ iff:*
the sequent $\vdash ?\Delta, i_m^{\perp}; \ldots; i_1^{\perp}, ((l_1 \bullet \ldots \bullet (l_{n-1} \bullet l_n)\ldots)$ is provable in $\mathcal{NCL}_{ocr}$.

The proof of the lemma-2 is obvious, so we only focus on the proof of the lemma-1.

1. **Necessary Condition** Let us suppose that there exists a cut-free proof of the sequent $\vdash ?\Delta, i_m^{\perp}; \ldots; i_1^{\perp}, ((l_1 \bullet \ldots \bullet (l_{n-1} \bullet l_n)\ldots)$; let us show that the conditions 1) and 2) are then satisfied. We essentially use the induction principle applied to proof-trees.

a) **Basic case**: the proof-tree is a one-node tree. The sequent $\vdash ?\Delta, i_{\bar{k}}^{\perp}, l_k$, is an axiom of $\mathcal{NCL}_{ocr}$ after applying weakening rules (w) to ?ed formulas. The conditions i) and ii) are satisfied.

b) **General case**: let T be the proof-tree of the sequent $\vdash ?\Delta, i_{\bar{m}}^{\perp}; ...; i_{\bar{1}}^{\perp}, ((l_1 \bullet ... \bullet (l_{n-1} \bullet l_n)...))$. The last rule applied may be (d), (w), (c), a structural rule (*entropy, seesaw, center1, center2*), $\otimes$ (*times*) or the rule $\bullet$ (*next*). The rules (d), (w), (c), $(center1)$, $(center2)$ only involve ?ed formulas. The (*entropy*) or (*seesaw*) rules lead to cyclic exchanges. So, the induction hypothesis holds and we can conclude easily in one of these cases.

   i. **the last rule applied is** $\otimes$: this connective only involves a formula of Δ, let $(p^{\perp} \otimes q)$ be this formula; the proof-tree T is then the following one:

$$\frac{T_1 \qquad T_2}{\vdash \Gamma_1, (p^{\perp} \otimes q), \Gamma_2} \otimes$$

   A. **proof of the condition 1)** T_1 is the proof-tree of the sequent $\vdash \Gamma_1, p^{\perp}$. By the induction hypothesis, we have $m_1 = n_1$ and the sequents $\vdash ?\Delta, i_{\bar{k1}}^{\perp}, l_{k1}$ are provable for every k1 : $1 \leq k1 \leq m_1$. In particular, we can isolate the sequent $\vdash ?\Delta, p, p^{\perp}$, which is provable too. This result also holds for T_2 : we have $m_2 = n_2$, and the sequents $\vdash ?\Delta, i_{\bar{k2}}^{\perp}, l_{k2}$ are provable for every k2 : $1 \leq k2 \leq m_2$. In particular, the sequent $\vdash q, q^{\perp}, \Delta$ (obtained by exchanging) is provable too. As the connective juxtaposes the contexts, we are left with the following results: $m = m_1 + m_2 - 1 = n_1 + n_2 - 1 = n$; then the condition i) is satisfied.

   B. **proof of the condition 2)** We know that the sequents $\vdash ?\Delta, i_{\bar{k1}}^{\perp}, l_{k1}$ and $\vdash ?\Delta, i_{\bar{k2}}^{\perp}, l_{k2}$ are provable for every k1 : $1 \leq k1 \leq m - 1$, and for every k2 : $1 \leq k2 \leq m - 1$. Let us focus on the pair $\vdash ?\Delta, p, p^{\perp}$ and $\vdash q, q^{\perp}, ?\Delta$. As they are provable we can apply the $\otimes$ rule to prove the sequent $\vdash ?\Delta, p, (p^{\perp} \otimes q), q^{\perp}, ?\Delta$. By iterating the contraction rule on ?ed formulas, we get the proof of the sequent $\vdash ?\Delta, p, (p^{\perp} \otimes q), q^{\perp}$; by applying the (d) rule, we prove the sequent $\vdash ?\Delta, p, ?(p^{\perp} \otimes q), q^{\perp}$, and at last, by applying the contraction rule on the formula $?(p^{\perp} \otimes q)$ (which also has an occurrence in $?\Delta$), we satisfy the condition 2).

   ii. **the last rule applied is** $\bullet$ **(next)** we can proceed in the same manner because the contexts are also juxtaposed. So, from $m_1 = n_1$ and $m_2 = n_2$, we can conclude that $m = m_1 + n_1 = m_2 + n_2$, and for every k: $1 \leq k \leq m$, the sequents $\vdash ?\Delta, i_{\bar{k}}^{\perp}, l_k$ are obviously provable by using the induction hypothesis.

2. **Sufficient Condition:** we just give an idea of the proof of the sequent $\vdash ?\Delta, i_{\bar{m}}^{\perp}; ...; i_{\bar{1}}^{\perp}, ((l_1 \bullet ... \bullet (l_{n-1} \bullet l_n)...))$. Let us suppose that the two conditions 1) and 2) are satisfied. From the sequents $\vdash ?\Delta, i_{\bar{k}}^{\perp}, l_k$, which are provable for every i: $1 \leq k \leq m$, the proof consists in iterating the application of the

(next) rule, followed by some contractions on ?ed formulas combined with cyclic exchanges.

7   What Is Gained by Carrying Out Proofs in $\mathcal{NCL}_{ocr}$

What is learnt from our formal analysis can be outlined now. First, we are able to evaluate accurately the complexity of our method. This point is not clear at all for the approaches refered to in section 2. Second, we will show that the word recognition problem is as difficult to solve as the reachability problem for a Petri net [13]; so, it still remains a challenge. We also will prove that segmenting a word consists in either duplicating some stroke or removing another one in this word. $\mathcal{NCL}_{ocr}$ is an appropriate framework to study the problem of word segmentation.

7.1   Complexity of Our Approach

As our recognition process can be fully captured by means of proofs in $\mathcal{NCL}_{ocr}$ of the sequents of the form: $\vdash ?\Delta, i_m^\perp; ...; i_1^\perp, ((l_1 \bullet ... \bullet (l_{n-1} \bullet l_n)...)$, the two following points are straigthforward.

1. From the condition 1) m = n of the lemma-1, we can conclude that the strings manipulated in our system of recognition **in the context of the rewriting rules E** must have the same **length**. This result justifies our choice to normalize all the input characters by giving them the same size (set to 22).
2. From the condition 2) for every $k : 1 \le k \le m$, the minimal number of times we apply the $\otimes$ rule in the proofs of the sequents $\vdash ?\Delta, i_k^\perp, l_k$ can be used to compute the distance between the two graphic primitives $|i_k, l_k|$ (see the definitions in section 3, and the Example-9). So, the distance between two normalized strings can also be computed by means of depths of proof-trees. As the formulas of $?\Delta$ are restricted to atoms of the language, the recognition process can be achieved in **linear time**, depending only on the length m of strings, and on the number of rewriting rules involved in the proofs of the sequents $\vdash ?\Delta, i_k^\perp, l_k)$.

    Moreover, we can carry out proofs faster with the help of our symbolic learning algorithm because some sequents of the form $\vdash ?\Delta, i_k^\perp, \top$ allow us to conclude at once by applying the Constant rule.

7.2   Segmentation and Recognition

To illustrate the segmentation problem, we give an example of cursive writing in which a number of consecutive characters may be written between the two marks of pen-up and pen-down.

Fig. 4. The word "tall" in cursive style

Example 10. The word "tall" may have the form given in Fig.4. The first character "t" can be isolated in a box from the rest of the word because the writer explicitly has introduced the marks of pen-up and pen-down when he added the cross of the letter "t". But the consecutive characters "all"are written using an unbroken stroke. So, we cannot count on the beginning and ending marks to define possible character segmentation. Let us formalise this point.

In $\mathcal{NCL}_{ocr}$, the word "tall" is represented by the formula $W = (f \bullet g \bullet v \bullet f \bullet e \bullet h) \bullet e \bullet g \bullet d \bullet f \bullet d \bullet f \bullet t \bullet e \bullet v \bullet d \bullet f \bullet e \bullet v \bullet h \bullet f$.

A correct recognition of the word "tall" supposes that some strategy is available to stress the frontier of the characters which are hidden in the word "tall", i.e, to transform the formula W into the formula $WS = (f \bullet g \bullet v \bullet f \bullet e \bullet h) \bullet (e \bullet g \bullet d \bullet f \bullet d \bullet f \bullet)(f \bullet t \bullet e \bullet v \bullet d \bullet f) \bullet (f \bullet e \bullet v \bullet h \bullet f)$, where the strokes at the frontier of characters are duplicated. All the characters contained in the word W are isolated now, and can be submitted to our character recognition procedure. A major difficulty to deal with is to locate the segmentation point.

Complexity of Word Recognition. If we extend roughly our approach to word recognition, as we don't know the number of characters which appear in a word, we can no longer normalize them.

Let us consider this general case, where the input I is given by the formula $I = (i_1 \bullet ... \bullet i_{m-1} \bullet i_m...)$, and some chain of character models is represented by the formula $L = (l_1 \bullet ...) \bullet (... \bullet l_{n-1} \bullet l_n)...)$ have different lengths : $m \neq n$. We know from the lemma-1, that we have to improve the expressive power of the set E in order to carry out a proof of the sequent $!E, I \vdash L$. So, let us now suppose that the rewriting rules of E are of the form: $(p_1 \bullet ... \bullet (p_{r-1} \bullet p_r)...) \multimap (q_1 \bullet ... \bullet (q_{s-1} \bullet q_s)...)$, where $r \neq s$, the premise and the conclusion are no longer limited to atomic formulas. We are left with the following sequents $!E, I \vdash L$, to be proved in a more general context.

It is likely that our problem is very complex; we can compare it to the reachability problem for a Petri net, see [8,13], problem which can be stated as follows :

Given a Petri net R (R is a tuple $(\mathcal{S}, \mathcal{T})$, where $\mathcal{S}$ is the set of places, and $\mathcal{T}$ is a set of transitions), and two markings m_i and m_f find an algorithm to decide if there exists a sequence of transitions such that m_f can be reached from m_i.

In our application the markings are given by the atoms which appear in formulas I and L, the sequence of transitions are the rewriting rules of the set E.

A rough approach to word recognition is likely to be of exponential complexity, depending on the number of variables in $\mathcal{NCL}_{ocr}$ and their interconnections to the linear implications of the set E.

First Investigation in Word Recognition. Consequently, we have investigated a new solution by restricting the segmentation hypotheses. We propose to proceed as follows:

1. First Step: Let I be the input formula. Let $\mathcal{L}$ be a (non normalized) character model of length m in the data base; we match $\mathcal{L}$ to I from the position 1 to m. The predicate i_m is duplicated, and we submit the subformula of I, $I_1 = i_1 \bullet ... \bullet i_m$ to our recognition process. We repeat this step for all the character models of different lengths in the data base.
2. Second Step: We select the best match (see section 4). Then, we iterate again the process on the rest of the word.

We run this experimentation on small test sets containing 120 words from a total word lexicon of 400 (usual english) words. The word recognition rate is 80% for unconstrained cursive style, and can go up to 90% for limited cursive style.

8   Conclusion

From our point of view it is beneficial to analyse recognition system with the help of $\mathcal{NCL}_{ocr}$. The connective $\bullet$(next) has the good "properties" to capture knowledge in data base. When structural rules - contraction, weakening and exchange- are used, their role appears clearly. Inferences are carried out step by step, and so, computational complexity can be accurately evaluated. We also learn from this formal analysis, that we need appropriate strategies to limit the complexity of the word recognition problem.

Acknowledgements. I wish to thank Jacqueline Vauzeilles for helpful comments on the contents of this paper. I would also like to thank Pierre Brézellec for coming to my assistance when I used his learning algorithm. Finally, I thank the AISC'02 reviewers for their helpful reviews.

References

1. Abrusci, V.M., Ruet, P.: Non-Commutative Logic I : the Multiplicative Fragment, Annals Pure Appl. Logic, 1998.
2. Brézellec, P., Soldano, H.: ÉLÉNA: A Bottom-Up Learning Method, Proceedings of the Tenth International Conference on Machine Learning, Ahmerst 93, Morgan Kaufmann, pp 9–16.
3. Castaing, J., Brézellec, P.: On-Line Chinese Characters Recognition by means of Symbolic Learning, Proceedings of the International Conference on Chinese Computing'96 June 4-7 Univ. of Singapor
4. Castaing, J., Brézellec, P.: Une Méthode Symbolique pour la Reconnaissance de l'écriture Manuelle en Ligne, RFIA 96 Rennes
5. Chan, K.-F., Yeung, D.-Y.: Elastic Structural Matching for On-Line Handwritten Alphanumeric Character Recognition, Proceedings 14th Int. Conf. Pattern Recognition vol.2, Brisbane, Australia, pp. 1508–1511

6. Connell, S.: On-Line Handwriting Recognition using Multiple Pattern Class Models, submitted to Michigan State University in partial fulfillment of the requirements for the degree Doctor of Pilosophy Department of Computer Science and Engineering 2000
7. Frey, P. , Slate, D.: Letter Recognition Using Hollad-Style Adaptive Classifiers, Machine Learning Vol.6 num.2, Kluwer Academic Publishers pp. 161–182
8. Kanovitch, M.I.: Linear Logic as a Logic of Computations, Proceedings 7-th Annual IEEE Symposium on Logic in Computer science, Santa Cruz, pp 200–210, 1992
9. Girard, J.-Y,: Linear Logic, Theoret.Comp. Sci., 50(1) :1–102, 1987
10. Hu, J., Turin, W.: HMM Based On-Line Handwriting Recognition, IEEE Trans. Pattern Analysis and Machine Intelligence, vol.18, n0. 10, pp. 1039–1045, Oct. 1996.
11. Manke, S., Finke, M., Waibel A.: The Use of Dynamic Writing Information in a Connectionist On-Line Cursive Handwriting Recognition System, Neural Information Processing System NIPS 94, pp 1093–1100
12. Nathan, K.S., Bellegarda, J.R., Nahamou, D., Bellegarda, E.J. :On-Line Handwriting Recognition Using Continuous Parameter Hidden Markov Models, Proc. ICASSP'93, vol.5, Minneapolis, MN, pp. 121–124, May 1993.
13. Peterson, J.L,: Computation sequence sets, J.Comput. System Sci.13 1–24 1976
14. Prevost, L., Milgram, M.: Automatic Allograph Selection and Multiple Classification for Totally UnconstrainedHandwritten Character Recognition, Proceedings 14th Int. Conf. Pattern Recognition vol.2, Brisbane, Australia, pp 381–383
15. Rabiner, L.R: A tutorial on Hidden Markov Models and Selected Application in Speech Recognition, Proceedings of IEEE, 77(2) 1989
16. Rigoll, G., Kosmala, A., Willet, D.: A New Hybrid Approach to Large Vocabulary Cursive handwriting Recognition, Proc. 14th Int. Conf. on Pattern Recognition, Brisbane, Australia, pp. 1512–1514, Aug. 1998.
17. Robinson, J.A.: A Machine Oriented Logic Basedon the Resolution Principle J.ACM 12(1), 23–41
18. Ruet, P.: Non-commutative logic II : sequent calculus and phase semantics, to appear in Mathematical Structure in Computer Science
19. Scattolin, P., Krzyzak, A.: Weighted Elastic Matching Method for Recognition of Handwritten Numerals in Vision Interface'94, pp178–185, 1994
20. Editor Wang, P.S.P: Characters & Handwriting Recognition: Expanding Frontiers, World Scientific Series 1991
21. Yanikoglu, B.A, Sandon, P.A: Recognizing Off-Line Cursive Handwriting, IEEE Computer Society Conference On Computer Vision and Pattern Recognition, CVPR'94, pp397–403

From Numerical to Symbolic Data during the Recognition of Scenarii

S. Loriette-Rougegrez

University of Technology of Troyes, Laboratory LM2S,
BP 2060, 10010 Troyes, France,
loriette@utt.fr

Abstract. The objective of this paper is to present a system that is able to recognize the occurrence of a scenario evolving over time and space. Scenarii are considered to be made up of several stages. The transition from a stage to another one requires the satisfaction of conditions. These features have led us to the construction of a graph which is run by means of a rule-based system. Transitions are validated with the transformation of numerical data into symbolic ones. Data's uncertainty is considered by means of the computation of an evidence's mass for each transition. The system described in this paper is applied to the recognition of maneuvers performed by a car driver.

1 Introduction

Our work takes place inside the CASSICE project. It aims at building a system permitting to acquire the description of driving situations, and facilitate their analyzis. These situations are assumed to take place on a straight urban motorway. CASSICE includes several components. We focus in this paper on one of them, the DSRC[1] system, which role is the recognition of the performed maneuvers, namely the overtakings.

Several data inform the system DSRC about the respective positions of the considered cars. The recognition of the overtaking maneuver is fulfilled by means of several steps. In the next part, we will present the CASSICE project. We will carry on in section 3 with the description of the input data, permitting the recognition of the performed maneuver. Then the system DSRC will be presented: the principles underlying the development of the system will be illustrated inside DSRC version 1 in section 4, and a second version taking into account the uncertainty of data will be presented in section 5. Knowledge representation used in DSRC v2 will be detailed in section 6. We will conclude with results and perspectives in part 7.

2 The CASSICE Project

Over the last years, several driving assistance systems have been developed, in order to increase the security of the car driver or facilitate his task. In order

---

[1] Acronym for Driving Situation ReCognition.

J. Calmet et al. (Eds.): AISC-Calculemus 2002, LNAI 2385, pp. 154–167, 2002.

to improve these systems' quality, it is necessary to evaluate their impact on the driver's behavior. To do so, comparisons are fulfilled between the driving activity *with assistance system(s)* and the driving activity *without assistance system(s)* [1]. Now, this activity can not be evaluated independently from its context: infrastructure, driver's actions, social data (driver's experience, gender, driver's knowledge of the way, etc.).

The methodology used consists in observing and video-recording the driver's behavior inside an equipped vehicle [2]. They allow to conduct interviews after the journey so as to identify his (resp. her) objectives and strategies, his (resp. her) knowledge and representations of the studied driving situation. Several kinds of data are acquired. Some of them are context-dependent (e.g., social data). Other ones are time-dependent. These last ones, except for a few ones[2], allow first to fill in a grid of a spreadsheet. They are time-stamped. They describe what the driver may observe or do: the motorway infrastructure (traffic entry zones, directional zones, traffic joining zones, etc.), the intensity of traffic, and several features of the considered car (speed, action on the brake, lane changing, distance to the followed vehicle), or of the car followed (speed, lane occupied). Different profiles of driver, defined by different experiences, etc. lead to the filling of different grids, which will be analyzed separately.

Each grid allows afterwards the psychologist to compute the answer to questions such as: *how many seconds does the driver require in order to perform an overtaking?. Between a left changing lane and the corresponding right changing lane, calculate the number of vehicles that have been overtaken by the equipped car*, etc. Such questions, and many other ones, permit to the psychologist to model the driver's behavior.

The interest of the CASSICE project, and its originality, is to automatize this analysis. The objective is indeed to build an integrated system, made up of two parts [3]:

– an equipped vehicle (EV) to record automatically a physical description of the driving situation,
– a software to build a database of driving situations, and automatically answer questions such as the ones mentioned above.

The first part of CASSICE manipulates raw numerical data, whereas the second one requires a symbolic description of the situations. A transformation's step of numerical data into symbolic ones is then necessary.

3 Input Data

The different layers of CASSICE are fulfilled at the same time by different researchers. Especially, the experimental vehicle has just been equipped with differ-

---

[2] e.g., the flashes of the left turn signal of the vehicle to be overtaken. This data is considered during a later analysis.

ent sensors [4]. That's the reason why DSRC has been tested on simulated data
that we present next. The java-written simulator permits to enter the definition
of road scenarii, to visualize them graphically, and to generate corresponding
sensors data [5]. We interested only in the overtaking maneuver. This maneuver
will be presented afterwards.

Table 1. Simulated data.

Time	X	Y	V	θ	φ	Rg	Rd
0.01	-32.00	0	15	0	0	-3.50	1.50
0.02	-31.85	0	15	0	0	-3.50	1.50
0.03	-31.70	0	15	0	0	-3.50	1.50
...	...	...	...	...	...	...	...
1.12	-15.52	-2.01	15	-9.91	3	-1.46	3.54
1.13	-15.37	-2.04	15	-9.68	3	-1.44	3.56
1.14	-15.22	-2.06	15	-9.46	3	-1.41	3.59

Table 2. Data's meaning.

Acc	Acceleration of EV relative to TV (m²/s)
φ (phi)	Front wheel angle of EV (in degrees)
Rd	Position of EV relatively to the right side of the road(m)
Rg	Position of EV relatively to the left side of the road (m)
θ (theta)	Angle that the target TV forms with the direction indicated by the road (degrees)
V	Speed of EV relative to TV (m/s)
X	Position on the x's axis of TV against EV (m)
Y	Position on the y's axis of EV against EV (m)

3.1   Raw Data Acquired

The maneuver recognition needs first to measure a lot of parameters that characterize the driving situation. The simulated data used are presented in table
1. The meaning of the different variables is presented in table 2. They refer to
a *target vehicle* (TV). It is the vehicle which is assumed to be followed, then
overtaken, by the driver of the *equipped vehicle* (EV).

Simulated data, as well as the future real data, are time-stamped.

3.2   Overtaking Maneuver Example

The overtaking maneuver, that one has to recognize inside the data of table 1,
may be simply described in this way: an experimental vehicle (EV) goes to the

right lane of the highway. It catches a target vehicle (TV) running on the same lane with a lower speed. The EV is beginning an overtaking of the TV. It begins going to left for a lane changing, then it is running straight forward. When EV has overtaken TV, it goes right to the right lane.

Each step needs to consider several variables. For instance, to detect a change of lane inside an overtaking, one has to compare the variable Y with 0, consider the speed of EV, etc. Each step of the maneuver depends on a configuration of certain variables' values. It depends too on the stage of recognition of the maneuver. Indeed, a given configuration of variables' values will not always have the same signification during the recognition process. For instance, we have to detect that the four wheels of the EV are on the same lane. This condition is useful for 3 different stages in the recognition process: before EV begins its overtaking, when EV is passing, when EV has completed its maneuver. The recognition process is thus context-dependent.

4   Recognition of a Maneuver by Means of a Graph

The last remark has led us to this conclusion: to recognize a maneuver performance, one needs to take into account:

- the variables relative to the position of EV towards TV,
- the actions of the EV's driver,
- the stage of recognition of the maneuver.

We chose to model a manoeuver with a graph. An extract is described in figure 1. A *state* corresponds to a stage of fulfillment of the maneuver, some of them are optional. For instance, if we consider the overtaking maneuver, several steps may be identified: indication of the intention of overtaking, crossing of the left continuous line, etc. The first is optional whereas the second is obligatory. *Transitions* between states are fulfilled by means of the occurrence of precise events or special configurations of the location of EV towards TV. Events or configurations will be uniformly called in the follow-up *conditions*. *Wait for overtaking, turn of the steering wheel, EV behind TV*, are instances of such conditions. These conditions are potentially numerous.

In DSRC v1 [6], the maneuver is considered recognized if the last state is reached. In DSRC v2, things are slightly different because an evidence's mass should be associated with each state. This problem will be considered in section 5.

4.1   A Rule-Based Recognition

Input data are processed in batch mode. Lines of table 1 representing simulated data, are processed line by line. The conditions' satisfaction evoked in paragraph 3.2, such as *four wheels on the same lane* is detected by means of tests on a subset of variables' values in 2 or 3 successive lines. Once the satisfied conditions have been detected, we have to search for the transitions of the graph that these conditions permit to validate. They will permit in turn to validate states

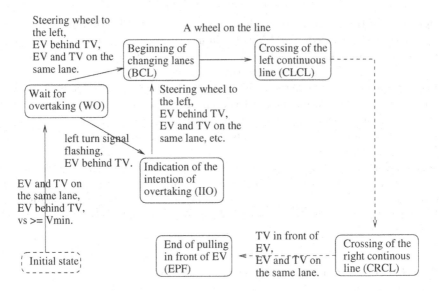

Fig. 1. States making up the overtaking maneuver

of the graph. The data processing relies at the beginning on the detection of variables' values' combinations, such as X, Y, etc. That's the reason why we chose to handle this problem by means of a rule-based approach.

The DSRC system is made up of two levels of rules. We have used the CLIPS formalism [7] [8]. The first-level's rules use raw data and they have to detect the satisfaction of the transitions' conditions. Subsets of conditions recognized make up the transitions validated in the graph. Validated transitions allow in turn the second-level of rules to generate hypotheses about the stage of fulfillment of the maneuver, that is saying to identify the current state.

Table 3. First-level rules in DSRC.

If at time t, EV has a negative value for X then EV is behind TV
If, at time t, Y is in [-1.00, +1.00] then both vehicles are on the same lane

First-Level Rules. The condition part of this kind of rule evaluates the variables' values relating to the detection of a transition's condition. The values are considered over a time period of 2 or 3 ms. This rule-set contains about 20 rules. Two rules are presented in table 3.

Second-Level Rules. They allow to run through the graph. For a couple of states E_i, E_j, and a transition noted "A" between both states, we define the following rule: if, at instant t, state E_i is the current state, and condition(s) in A is (resp. are) detected then the current state becomes E_j. Two other rules are used, of which a rule allowing to initialize the maneuver's recognition.

Rule basis have been used on several occasions in the CASSICE project. The system IDRES [3] has to recognize maneuvers performed too. It is independent from any graph. Input data are directly transformed into states, then states' sequences are pieced together to match maneuver models consisting of a states' succession.

4.2   Results

DSRC has been experimented on simulated data. These data contain about 600 descriptions of variables' values descriptions, with an interval of length 0.01 s between 2 descriptions. The fulfilled recognition presented in figure 2 shows the time periods during which each condition associated with one or several transitions' graph is recognized. On the top of the figure, conditions are associated with intervals during which they are validated. To go from a state to another one at a given time point, all the conditions associated with the corresponding transition need to be validated. We show in the bottom of the figure a succession of time points at which each state is validated[3]. These results permit to conclude that an overtaking maneuver has been recognized inside the considered set of data.

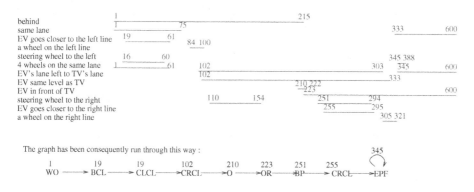

Fig. 2. DSRC v1 results

In this version of DSRC, the recognition of conditions, then transitions and states, is boolean. A state/condition/transition is true or false. In the reality,

---

[3] Time-stamps are the earliest time-points at which the states are validated. States are abbreviated. Abbreviations appear in figure 1, except for "O", "OR", "BP", which mean respectively *overtaking, overtaking fulfilled, beginning of pulling in.*

things are not so clear. For instance, to recognize that the car driver has turned the steering wheel, which rotation is considered significative?. In fact, almost all the conditions of the graph's transitions are concerned about this problem. The second problem is how to run through the graph. If true conditions at a given time point allow the recognition process to remain in the same state or to go to the following, the software will choose to go in the next state. We would like to introduce in DSRC a choice based on solid criteria, such as a confidence in the next state for instance.

5   Taking into Account the Data's Uncertainty

If we consider the following condition: *The car driver has turned the steering wheel*, we need to know the value of the rotation's angle, represented with the variable ϕ. ϕ takes values in [-3,3]. It should have a negative value so as to satisfy the condition. If the value is -3, we may consider that the driver has *intentionally* turned to the left. If the value is -0.5, we may consider that the car driver has *slightly*, perhaps accidentally, turned the steering wheel. We can tell that this action's detection is uncertain. We propose to associate with each transition's condition, a belief into the condition's fulfillment.

We suggest to apply the Dempster-Shafer theory [9]. This theory's interest is that it allows to associate an evidence's mass with a set of hypotheses. We can model a basic belief assignment by a distribution of evidence's masses m on the propositions A, subsets of the hypotheses' set $\Omega = \{H_i, H_j, H_k...\}$. A can be a singleton (or single) proposition such as $\{H_i\}$ but also a composed proposition such as $\{H_i, H_j\}$ for instance. The mass' distribution takes values in 2^Ω, the power set of Ω. The sum of the masses over 2^Ω is 1.

$$m : 2^\Omega \to [0, 1]$$
$$A \to m(A)$$

In the following section, we will explain how we translate numerical variables' values into an evidence's mass. We then obtain a set of conditions associated with a belief in their fulfillment. We will then explain how we work out a confidence in the transitions.

5.1   From Values of Variables to Conditions

Conditions associated with transitions, such as "EV is behind TV", requires the consideration of one or more input data variables. Inversely, each variable will allow to validate one or more conditions. We choose to evaluate for each condition C a confidence in the following statements: "C is true", "C is false", "C is true or false". $\Omega_C = \{C_{true}, C_{false}\}$ makes then up our discernment's space towards the Dempster-Shafer theory.
We have used the fuzzy sets theory [10]. We have defined the DSRC's fuzzy sets

by means of a comparison between simulated data and our knowledge of the maneuver's progress. It will be interesting later to use the same fuzzy sets with real data and to compare the results.

Let's consider the fuzzy sets of figure 3. They gather information about the conditions *steering wheel to the left* (SWL) and *steering wheel to the right* (SWR), which validation depends on phi. We interest here in SWL. We define $\Omega_{SWL} = \{SWL_{true}, SWL_{false}\}$. Figure 3 indicates that this condition is *true* in $]-\infty, -1]$, that it is *true* or *false* in $]-1, 0]$, and that it is *false* in $[0, +\infty[$. The confidences in the statement SWL are summarized in table 4. This stage in the data processing is fulfilled with an adapted version of first-level rules of paragraph 4.1.

SWL is a condition depending on one single variable. It means that in the condition part of the first-level rule permitting the calculation of the confidence into its fulfillment, one has only to read the associated fuzzy set to extract from it the value in $[0, 1]$. Certain conditions are related, not to one, but to 2 variables. We obtain then from the relating fuzzy sets' reading, 2 values in $[0, 1]$. We have to combine them. We then use the combination-rule of Dempster-Shafer described in the next paragraph.

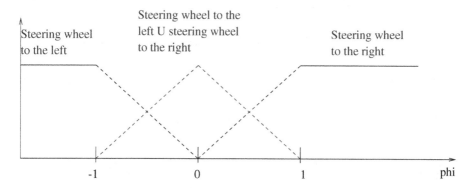

Fig. 3. Belief functions for the conditions steering wheel to the left (resp. right)

Table 4. Confidence in the condition "steering wheel to the left".

Truth value/interval	$]-\infty, -1]$	$]-1, 0]$	$[0, +\infty[$
true	1	0	0
false	0	0	1
true or *false*	0	1	0

5.2   From Conditions to Transitions

The result of the last paragraph is to associate a condition with a masses' distribution over the truth-values *true*, *false*, *true* or *false*. A transition is a disjunction of conditions for which we have to evaluate a confidence in its validation.

A condition may be viewed as a source of information for one or more transitions. A transition is then associated with several sources of information: all of them contribute to the truthfulness or the falsity of the transition's validation. Dempster-Shafer's theory allows to model this situation by the combination-rule. Let's consider m_1 and m_2 the distribution of masses associated with 2 information' sources S_1 and S_2. Let's consider X, an hypothesis, which we wish to compute its evidence's mass. X is given by:

$$m(X) = \sum_{i,j,A_i \bigcap A_j = X} m^{S_1}(A_i) * m^{S_2}(B_j)$$

in which A_i and B_j represent hypotheses coming out from S_1 and S_2.
We calculate now the evidence's mass associated with a transition T, assumed to be dependent on 2 conditions $cond_i$ and $cond_j$, that is to say $T = cond_i \bigwedge cond_j$.

Notations. We use the following notations:

1. T is the transition considered,
2. $\Omega_T = \{T_{true}, T_{false}\}$ is the discernment's space of T,
3. $\Omega_{cond_i} = \{cond_i^{true}, cond_i^{false}\}$ is the discernment's space of $cond_i$,
4. Ω_{cond_j} is the discernment's space of $cond_j$ with $\Omega_{cond_j} = \{cond_j^{true} cond_j^{false}\}$,
5. m_i is a distribution of mass associated with $cond_i$,
6. m_j is a distribution of mass associated with $cond_j$.

We have to compute $m_T = m_i \oplus m_j$.

Computing of the Evidence's mass of a Transition. Table 5 indicates in its first line and first column the values of the mass' distributions m_i and m_j, computed in the paragraph 5.1. Each square contains the product of the evidence's masses of the corresponding line and column, which by application of the Dempster-Shafer theory's combination-rule, will contribute to the calculation of the mass' distribution of m_T. For instance, the table 5 indicates how to compute $m_T^{true\ or\ false}$:

$$m_T^{true\ or\ false} = m_i^{true} \times m_j^{true\ or\ false} + m_i^{true\ or\ false} \times m_j^{true}$$
$$+ m_i^{true\ or\ false} \times m_j^{true\ or\ false}$$

A transition may be associated with more than 2 conditions. The algorithm of figure 4 considers this case.

Table 5. Computing of the mass of evidence of a disjunction of conditions.

Masses	m_i^{true}	m_i^{false}	$m_i^{true\ or\ false}$
m_j^{true}	m_T^{true}	m_T^{false}	$m_T^{true\ or\ false}$
m_j^{false}	m_T^{false}	m_T^{false}	m_T^{false}
$m_j^{true\ or\ false}$	$m_T^{true\ or\ false}$	m_T^{false}	$m_T^{true\ or\ false}$

List <- list of conditions relative to the transition T considered
cond_i <- 1st condition of List; List <- rest(List)

If the number of conditions associated with T = 1 then
        mass(T) = mass(cond_i)
else
        while all the conditions have not been processed do
                cond_j <- 1st condition of List
                List <- rest(List)
                mass(T) = cond_i $\oplus$ cond_j
                cond_i <- cond_j
        End while
End if
Return mass(T)

Fig. 4. Computation of a transition's mass

6 Knowledge Representation

In this section, we describe the knowledge representation chosen. The constraints are the following:

- the description of input data is a list of *lines*. A line associates a time-point with several variables' values, as described in table 1,
- each variable is likely to validate one or more conditions of the graph. Inversely, a condition may be validated with one or several variables, as told about in paragraph 5.1,
- a condition is likely to validate one or more transitions. Inversely, a transition is associated with a conditions' disjunction, as described in paragraph 5.2,
- a first-level rule, in order to compute the evidence's mass of a given condition, has to quickly examine the fuzzy set associated with a variable for this condition. This has been presented in section 5.1.

We have used the object-oriented language (COOL) which is part of CLIPS. The main classes defined are: $C - maneuver - t$, $C - interval - condition$. The graph has an object-oriented representation too. It gathers transitions, considered too as objects. An object *transition* is associated with a transition of a graph, a time-point and a mass' distribution, such as the one talked about in the next paragraph.

Other knowledge structures are used, in order to link transitions to conditions, first-level rules to considered variables, etc.

$C - maneuver - t.$ This class permits to describe at a given time point the mass' distribution associated with each condition of the graph. There is a slot for the time, and a slot for each condition considered. To associate a mass' distribution with each condition, we have defined another class $C - condition - mass$ with the slots *true, false, true or false*. This class is useful for the action part of the first-level rules. It allows indeed to store the evidence's masses computed for all the conditions. Some conditions need the firing of several first-level rules to compute their mass' distribution. That's the case for all the conditions in which several variables are implied. The instances of the class $C - maneuver - t$ allows then to temporarily store the results of the rules calculations. The $C - maneuver - t$ is partially described below:

```
(defclass C_maneuver_t (is-a  C_conditions)
(role concrete)
(pattern-match reactive)
   (slot time (default-dynamic
              (new_C_masse_condition))
              (create-accessor read-write))
   (slot meme_file (default-dynamic
              (new_C_masse_condition))
              (create-accessor read-write))
   (slot vitesse_sup (default-dynamic
              (new_C_masse_condition))
              (create-accessor read-write))...
   (slot vol_a_G (default-dynamic
              (new_C_masse_condition))
              (create-accessor read-write)))
```

$C - interval - condition.$ This class permits to associate with each couple ⟨variable, condition⟩, the list of values' intervals of the *variable*, where the evidence's mass of the *condition* may be different of 0. This class is very useful for the condition part of the first-level rules. Each interval is associated with an interval of *variable*'s values, and an affine function allowing to know the evidence's mass associated with it.

```
(defclass C_interval_condition (is-a  USER)
   (slot inf (default U)
              (create-accessor read-write))
   (slot sup
```

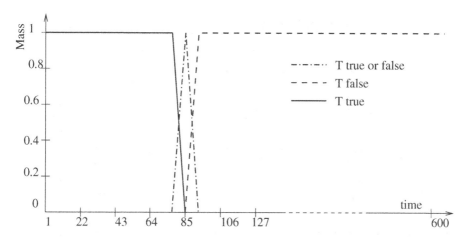

Fig. 5. Mass of the source transition

```
            (default U)
            (create-accessor read-write))
    (multislot listeI
            (default U)
            (create-accessor read-write)))
```

One of its instances is partially described below:

```
(IVC_phi_vol_a_G of C_intervalle_condition
    (inf #-infini)
    (sup 0)
    (listeI
        (create$ -1 (affine phi_vol_a_G_1 -2 1 -1 1)
                 0
                 (affine phi_vol_a_G_2 -1 1 0 0)))))
```

The syntax of the instances' names[4], such as *IVC-phi-vol-a-G*, has been chosen so as to recognize immediately the variables and the condition concerned by it. It allows the first-level rules too to quickly identify the instance to which it has to send a message in order to ask it to return an evidence's mass. The instance above indicates that the condition *vol-a-G* (steering wheel to the left), which depends on ϕ may be associated with a non zero evidence's mass in $]-\infty, 0]$. It details afterwards the values of the mass in subintervals. Each subinterval is associated with an affine function which will automatically be generated from the description, in the instance, of the coordinates (X, Y) of 2 points of this subinterval. X concerns the value of ϕ. Y concerns the evidence's mass associated with X.

---

[4] $< IVC > - < variable\ considered > - < condition\ considered >$, IVC standing for Interval of Validity of Conditions.

7 Results and Perspectives

We have presented a system, DSRC, which, from a graph-based representation of a maneuver, computes the conditions' validity and then the transitions' validity, so as to recognize a fulfillment of the maneuver. The process is 2-steps. A first step consists of transforming, by means of a rule-basis, raw numerical data into symbolic ones, i.e. conditions. DSRC v2 furthermore associates with each condition a confidence into its recognition. The second step combines the confidences associated with each transition's conditions, in order to obtain a single confidence. Fuzzy sets theory and Dempster-Shafer theory seemed to be very suitable to us. The computation of an evidence's mass for each transition is indeed important for the recognition of the overall maneuver. Our work is to our knowledge the first one that uses the Dempster-Shafer theory for the recognition of the validity of transitions inside a graph.

We have next to choose the criteria that will permit to consider that the maneuver is fulfilled. This task is not simple. Transitions are validated upon intervals and they are associated with a confidence. If a transition T_k between states E_i and E_j is validated during an interval $[t_1, t_2]$, when will E_j be validated?.

From the works fulfilled on Petri nets [11], namely the consideration of uncertainty with fuzzy sets on the transitions, we may by means of the combination-rule of Dempster-Shafer theory compute an evidence's mass for each state, and for each time point, like in [12]. We may then simply consider that the maneuver is recognized if states are validated one after the other, or if the final state has been reached etc. But for the realism of the recognition, shouldn't we consider that the recognition process should remain in each state during a minimal time, to define?. We will then have to introduce temporal constraints [13].

We then have to pursue our research in the direction of the graphs and/or Petri nets. We do not forget that we have worked until now with simulated data. One of our future perspectives is the use of real data.

References

1. Saad, F., Villame, T.: Intégration d'un nouveau système d'assistance dans l'activité des conducteurs d'automobile. In Ganascia, J., ed.: Sécurité et cognition, Hermès France (1999) 105–114 chap. 9.
2. Saad, F.: Driver strategies in car-following situations. In et al., A.G., ed.: Vision in Vehicles, Elsevier Science B.V. (1996) 61–70
3. Nigro, J., Loriette-Rougegrez, S.: Characterization of driving situation. In: International Conference on Modelling and Simulation, MS'99, Santiago de Compostela, Spain (1999) 287–297
4. Shawky, M., Crubille, P., Bonnifait, P.: Archiving and indexing of large volume sensor data of an equipped vehicle. In: DriiVE - Driving research in instrumented vehicles workshop, Helsinki, Finland (1999)
5. Simulator: (http://www.hds.utc.fr/ crubille/web/simulateur/)

6. Loriette-Rougegrez, S., Nigro, J.M., Jarkass, I.: Rule-based approaches for the recognition of driving maneuvers. In: AISTA'2000 (International Conference on Advances in Intelligent Systems: Theory and Applications), Canberra (Australia) (2000)
7. Giarratano, Riley: Expert systems: Principles and programming. (ISBN 0-534-95053-1)
8. Clips: (http://www.ghg.net/clips/download/documentation/)
9. Shafer, G.: A mathematical theory of evidence. Volume 2702. Princeton University Press (1976)
10. Bouchon-Meunier, B.: La logique floue. In: Que sais-je? Volume 2702. Presse universitaire de France (1993)
11. David, R., Alla, H.: Du Grafcet aux réseaux de Petri. 2ième edn. Traité des nouvelles technologies. Série automatique. Hermès (1992)
12. Jarkass, I., Rombaut, M.: Reconnaissance de séquences temporelles à l'aide de réseau de petri crédibiliste. In Hermès, ed.: Journal Européen des Systèmes Automatisés (APII-JESA). Volume 32. (1998)
13. Fontaine, D.: Une approche par graphes pour la reconnaissance de scenarios temporels. Revue d'intelligence artificielle 10 (1996) 439–468

On Mathematical Modeling of Networks and Implementation Aspects

Regina Bernhaupt and Jochen Pfalzgraf

Department of Computer Science
University of Salzburg
Jakob-Haringer-Str. 2
A-5020 Salzburg/Austria
{rbern, jpfalz}@cosy.sbg.ac.at

Abstract. Based on existing work where categorical and geometrical methods were used to establish a mathematical model of neural net structures, we develop a new very general model for artificial neural networks (ANN), where all basic components of a network are described abstractly. This mathematical model serves as a guideline for design and implementation of a new ANN-simulator. The proposed model of neuron types will be illustrated by the discussion of an example, using the Single Spiking Neuron Model (SSM). The main building blocks of the simulation tool, a new construction principle for ANN, and abstract modeling of connection weights are presented.

1 Introduction

Currently, there is a large number of software packages in use for simulations of artificial neural networks (ANN), ranging from general simulators to very special ones. A general simulator used in Europe is the SNNS [Zel94]. On the scale between general and specialized simulators there are many software packages e. g. SpikeNet in use for large networks with specialized neuron types [DGRT99]. A special software package is the sophisticated program GENESIS performing detailed physiological simulations [BB98]. All these simulators are lacking the ability to use special connection structures between neurons. In addition they can hardly be linked to other software modules to allow hybrid approaches, like the combination of fuzzy-logic, multi agent systems, and ANN.

We are currently working on a new connectionist simulator which was inspired as described subsequently. First the fruitful cooperation of the second author with H. Geiger let to the foundation of the mathematical framework presented in [Pfa01], [Pfa02]. H. Geiger is working with ANN in industrial applications [GP95], [Gei94]. Second the working group of the second author at RISC-Linz started to implement a simulation tool with a special command language for network simulation [Six94]. Later this tool was extended at the University of Salzburg. The results of these cooperations can be found in several diploma and doctoral theses [Lan01], [Six94].

J. Calmet et al. (Eds.): AISC-Calculemus 2002, LNAI 2385, pp. 168–180, 2002.

Using these contributions and ideas we are now designing and implementing a new improved simulation tool. The previously mentioned mathematical framework provides the guideline for implementing and simulating a large class of neural network structures in a flexible way. The emphasis of the mathematical approach is on modeling the neural network structures in the interpretation of geometric nets. Initially it was established to represent the structure of Geiger's network paradigm, but our new general mathematical definition of a network goes far beyond.

2   Towards a General Definition for ANN

As already mentioned before, very important for our construction of the simulator is the formulation of a general mathematical framework. In the past we observed that the analysis of network structures showed the possibility to introduce geometric and categorical modeling approaches. As pointed out in [Pfa01], [Pfa02] to any given ANN a geometric net can be associated. Since geometric spaces form a category, one obtains a category of geometric nets with a suitable notion of morphism. Neurons and connection structures can now be described using these formalisms. For a suitable implementation the mathematical description has to comprise further aspects, as described below.

- Implementing ANN is always based on the possibilities to simulate neurons and their connections. An important aspect is the description of a layer. In practice layers are named e. g. input, hidden or output layer.
- The possibility of describing network structures must be given. When do we call two neurons connected? And how can the neuron model be described using these structure descriptions? Do there exist curcuits (feedback loops)?
- Is the mathematical description of a neuron type a good basis for the implementation? A differential equation describing a neuron type may be precise, but can hardly be implemented without further knowledge. We have to find a way to describe neurons more generally, easy to understand for the user and still easy to implement.
- Neurons and their models are biologically oriented. In hybrid approaches these neurons have to communicate with other processing entities or modules. In image processing fast algorithms are used for preprocessing images. In general, the notion of a neuron may not sufficiently describe this behaviour. A new concept with a generic description can solve this problem.
- The general mathematical model of an ANN as proposed below shall allow flexible construction of ANN. Any neural network can be connected to another (sub-) network without further definitions, despite the specialized connection structure.

With these introductory and motivating remarks we start to prepare our general definition as previously announced.

2.1   Mathematical Definition of ANN

The following definition of an artificial neural network forms the basis for the implementation of the new simulator.

The elementary basis of the general ANN model that we are developing is the notion of a geometric net as introduced in [Pfa01], [Pfa02], where we mention the category **GeoNET** induced by the category of noncommutative geometric spaces **NCG**. We briefly recall: let X be a set of points and R a set, then a geometric space (i. e. object in **NCG**) is given by a so-called parallel map (parallelism structure) $<,>: X \times X \to R$.

Interpreted in terms of a geometric net (object in **GeoNET**) this map describes a net with X as a set of nodes, $X^2 = X \times X$ the set of all possible directed edges and $<,>: X^2 \to R$ is the coloring (weighting) of edges. We say R is the set of weights and for $x, y \in X$ the element $<x, y>$ is the weight of the edge (x, y).

For practical purposes we want to be able to model (feedback) loops in an ANN. Mathematically, we define a loop as a closed path (circuit) in the underlying geometric net. More specifically: a loop of length n in a geometric net consists of a sequence of consecutively connected nodes $x = x_0, x_1, x_2, ..., x_{n-1}, x_n = x$, where (x_{i-1}, x_i) are edges of the net with weights $<x_{i-1}, x_i>$, for $i = 1, ..., n$.

In our general definition below it is necessary to distinguish between different types of nodes. We will use the following notation. Very generally spoken, the types of nodes will be expressed by a set $\mathcal{T}(X)$ (the 'names' of the types) and an element $t \in \mathcal{T}(X)$ will be called a (specific) type. Let $X[t]$ denote the set of nodes of type t, then X is partitioned by all $X[t], t \in \mathcal{T}(X)$, i. e. $X = \coprod_{t \in \mathcal{T}(X)} X[t]$ (disjoint union).

We can express this equivalently by introducing the 'type function' $\tau(X)$: $X \to \mathcal{T}(X)$ assigning to a node $x \in X$ its type $\tau(x) \in \mathcal{T}(X)$. Thus we obtain $X[t] = \tau^{-1}(t)$ $(= \{x \in X | \tau(x) = t\}$ the preimage set of type $t)$.

Analogously, we will speak of different types of weights expressed by a corresponding type function $\tau(R) : R \to \mathcal{T}(R)$. Now we come to the introduction of our general model of an ANN.

Definition 1 *An artificial neural net is defined in the following way:*

(i)   *A set of nodes (neurons) X, a set of weights R, a weight mapping $<,>$: $X^2 \to R$, assigning to a directed edge (x,y) the (abstract) weight $<x, y>$. Sometimes we symbolize it by the triple $(X, <, >, R)$.*

(ii)   *With corresponding type functions $\tau(X)$ and $\tau(R)$ we define the occurring types of neurons and weights respectively.*

(iii)   *There is a partitioning of X into disjoint subsets $X_1, ..., X_n$ called the layers of the ANN. If $<, >_l: X_l^2 \to R_l$ denotes the restriction of $<, >$ to layer X_l, with corresponding set of weights R_l, then we can interpret this structure as a 'subnet' $(X_l, <, >_l, R_l)$ of $(X, <, >, R)$.*

(iv)   *As mentioned, $<x, y>$ is the weight of the directed edge $(x, y) \in X^2$. If there is no 'synaptic connection' between two neurons u, v, i. e. if $u, v \in X$*

are not connected by an edge in the underlying directed graph of the ANN, then we can define $< u, v > := \infty$, for a distinguished symbol ∞ added to R. This can be interpreted, for example, as infinite resistors, symbol of no information flow, no connection.

(v)   *In practical applications, if it is necessary to model feedback loops, we use the notion of a circuit as defined above.*

(vi) *For the moment in our definition the type of a neuron $t \in \mathcal{T}$ is determined by parameter sets $P_{stat}(t)$ and $P_{dyn}(t)$ and a set of mathematical equations (calculation rules) $F(t)$.*

This definition evolved through various stages. A first version is mentioned in [Pfa02]. A new aspect is the interpretation of the partitioning of X not only into layers but into sub nets and the notion of the preprocessor. These interpretations shall lead to a completely new implementation of the simulator. The definition of neuron types as implementational relevant functions is a new aspect in the mathematically oriented net community and shall lead to comparable neuron type implementations. To make the notation of a neuron type more precise and comparable, we describe a neuron type using the implementation relevant functions (cf. neuron type implementation).

We will now turn to the description of a special single spiking neuron type and its defining sets $P_{stat}(t)$, $P_{dyn}(t)$ and $F(t)$ to make the above definition of node (neuron) types more concrete.

3   Single Spiking Neuron Model

The Single Spiking Neuron Model (SSM) was first used by H. Geiger and coworkers. We will use this biologically oriented model to instantiate the general definition of an artificial neural network and its neuron types. Related work in the area of time coded neuron types mainly deals with some specialized image processing problems. The SSM is a central point in the work of the first author. The main idea is solving the recognition of several objects within a scene using different time stamps for differing objects. The construction of the ANN shall consist of several ANN-subnets (e.g. subnets are called V1, V2, V4). Each subnet is capable of the specialized information processing (close to biology).

In our framework a neuron type t is described using the 'tupel' $(P_{stat}(t), P_{dyn}(t), F(t))$. This 'tupel' can always be completed by further mathematical descriptions needed to make the functionality of the neuron type more precise. We use the abstract and general mathematical model as a guideline for our implementation of the simulation tool.

Corresponding to this, the mathematical description of the SSM can be expressed in the following way: The Single Spiking Neuron Model (SSM) is a specialized model of a compartmental neuron model. For a mathematical description see [BP01]. The mathematical description can be summed up as follows: The spike train of a neuron i is described as a series of δ-functions. The firing times t_i^f of a neuron i are labeled by an upper index f. To simplify the formulas subsequently presented, we simply replace t^f by t (not to be confused with the same symbol used for neuron type t).

The membrane potential $V_m(t)$ is given by

$$V_m(t) = \frac{\sum_k g_i^{(k)}(t) E_i^k}{g_{total}} (1 - e^{-g_{total} \cdot C_m(t)}) \tag{1}$$

the discrete representation of the solved differential equation (using a fixed time window) [Lan01]. The membrane potential depends on the typical conductances of the various synapse types g_i (they may be different for the branches k of each synapse type), the synaptic potentials E_i are summed up over all branches k, the capacity of the cell ($C_m(t)$) and the sum of all conductances (g_{total}).

The refractory function is denoted by $\eta(t)$ which describes the refractory period of the neuron after a spike. The membrane potential at time t may thus be expressed by $F_{sum}(t) = V_m(t) + \eta(t)$.

$F_{threshold}$ is used to compare the membrane potential with the given threshold θ, and to compute the difference between them.

$$F_{prob} = \frac{1}{2 * (1 + e^{\beta * \frac{diff}{noise}})} \tag{2}$$

describes the probability function used to determine when the neuron fires. β, $noise$ are constants, $diff = \theta - V_m(t)$ and θ is the threshold.

If the neuron fires (randomly generated number less than F_{prob}), we will use a norming-function (F_{norm}) to simply add the time stamp to the firing time array (t_i^f).

Summarizing, the SSM can be described following the above notation of a neuron type t consisting of $(P_{stat}(t), P_{dyn}(t), F(t))$. Subsequently, we suppress the corresponding neuron type letter t.

$$F = \{F_{sum}, F_{threshold}, F_{prob}, F_{norm}\} \tag{3}$$

$$P_{dyn} = \{V_m(t), C_m(t), g_i^{(k)}, E_i^k\} \tag{4}$$

$$P_{stat} = \{\eta, \theta, \beta, noise\} \tag{5}$$

To summarize the above statements: first we start with the mathematical description of the neuron type and the neuron states and second the computation of the neuron state is decomposed into several functions, close to the implementation. These functions and their dynamical and static parameter sets may be used (implemented) directly in any simulator. The description is easy comparable to other implemented neuron types, since the biologically oriented processing steps $F = \{F_{sum}, F_{threshold}, F_{prob}, F_{norm}\}$ can be the basis for any comparison with other implemented neuron types.

The behaviour of the SSM is similar to the Spike Response Model (SRM) [MB99]. The significant difference is the use of the capacitor and the possibility to describe several synapse types (besides inhibitory and excitatory connections) in our model. While the SRM is closely related to the integrate and fire model [Ger01], the SSM is a specialized type of a compartmental model. It would be a very interesting topic of future research to represent SRM in a similar way as the SSM in terms of suitable functions. As far as we know this has not been done yet.

Besides biological oriented neuron types we use the notion of a preprocessor. Any combination of mathematical functions can be interpreted as a neuron type.

4   The New Simulation Tool

The new tool for the simulation of artificial neural networks is under construction in our working group. Having started in August 2001, we expect the first release by the end of june 2002. The main focus is on the realization of the special connection structures and processing entities described in definition 1. The ANN-Simulator will be used in image and speech processing and in the field of multi agent systems (MAS) and in combination with fuzzy-logic modules and computer algebra systems. Therefore an interface to other modules like fuzzy-logic module, genetic algorithms and symbolic computation modules has to be specified.

Both architecture and design are object oriented, using UML for design and C++ for implementation. From the architecture point of view the simulator itself is decomposed into several modules. The main module, called kernel, implements processing entities, net structure and learning algorithms. The kernel with its net components and structures will be explained in detail in the following section. From the software engineering point of view, the usage of linear lists as data structures to reduce memory load and to speed up the system is of interest.

Besides the kernel, the simulator further consists of an XML-handling module, which mainly deals with load and save operations based on the XML-standards (e.g. saving the neural networks, training patterns, weights and even user defined workspace settings). The training- and simulation module will be used to build the training- and simulation sets. Combined with the GUI-module Version 1.0 is completely described.

Comparing the new simulation tool with other ANN-Tools we introduce the following new concepts and possibilities, which can not be achieved using either SNNS or GENESIS [Zel94], [DGRT99]:

- A new net construction principle, allowing the flexible combination of neural networks with other neural networks, the usage of several different neuron types within these neural networks, and the usage of so called preprocessors. We invented a new embedding principle for the recursive neural net construction (figure 1 gives a basic idea).
- We use the XML standard for saving and loading of neural networks. The ANN description can thus be edited by hand (scripting language).
- Based on the mathematical framework we are able to implement a flexible connection structure (called projection) to model structure in ANN's.

- We model abstract connection weights (e.g. logical values), which can serve as a possible connection to existing multi agent system implementations.
- Neuron types are added dynamically using dynamic link libraries (dll) during runtime.

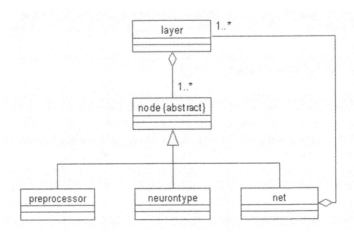

Fig. 1. A layer is of abstract nodetype (a layer has a nodetype). Using inheritance we get the class of a preprocessor, a neurontype or the abstract nodetype represents a whole (sub-)net. The (sub-) net has several layers. A neural net can be constructed recursively adding several (sub-) nets, which are 'represented' by an abstract node.

4.1   Building a Neural Net

Net construction. A neural network is described in the simulator via a recursive construction principle. The net consists of at least one layer. Each layer is assigned to a special node type. Figure 1 shows the description of an abstract node. We inherit from the abstract nodetype the classes preprocessor, neuron type, and net. Net can be again a whole subnet. Thus we are able to define nets recursively. This recursive inheritance principle allows us a special kind of ANN construction. To make the usage more precise: Starting with an ANN consisting of 3 layers (input layer, hidden layer, output layer) all of neuron type SSM, we can add a whole subnet easily. We define a new layer of type sub net. The position of the layer is described (e. g. between hidden layer and output layer) and then the connection structures is defined. The nodes of the hidden layer are connected to the subnet as described in the connection structure e. g. the nodes of the hidden layer are connected to the designated input layer of the subnet and the nodes of the output layer of the subnet are connected to the output layer of the original net. For simplicity we choose the function 'all' (all nodes are

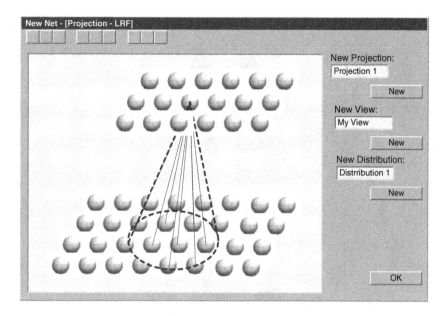

Fig. 2. Dialog for the definition of a special large receptive field (LRF) projection

connected to all other nodes) between these layers. Connections may be specified in detail using the connection structure features (cf.).

Consequently the instance of a layer in the simulation tool will consist of abstract nodes of one and the same type. For real applications this yields great flexibility in constructing ANN having processing entities of various types. The partitioning of the global net into layers corresponding to the description in our general definition is a separate process (currently done by the user while construction the network).

The internal structure of a neural net consists of net components (layers of neurons or preprocessors or subnets). The data structure used for the representation of the internal net structure is independent from the connection structure between processing entities (nodes) or subnets. All processing entities (layers) are linked using linear lists. The structure of the network (in the sense of graph theory, i. e. its topology) will be specified separately following our mathematical model.

Realization of net structures. Connections between ANN components are stored in a second step. The data structures used for the locally regularly modeled connection structures are multi-dimensional linear lists. To make the usage of these structures more clear we start with a description of the user interaction. (Figure 2 shows the prototype GUI-dialogue for structure modeling.)

From the user point of view a locally regular structure is modeled in 4 steps.

– Selection of a reference point in the postsynaptic layer.

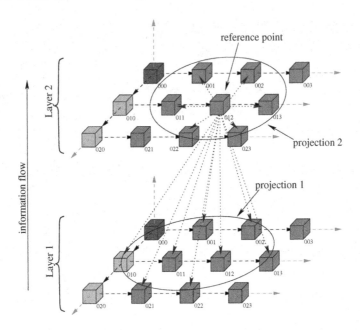

Fig. 3. Visualization of an internal structure of two layers and their connections. The connections are defined by two projection, projection 1 between layer 1 and layer 2, projection 2 within layer2.

- Determination of the visible area in the presynaptic layer, this area is called view.
- Determination of the distribution of the connections to the presynaptic layer. The distribution is the number of neurons connected with the reference point. A projection is defined as the collection of a reference point, distribution and view.
- Layers can be different in size (in the amount of neurons). To make the automatic usage of structure more convenient, we use a so-called step size. E. g. given a retina layer of 256 x 256 neurons and an edge detection layer consisting of 64 x 64 neurons a step size of 8 is used. If the reference point is shifted to a neighboring neuron in the edge detection layer, the view is shifted 8 neurons. Additionally the step size is dependent on the view. In this example the receptive field spans an area of at least 8 neurons.

To ease the use of the simulator, the user may define several projections. We noticed that in practice the user can easier define connection structures within a layer (lateral connections) in contrast to generating interconnections between different layers. The implementation enables the combination of several (user friendly) projections. Internally we exploit the mathematical concept of pointed spaces [Pfa01], [Pfa02]. Figure 3 shows an example of a projection between two layers.

Modeling abstract connection weights. Following our general definition connection weights can be of abstract type. For concrete applications we have to specify the types. A possible connection type can be shunting inhibition being modeled as a simple data structure. Invisible for the user the two possibilities are handled differently in the implementation. In the classical case where connection weights are real values, they are modeled as simple data structures. In contrast to that, where we have to represent more complex weights (abstract weights), the connection between two neurons is decomposed into a sequence of two arrows as shown in figure 4, using a preprocessor as an intermediate node between two neurons under consideration. Biologically the preprocessor models the synaptic behaviour and simulates the electro-chemical processes between the two connected neurons.

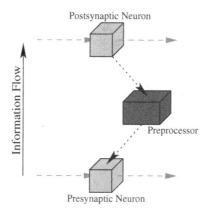

Fig. 4. Internal representation of specialized weights / synapses

Neuron type implementation. A neuron type can be defined by the user. In a first step a neuron base class must be chosen. Then the user has to choose from a set of functions his neuron type defining set F (see definition 1). The current version allows to choose up to 4 functions. Neuron types can be added dynamically during runtime.

As an example we choose the class rate coded neuron and the neuron type defining set of functions $F = \{f_{sum}, f_{thres}\}$, where f_{sum} represents the forming of weighted input sums and f_{thres} the step function. The user has to decide whether the neurons have constant threshold or variable thresholds. Depending on this specification we obtain $P_{stat} = \{\theta_{fix}\}$ and $P_{dyn} = \{activity\}$, where *activity* is the variable (in the software engineering sense) representing the neural activation. Concrete instances of this variable are calculated by f_{sum}. In the specific case of a classical McCulloch-Pitts neuron the variable activity has values $\{0, 1\}$.

Considering a standard perceptron model where the threshold varies, $P_{stat} = \{\}$ and $P_{dyn} = \{\theta, activation\}$, in a concrete application the values of θ are determined by the used learning procedure and *active* by evaluating the weighted sum and the comparison with the current threshold (i. e. the application of the actual f_{thres}).

4.2   Current State of Work

Currently we are improving the basic modules, like the kernel, training and simulation handling, XML-handling and we are implementing the ANN-simulator interface. The number of available functions constituting the neuron types are limited in the current version. The available functions are rate coded neuron types (the perceptron and the conductivity coded neuron model). Next steps are the implementation of the SSM and the implementations of further modules (e. g. additional learning algorithms). Our goal is to keep the interface of the ANN-simulator as 'open' and flexible as possible. An additional module may complete the ANN-simulator to act as an ANN-simulator-server, accessible for clients over any net via a special command language. Using a GUI-module a user-friendly version of the simulator is given for teaching purposes. We already implemented a prototype- GUI. The main goal in the GUI design is a user-friendly interface, capable of several visualization features (e. g. a 3D-engine for the visualization of structures and weight changes during learning steps).

5   Summary and Prospects

This contribution deals with the development of a generic mathematical model for neural networks. It is a formal basis for design and implementation of a new neural network simulator where non standard network paradigms can be simulated. A major aspect of our mathematical approach is the representation of the network structures (geometric and categorical methods can be applied and successfully exploited). The main components of the new simulation tool are described in the second part of the article. A variety of problems remains providing interesting topics for future work.

Among others we mention the following subjects. As pointed out in [Pfa02] group operations play a basic role in the problem field of network structuring. With a view to the new simulator it will be relevant to extend the tool by a module (interface) to computer algebra systems for group theory to handle the corresponding groups represented by generators and relations. Such a module will be useful with respect to automatic generation of network structures.

During the implementation process we have learned that flexible design of ANN, especially when specialized connection structures are used, must lead to a new insight on learning and learning algorithms. At the moment we are working on the possibilities to assign each connection to a learning algorithm, not a whole layer or even subnet. Using different neuron types leads to an extensive use of preprocessors e. g. between time coded and rate coded neuron types. We are seeking to find new algorithms to close this gap, to easily transform rate coded information into time coded information and vice versa.

Further modules extending the simulator concern fuzzy-techniques, genetic algorithms, robot simulation, and multi agent system (MAS) techniques. In our group, a new area of activity deals with the combination of MAS and ANN approaches. Intended applications, among others, will be in the realm of search engines. A large area of future research concerns homomorphic learning.

Concerning the promising combination of MAS techniques and ANN we aim at modeling learning agents (and learning MAS). The simulator with its special network structure may simulate several subnets, representing the agents, which all use the same input layer (database of the search engine). Logic modeling (especially so called logical fiberings) will also play a basic role [PM00], [Pfa00].

References

[BB98]   J. M. Bower and D. Beeman. *The book of GENESIS: Exploring realistic neural models with the GEnereal SImulation System.* Springer-Verlag, New York, 1998.

[BP01]   R. Bernhaupt and J. Pfalzgraf. *Time Coded Neurons, Geometric Networks and Homomorphic Learning In: Advances in Neural Networks and Applications,* pages 268–273. World SES Press, 2001.

[DGRT99]   A. Delorme, J. Gautrais, R. Rullen, and S. Thorpe. SpikeNET: a simulator for modeling large networks of integrate and fire neurons. *Neurocomputing,* 26-27:989–996, 1999.

[Gei94]   H. Geiger. Optical quality control with selflearning systems using a combination of algorithmic and neural network approaches. Proceedings of the Second European Congress on Intelligent Techniques and Soft Computing, EUFIT'94, Aachen, September 20-23, 1994.

[Ger01]   W. Gerstner. *The Handbook of Biological Physics,* volume 4, chapter 12, pages 447–494. Elsevier Science, 2001.

[GP95]   H. Geiger and J. Pfalzgraf. Quality control connectionist networks supported by a mathematical model. Proceedings of the International Conference on Engineering Applications of Artificial Neural Networks (EANN'95), 21-23 August 1995, Helsinki. A.B.Bulsari, S.Kallio (Editors), Finnish AI Society, 1995.

[Lan01]   K. Lang. Single spiking neuronal networks for object recognition. Master's thesis, Universität Salzburg, Institut für Computerwissenschaften, 2001.

[MB99]   W. Maass and C. M. Bishop. *Pulsed Neural Networks.* MIT Press, 1999.

[Pfa00]   J. Pfalzgraf. The concept of logical fiberings and fibered logical controllers. Proceedings CASYS 2000, August 2000, Liege, Belgium, American Institute of Physics, D. M. Dubois (ed.), 2000.

[Pfa01]   J. Pfalzgraf. A note on modeling connectionist network structures: geometric and categorical aspects. In *Proceedings Artificial Intelligence and Symbolic Computation, AISC'2000, July 17-19, 2000, Madrid. Springer Lecture Notes in AI, vol.1930. J.Calmet, E.Roanes (Eds.),* 2001.

[Pfa02]   J. Pfalzgraf. Modeling Connectionist Network Structures: Some Geometric and Categorical Aspects. *Annals of Mathematics and AI (to appear),* 2002.

[PM00]   J. Pfalzgraf and W. Meixl. A logical approach to model concurrency in multi agent systems. Proceedings EMCSR 2000, April 25 - 28, Vienna, 2000.

[Six94]   J. Sixt. Design of an artificial neural network simulator and its integration with a robot simulation environment. Master's thesis, Johannes Kepler University Linz, Institute of Mathematics, 1994.

[Zel94]   A. Zell. *Simulation Neuronaler Netze*. Addison-Wesley, Bonn, 1994.

Continuous First-Order Constraint Satisfaction

Stefan Ratschan

Institut d'Informatica i Aplicacions, Universitat de Girona, Spain,
`stefan.ratschan@risc.uni-linz.ac.at`

Abstract. This paper shows how to use constraint programming techniques for solving first-order constraints over the reals (i.e., formulas in the first-order predicate language over the structure of the real numbers). More specifically, based on a narrowing operator that implements an arbitrary notion of consistency for atomic constraints over the reals (e.g., box-consistency), the paper provides a narrowing operator for first-order constraints that implements a corresponding notion of first-order consistency, and a solver based on such a narrowing operator. As a consequence, this solver can take over various favorable properties from the field of constraint programming.

Keywords: Constraint Programming, Reasoning

1 Introduction

The problem of solving first-order constraints over the reals has numerous applications (we have created a web-page that lists more than fifty references [23]). This paper shows how to solve this problem, based on techniques from the field of constraint programming [5,6,29,12]. The basic idea of these techniques is to reduce the average run-time of algorithms for computationally hard problems by replacing expensive exhaustive search as much as possible by methods for pruning elements from the search space for which it is easy to show that they do not contain solutions. In this paper we extend this idea to first-order constraints: For proving existential quantifiers, one has to search for true elements; for disproving universal quantifiers, one has to search for false elements; as usual for continuous domains, search means branching here, and in our case, replacing sub-constraints of the form $\forall x \in I \ \phi$ by $\forall x \in I_1 \ \phi \ \wedge \ \forall x \in I_2 \ \phi$ where $I = I_1 \cup I_2$ (and the corresponding existential case). We try to avoid branching as much as possible, by first extracting elements for which it is easy to compute that they are (are not) a solution, that is, by replacing a sub-constraint of the form $\forall x \in I \ \phi$ by $\forall x \in I' \ \phi$ where $I' \subset I$ (and the corresponding existential case).

The structure of the proposed solution is parametric in the sense that it takes as input theory and algorithms from constraint programming, and provides as output corresponding new theory and algorithms for solving first-order constraints. More specifically, it takes at input: A specification describing a consistency notion for atomic constraints (e.g., box-consistency [5]), and a narrowing operator that implements this specification. It provides as output: A specification describing a corresponding consistency notion for first-order constraints, a

J. Calmet et al. (Eds.): AISC-Calculemus 2002, LNAI 2385, pp. 181–195, 2002.

narrowing operator that implements this specification, a branch and prune algorithm for computing approximate solutions of first-order constraints over the reals that uses this narrowing operator for pruning. These outputs are accompanied with proofs of their usefulness/optimality. The advantage of using such a parametric structure is that it provides a clear separation between dealing with quantifiers and narrowing of atomic constraints. This allows the use and combination of different implementations of narrowing operators [12,4,14] for atomic constraints. As a consequence one can directly benefit from further progress in continuous (atomic) constraint satisfaction.

As the emphasis of this paper is a general framework on which one can soundly base various algorithms and further developments, it strives for ease of reasoning, elegance, and foundational results, instead of detailed efficiency improvements. However, existing similar algorithms and special cases of our framework [3,25], give strong evidence for the efficiency of algorithms based on our approach.

Up to recently, all algorithms for dealing with this problem, have been based on computer algebra methods [8,7], which resulted in certain drawbacks (e.g., efficiency, restriction to polynomials). In an earlier paper [25] the author of this paper proposed a scheme for solving first-order constraints approximately that followed the idea of quantifier elimination by cylindrical algebraic composition [8, 7], but decomposed space into floating-point boxes instead of semi-algebraic cells. This approach was successful in showing that one can efficiently compute approximate solutions of first-order constraints using interval methods. However it still had several drawbacks. Especially, it was not clear when and how to optimize box splitting, because the algorithm was not separated into (inherently exponential) search, and pruning. The current paper provides a solution to this, and other, problems of the older approach.

To our knowledge, other authors have applied constraint programming techniques only to the special case of first-order constraints with one universally quantified variable [3], or to the special case of disjunctive constraints [16,13, 19,11]. Up to now it was unclear how to extend these approaches to first-order constraints, and which properties such an extension would have.

The content of the paper is as follows: Section 2 gives various preliminaries. Section 3 introduces the notion of narrowing operator for first-order constraints. Section 4 describes an according notion of first-order consistency that specifies the pruning power of narrowing operators. Section 5 gives a generic algorithm that implements a first-order narrowing operator ensuring first-order consistency. Section 6 bases an according branch-and-prune solver for first-order constraints on this narrowing algorithm. Section 7 discusses the relation of the results to classical decision algorithms, and Section 8 concludes the paper.

2   Preliminaries

We fix a set V of variables. A first-order constraint is a formula in the first-order predicate language over the reals with predicate and function symbols interpreted as suitable relations and functions, and with variables in V. In this paper we restrict ourselves to the predicate symbols $<$, $>$, $\leq$, and $\geq$, and assume that equalities are expressed by inequalities on the residual (i.e., $f = 0$ as $|f| \leq \varepsilon$ or $f^2 \leq \varepsilon$, where ε is a small positive real constant[1]). Furthermore we only deal with first-order constraints without negation symbols because one can easily eliminate negation symbols from first-order constraints by pushing them down, and replacing atomic constraints of the form $\neg(f \leq g)$ by $f > g$, and $\neg(f < g)$ by $f \geq g$, respectively. As a slight modification to the usual syntax in logic we require that every quantifier be bounded by an associated *quantifier bound*. This means that quantifiers are of the form $\exists x \in I$ or $\forall x \in I$, where I is a closed interval.

A *variable assignment* is a function from V to $\mathbb{R}$. We denote the semantics of a constraint ϕ, the set of variables assignments that make ϕ true, by $[\![\phi]\!]$. For any variable assignment d, variable $v \in V$ and real number r, we denote by d_v^r the variable assignment that is equal to d except that it assigns r to v. The semantics of a constraint of the form $\exists x \in I\ \phi$ is equal to the semantics of $\exists x\ x \in I \wedge \phi$, and the semantics of a constraint of the form $\forall x \in I\ \phi$ is equal to the semantics of $\forall x\ x \in I \rightarrow \phi$.

Let $\mathbb{I}$ be the set of closed real intervals. We denote by $I_1 \uplus I_2$ the smallest interval containing both intervals I_1 and I_2. A *box assignment* is a set of variable assignments that can be represented by functions from V to $\mathbb{I}$; that is, for every box assignment B, there is a function $D : V \rightarrow \mathbb{I}$ such that B is the set of all d such that for all $v \in V$, $d(v) \in D(v)$. From now on we will use a box assignment and its interval function representation interchangeably. For any box assignment B, variable $v \in V$ and interval I, we denote by B_v^I the box assignment that is equal to B except that it assigns I to v.

The notation $\{x \mapsto [-1, 1], y \mapsto\}$ denotes a box assignment that assigns the interval $[-1, 1]$ to the variable x and an arbitrary value to the variable y.

Traditionally, constraint programming techniques [5,6,29,12] use boxes (i.e., Cartesian products of intervals) instead of box assignments. However, when working with predicate logic, the additional flexibility of box assignments is very convenient in dealing with the scoping of variables. For efficiency reasons, an actual implementation might represent box assignments by boxes.

Note that set of box assignments is closed under the operations $\cap$ and $\uplus$. A box assignment B' is a *facet* of a box assignment B iff it results from B by replacing one of the assigned intervals $[\underline{a}, \overline{a}]$ by either $[\underline{a}, \underline{a}]$ or $[\overline{a}, \overline{a}]$. For any box assignment B and term t, interval evaluation [20] yields an overestimation of the range of t on B (i.e., a superset of all values of t under a variable assignment in B) and consequently for any atomic constraint ϕ an overestimation of the truth of ϕ on B.

We let fix : $((A \rightarrow A) \times A) \rightharpoonup A$ be a partial function such that for $f : A \rightarrow A$ and $a \in A$, if there is a positive integer n and a $b \in A$ such that for all $k > n$, $f^k(a) = b$ then fix$(f, a) = b$, and otherwise it is undefined.

---

[1] The constant ε needs to be non-zero because otherwise solutions would vanish under small perturbations of ε.

3   First-Order Narrowing

Let us call pairs (ϕ, B), where ϕ is a first-order constraint and B is a box assignment, *bounded constraints*. We call the second element of a bounded constraint its *free-variable bound*. Now suppose that for a given bounded constraint (ϕ, B) we want to remove elements from B that are guaranteed not to be within the solution set of ϕ. In a similar way we want to remove elements from B that are guaranteed to be within the solution set of ϕ.

Definition 1. *A* narrowing operator *is a function $N\updownarrow$ on bounded constraints such that for bounded constraints (ϕ, B), and (ϕ, B'), and for $N\updownarrow (\phi, B) = (\phi_N, B_N)$, and $N\updownarrow (\phi, B') = (\phi'_N, B'_N)$,*

- $B \supseteq B_N$ *(contractance),*
- $[\![\phi_N]\!] \cap B_N = [\![\phi]\!] \cap B_N$ *(soundness),*
- $B' \subseteq B$ *implies* $B'_N \subseteq B_N$ *(monotonicity), and*
- $N\updownarrow (N\updownarrow (\phi, B)) = N\updownarrow (\phi, B)$ *(idempotence).*

Note that, in contrast to narrowing operators for constraints without quantifiers [2,6], here a narrowing operator can also modify the constraint itself. Furthermore note that we use a soundness condition here, instead of a correctness condition: We require that the solution set of the resulting constraint be the same only on the *resulting* box, but not necessarily on the initial box. We will ensure full correctness by the next definition.

Constraint programming techniques for continuous domains traditionally compute outer approximations of the solution set. However, here we also want to compute inner approximations for three reasons: First, the solution set of first-order constraints with inequality predicates usually does not consist of singular points, but of sets that can have volume, and for many applications it is important to find points that are guaranteed to be within this solution set. Second, available inner approximations can speed up the computation of outer approximations and vice versa, because any element known to be within the solution set, or known to be not in the solution set, does not need to be inspected further. Third, inner estimations are needed for pruning the search space of universal quantifiers.

So we will allow two kinds of narrowing operators—one that only removes elements not in the solution set, and one that only removes elements in the solution set.

Definition 2. *An* up-narrowing operator *is a narrowing operator $N\uparrow$ such that for every bounded constraint (ϕ, B), for the free-variable bound B_N of $N\uparrow (\phi, B)$, $B_N \supseteq B \cap [\![\phi]\!]$. A* down-narrowing operator *is a narrowing operator $N\downarrow$ such that for every bounded constraint (ϕ, B), for the free-variable bound B_N of $N\downarrow (\phi, B)$, $B_N \supseteq B \setminus [\![\phi]\!]$.*

For any first-order constraint ϕ, let $\neg\phi$ (the *opposite of* ϕ) be the first-order constraints that results from $\neg\phi$ by eliminating the negation by pushing it down to the predicates. Now we have:

Theorem 1. *Let $N\!\uparrow$ be a function on bounded constraints and let $N\!\downarrow (\phi, B) := (\neg\phi_N, B_N)$ where $(\phi_N, B_N) = N\!\uparrow (\neg\phi, B)$. Then $N\!\uparrow$ is an up-narrowing operator iff $N\!\downarrow$ is a down-narrowing operator.*

Proof. Obviously $N\!\downarrow$ is a narrowing operator iff $N\!\uparrow$ is a narrowing operator. $N\!\uparrow$ is up-narrowing iff $N\!\downarrow$ is down-narrowing because for any bounded constraint (ϕ, B), $B \cap [\![\neg\phi]\!] = B \setminus [\![\phi]\!]$, and $B \setminus [\![\neg\phi]\!] = B \cap [\![\phi]\!]$. □

A similar observation has already been used for the special case of first-order constraints with one universal quantifier [3]. The above theorem allows us to concentrate on up-narrowing operators from now on. We get the corresponding down-narrowing operator for free by applying the up-narrowing operator on the opposite of the input.

4   First-Order Consistency

In constraint programming the notion of consistency is used to specify the pruning power of narrowing operators. In this section we generalize this approach to first-order constraints. We will call any predicate on bounded constraints that we will use for such specification purposes *consistency property*.

Example 1. For an atomic bounded constraint (ϕ, B), $BC(\phi, B)$ holds if there is no facet of B for which interval evaluation will prove that it contains no element of $[\![\phi]\!]$. In this case we say that ϕ is *box-consistent* wrt. B [5,12].

Note that the original definition of box consistency is slightly weaker in order to allow a floating-point implementation. However, in this paper we prefer to work with abstract, implementation-independent concepts.

The following is the strongest form of consistency that does not result in loss of information, that is, for which an up-narrowing operator exists.

Example 2. For an atomic bounded constraint (ϕ, B), $HC(\phi, B)$ holds if there is no box assignment B' such that $B' \subset B$ and $[\![\phi]\!] \cap B' = [\![\phi]\!] \cap B$. In this case we say that ϕ is *hull-consistent* wrt. B [6].

We can use consistency properties as specifications for the effectiveness of narrowing operators:

Definition 3. *Given a consistency property C, a narrowing operator $N\!\updownarrow$ ensures C iff for all bounded constraints (ϕ, B), $C(N\!\updownarrow (\phi, B))$ holds.*

Now we assume a certain consistency property on literals (i.e., atomic constraints and their negations) and give a corresponding consistency property on first-order constraints.

Definition 4. *Given a first-order constraint ϕ and a consistency property C on atomic constraints, let FOC^C be the following consistency property*

- if ϕ is atomic, then $FOC^C(\phi, B)$ iff $C(\phi, B)$
- if ϕ is of the form $\phi_1 \wedge \phi_2$ then $FOC^C(\phi, B)$ iff $FOC^C(\phi_1, B)$ and $FOC^C(\phi_2, B)$.
- if ϕ is of the form $\phi_1 \vee \phi_2$ then $FOC^C(\phi, B)$ iff $B = B_1 \uplus B_2$ where $FOC^C(\phi_1, B_1)$ and $FOC^C(\phi_2, B_2)$.
- if ϕ is of the form $Qy \in I' \, \phi'$, where Q is a quantifier, then $FOC^C(\phi, B)$ iff $FOC^C(\phi', B\frac{I'}{y})$.

If for a bounded constraint (ϕ, B), $FOC^C(\phi, B)$ holds, we say that (ϕ, B) is *first-order C-consistent* (FO^C-*consistent*). Note that, in the above definition, recursion for quantification puts the quantifier bound into the free-variable bound of the quantified constraint. This means that a narrowing operator also will have to modify the quantifier bounds in order to achieve first-order consistency.

Example 3. The bounded constraint $\left(\exists y \in [0, 1] \left[x^2 + y^2 \leq 1 \wedge y \geq 0 \right], \{x \mapsto [-1, 1], y \mapsto\} \right)$ is first-order hull consistent, and it will be the result of applying an according narrowing operator to an input such as $\left(\exists y \in [-2, 2] \left[x^2 + y^2 \leq 1 \wedge y \geq 0 \right], \{x \mapsto [-2, 2], y \mapsto\} \right)$.

Note that Definition 4 is compatible with the usual consistency notions for sets of constraints [5,4]. For example a set of atomic constraints $\{\phi_1, \ldots, \phi_n\}$ is box-consistent wrt. a box assignment B iff $(\phi_1 \wedge \cdots \wedge \phi_n, B)$ is FO^{BC}-consistent. In addition, the method for solving constraints with one universally quantified variable by Benhamou and Goualard [3] computes a special case of Definition 4.

In the following sense, our definition of FO^C-consistency is optimal (remember that hull consistency is the strongest possible consistency property when considering bounded constraints as atomic).

Theorem 2. *A FO^{HC}-consistent bounded constraint (ϕ, B), where ϕ neither contains conjunctions nor universal quantifiers, is HC-consistent.*

Proof. We proceed by induction over the structure of constraints. The atomic case trivially holds. Now assume constraints of the following types:

- For a FO^{HC}-consistent bounded constraint of the form $(\phi_1 \vee \phi_2, B)$, by definition, we have $B = B_1 \uplus B_2$ where $FOC^{HC}(\phi_1, B_1)$ and $FOC^{HC}(\phi_2, B_2)$. By the induction hypothesis both (ϕ_1, B_1) and (ϕ_2, B_2) are hull consistent. So for no box assignment $B_1' \subset B_1$, we have $B_1' \supseteq B_1 \cap [\![\phi_1]\!]$, and for no box assignment $B_2' \subset B_2$, we have $B_2' \supseteq B_2 \cap [\![\phi_2]\!]$. Thus also for no box assignment $B' \subset B_1 \uplus B_2$, we have $B' \supseteq B \cap ([\![\phi_1]\!] \cup [\![\phi_2]\!]) = B \cap [\![\phi_1 \vee \phi_2]\!]$.
- For a FO^{HC}-consistent bounded constraint of the form $(\exists y \in I' \, \phi', B)$, by definition $FO^{HC}(\phi', B\frac{I'}{y})$. Thus, by the induction hypothesis $(\phi', B\frac{I'}{y})$ is hull consistent. So for no box assignment $B_p' \subset B\frac{I'}{y}$, $B_p' \supseteq B\frac{I'}{y} \cap [\![\phi']\!]$. As a consequence also for no box assignment $B_p \subset B$, $B_p \supseteq B \cap [\![\exists y \in I' \, \phi']\!]$, and so $(\exists y \in I' \, \phi', B)$ is hull consistent.

$\square$

The fact that the above theorem does not hold for constraints with conjunctions is well known [6]. It is illustrated in Figure 1, where both ϕ_1 and ϕ_2 are hull consistent wrt. the box B (i.e. the larger box encloses the ellipses tightly), but $\phi_1 \wedge \phi_2$ is only hull consistent wrt. the smaller box B' (i.e., the smaller, but not the larger box, encloses the intersection of the ellipses tightly).

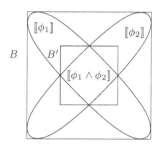

Fig. 1. Fully Pruned Conjunction

For universal quantification there is a similar problem: In Figure 2, ϕ is hull consistent wrt. the box B, but when considering $\forall y \in I \; \phi$ one can still narrow B horizontally. So any stronger consistency notion would have to treat universal quantification different from existential quantification.

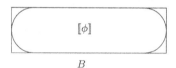

Fig. 2. Fully Pruned Universal Quantification

5   Narrowing Algorithm

Now assume that we have a consistency property C defined on atomic constraints and an up-narrowing operator $A{\uparrow}$ such that for all bounded constraints (ϕ, B), where ϕ is atomic, $C(A{\uparrow}\,(\phi, B))$ holds. From this we can construct a general up-narrowing operator that ensures $FO^C - consistency$.

Narrowing for conjunctive and disjunctive constraints is straightforward. For bounded constraints of the form $(Qx \in I \; \phi, B)$ the narrowing operator removes elements both from the free-variable bound B and the quantifier bound I. It removes those elements from B for which narrowing of $(\phi, B\frac{I}{x})$ showed that they are certainly not in the overall solution set. For existential quantification it additionally removes those elements from the quantifier bound for which narrowing showed that they are not in the solution set of ϕ. This does not change the solution set of the overall constraint (i.e., $[\![\exists x \in I \ \phi]\!] \cap B$), and thus keeps the soundness property of the narrowing operator. For universal quantification we do not remove elements from the quantifier bound, because here, whenever narrowing of the sub-constraint shows that certain elements of the quantifier bound are not in the solution set, we can immediately infer that the universal quantifier does not hold.

Definition 5.

- For atomic ϕ, $N\!\!\uparrow_A (\phi, B) = A\!\!\uparrow (\phi, B)$,
- $N\!\!\uparrow_A (\phi_1 \wedge \phi_2, B) = fix(N\!\!\uparrow'_A)(\phi_1 \wedge \phi_2, B)$
    where $N\!\!\uparrow'_A (\phi_1 \wedge \phi_2, B) = (\phi'_1 \wedge \phi'_2, B'_1 \cap B'_2)$
        where $(\phi'_1, B'_1) = N\!\!\uparrow_A (\phi_1, B)$ and $(\phi'_2, B'_2) = N\!\!\uparrow_A (\phi_2, B)$
- $N\!\!\uparrow_A (\phi_1 \vee \phi_2, B) = (\phi'_1 \vee \phi'_2, B'_1 \uplus B'_2)$
        where $(\phi'_1, B'_1) = N\!\!\uparrow_A (\phi_1, B)$ and $(\phi'_2, B'_2) = N\!\!\uparrow_A (\phi_2, B)$
- $N\!\!\uparrow_A (\exists x \in I \ \phi, B) = (\exists x \in B'(x) \ \phi', B')$,
        where $(\phi', B') = N\!\!\uparrow_A (\phi, B\frac{I}{x})$
- $N\!\!\uparrow_A (\forall x \in I \ \phi, B) = (\forall x \in I \ \phi', D)$,
        where $(\phi', B') = N\!\!\uparrow_A (\phi, B\frac{I}{x})$
        and $d \in D$ iff for all $r \in I$, $d\frac{r}{x} \in B'$

The reason why here, in contrast to the first-order consistency definition, we need different rules for existential and universal quantification, lies in the fact that for universal quantification we even have to remove these elements from the free-variable bound for which a single corresponding element of the bound variable has been removed by narrowing of the sub-constraint.

Example 4. For the input $\left(\exists y \in [-2, 2] \left[x^2 + y^2 \leq 1 \wedge y \geq 0 \right], \{x \mapsto [-2, 2], y \mapsto\} \right)$ already used in Example 3, a narrowing operator based on hull consistency applies itself recursively to $(x^2 + y^2 \leq 1 \wedge y \geq 0, \{x \mapsto [-2, 2], y \mapsto [-2, 2]\})$. Repeated applications of the atomic narrowing operator—until a fixpoint is reached—will create the constraint $(x^2 + y^2 \leq 1 \wedge y \geq 0, \{x \mapsto [-1, 1], y \mapsto [0, 1]\})$. As a final result we get $\left(\exists y \in [0, 1] \left[x^2 + y^2 \leq 1 \wedge y \geq 0 \right], \{x \mapsto [-1, 1], y \mapsto\} \right)$.

Example 5. For the input $(\forall x \in [-2, 2] \ x \geq 0, \{x \mapsto\})$, the algorithm will first narrow $(x \geq 0, \{x \mapsto [-2, 2]\})$ to $(x \geq 0, \{x \mapsto [0, 2]\})$, and then create $(\forall x \in [-2, 2] \ x \geq 0, \emptyset)$, indicating that the constraint is false.

Note that the fixed-point operator fix could result in a partial function, that is, the narrowing algorithm could fail to terminate. For ensuring termination, we require that narrowing on atomic constraints eventually terminates even when intermingled with shrinking of the free-variable bound by other operations:

Definition 6. *A narrowing operator $N_A^\uparrow$ is finitely contracting iff there is no infinite chain $(\phi_1, B_1), (\phi_2, B_2), \ldots$ for which for all $k \in \mathbb{N}$, for $(\phi', B') = N_A^\uparrow (\phi_k, B_k)$, $\phi_{k+1} = \phi'$ and B_{k+1} is a strict subset of B'.*

This property usually holds for practical implementations, because of the finiteness of floating point numbers.

Lemma 1. *If $A^\uparrow$ is a finitely contracting narrowing operator then $N^\uparrow_A$ is a total function.*

Proof. We assume that $A^\uparrow$ is finitely contracting but $N^\uparrow_A$ is not total. This can only happen if $\text{fix}(N^\uparrow_A')$ is undefined. Consider the chain $(\phi_1^1 \wedge \phi_2^1, B_1), (\phi_1^2 \wedge \phi_2^2, B_2), \ldots$ of bounded constraints created by repeated applications of $N^\uparrow_A'$. Here $(\phi_1^1, B_1), (\phi_1^2, B_2), \ldots$ is an infinite chain as in Definition 6. So $N^\uparrow_A$ is not finitely contracting, and by induction also $A^\uparrow$ is not finitely contracting—a contradiction. □

Theorem 3. *For every finitely contracting (atomic) up-narrowing operator $A^\uparrow$, $N^\uparrow_A$ is an up-narrowing operator.*

Proof. Contractance and idempotence hold by easy induction. For proving that $N^\uparrow_A$ is up-narrowing, sound and monotonic we proceed by induction. The ground case of atomic constraints holds by definition. Now we have:

- Obviously the composition of two narrowing operators is also a narrowing operator. So, for constraints of the form $\phi_1 \wedge \phi_2$ we just need to show that $N^\uparrow_A$ is up-narrowing. For $(\phi_1', B_1') = N^\uparrow (\phi_1, B)$ and $(\phi_2', B_2') = N^\uparrow (\phi_2, B)$, by the induction hypothesis $B_1' \supseteq B \cap [\![\phi_1]\!]$ and $B_2' \supseteq B \cap [\![\phi_2]\!]$. Thus also $B_1' \cap B_2' \supseteq B \cap [\![\phi_1]\!] \cap [\![\phi_2]\!] = B \cap [\![\phi_1 \wedge \phi_2]\!]$. The induction step for soundness and monotonicity is easy.
- For constraints of the form $\phi_1 \vee \phi_2$, for $(\phi_1', B) = N^\uparrow (\phi_1, B_1)$ and $(\phi_2', B_2') = N^\uparrow (\phi_2, B)$, by the induction hypothesis $B_1' \supseteq B \cap [\![\phi_1]\!]$ and $B_2' \supseteq B \cap [\![\phi_2]\!]$. Thus also $B_1' \uplus B_2' \supseteq B_1 \cup B_2' \supseteq B \cap ([\![\phi_1]\!] \cup [\![\phi_2]\!]) = B \cap [\![\phi_1 \vee \phi_2]\!]$. The induction step for soundness and monotonicity is easy.
- For constraints of the form $\exists x \in I\ \phi$, the induction step for up-narrowing is easy. For soundness we have to prove that $[\![\exists x \in I\ \phi]\!] \cap B' = [\![\exists x \in B'(x)\ \phi']\!] \cap B'$, where $B' = N^\uparrow_A (\phi, B\frac{I}{x})$. Now by the up-narrowing property $B' \supseteq B \cap [\![\phi]\!]$, and so $[\![\exists x \in I\ \phi]\!] \cap B' = [\![\exists x \in B'(x)\ \phi]\!] \cap B'$. This is equal to $[\![\exists x \in B'(x)\ \phi']\!] \cap B$, because by the induction hypothesis $[\![\phi']\!] \cap B' = [\![\phi]\!] \cap B'$.
- For constraints of the form $\forall x \in I\ \phi$, for up-narrowing we have to prove that $D \supseteq B \cap [\![\forall x \in I\ \phi]\!]$, where D is defined as in the corresponding rule of Definition 5. So we assume a variable assignment d that is both in B and $[\![\forall x \in I\ \phi]\!]$, and prove that $d \in D$. This means that we have to prove that for all $r \in I$, $d\frac{r}{x} \in B'$, where $(\phi', B') = N^\uparrow_A (\phi, B\frac{I}{x})$. This is clearly the case by the semantics of universal quantification and the induction hypothesis. For soundness we have to prove that $[\![\forall x \in I\ \phi]\!] \cap D = [\![\forall x \in I\ \phi']\!] \cap D$, where D is as above. By the quantifier semantics it suffices to prove that

$[\![\phi]\!] \cap D\frac{I}{x} = [\![\phi']\!] \cap D\frac{I}{x}$. This holds, because for all variable assignments $d \in D$, for all $r \in I$, $d\frac{r}{x} \in B'$, and moreover, by the induction hypothesis $[\![\phi']\!] \cap B' = [\![\phi]\!] \cap B'$.

$\square$

By easy induction we also get:

Theorem 4. $N\!\uparrow_A$ *ensures* FO^A-*consistency.*

By applying Theorem 1 we get a corresponding down-narrowing operator $N\!\downarrow_A$ from $N\!\uparrow_A$. Note, however, that $N\!\uparrow_A$ and $N\!\downarrow_A$ do not commute, and $N\!\uparrow_A \circ N\!\downarrow_A$ is not idempotent.

As for the classical conjunctive case, the complexity of the algorithm in a floating-point implementation is polynomial in the problem dimension (the number of floating point numbers that one can remove from the quantification bounds is polynomial). So, as desired, narrowing is efficient compared to expensive exhaustive search, and even more so compared to the doubly exponential complexity of symbolic solvers [8,7].

6   Solver

Now a solver can use narrowing operators to avoid computationally expensive search (i.e., branching) as much as possible. Algorithm 1 is such a solver for closed first-order constraints. A solver for open first-order constraints is an easy extension that would record the boxes that a narrowing operator proved to be in or out of the solution set.

Algorithm 1

Input: A closed first-order constraint ϕ
Output: The truth-value of ϕ
    $(\phi_F, L^F) \leftarrow N\!\uparrow (\phi, \mathbb{R}^{|V|})$
    $(\phi_T, L^T) \leftarrow N\!\uparrow (\neg\phi, \mathbb{R}^{|V|})$
    while L^F is not empty and L^T is not empty **do**
        $\phi_F \leftarrow \text{Branch}(\phi_F)$
        $\phi_T \leftarrow \text{Branch}(\phi_T)$
        $(\phi_F, L^F) \leftarrow N\!\uparrow (\phi_F, L^F)$
        $(\phi_T, L^T) \leftarrow N\!\uparrow (\phi_T, L^T)$
    if L^F is empty **then**
        return F
    else
        return T

Here the function Branch either replaces a sub-constraint of the form $\exists x \in I \; \phi$ by $\exists x \in I_1 \; \phi \; \vee \; \exists x \in I_2 \; \phi$, where $I_1 \cup I_2 = I$, or replaces a sub-constraint of the form $\forall x \in I \; \phi$ by $\forall x \in I_1 \; \phi \; \wedge \; \forall x \in I_2 \; \phi$, where $I_1 \cup I_2 = I$. We assume branching to be *fair*, in the sense that every quantifier will eventually be split (finding such a strategy is easy, but finding an optimal branching strategy is highly non-trivial).

For discussing termination of Algorithm 1 it is important to see, that the problem of computing truth-values/solution sets of first-order constraints can be numerically ill-posed [26]. An example is the first-order constraint $\exists x \in [-1,1] \ -x^2 \geq 0$ which is true, but becomes false under arbitrarily small positive perturbations of the constant 0. As a consequence, it is not possible to design an algorithm based on approximation that will always terminate (with a correct result). Note that this situation is similar for most computational problems of continuous mathematics (e.g., solving linear equations, solving differential equations). However, as in these cases, most inputs are still numerically well-posed (in fact, in a certain, realistic model this is the case with probability one [24]). One can even argue that, philosophically speaking, the well-posed problems are exactly the problems that model real-life problems in a meaningful way.

It is beyond the scope of this paper to present all the formal details for characterizing well-posed first-order constraints and we will introduce the necessary concepts in a semi-formal way. For this we replace the discrete notion of truth of a first-order constraint by a continuous notion [26]. We interpret universal quantifiers as infimum operators, existential quantifiers as supremum operators, conjunction as minimum, disjunction as maximum, atomic constraints of the form $f > g$ or $f \geq g$ as the function denoted by $f - g$, and atomic constraints of the form $f < g$ or $f \leq g$ as the function denoted by $g - f$. We call the result the *degree of truth* of a first-order constraint and denote it by $[\![\phi]\!]^\circ$ for any constraint ϕ. This function assigns to every variable assignment a real value that is independent of the variables that are not free in ϕ. The idea is that the degree of truth is greater or equal zero for variable assignments that make ϕ true, and less or equal zero for variable assignments that make ϕ false. One can prove [26] that the problem of computing the truth value of a closed first-order constraint is numerically well-posed iff its degree of truth is non-zero.

For giving a termination proof of Algorithm 1, we assume that the given narrowing operator for atomic constraints always succeeds for well-posed inputs:

Definition 7. *An up-narrowing operator $A\!\uparrow$ is* converging *iff for all atomic constraints ϕ and chains $B^0 \supseteq B^1 \supseteq \ldots$ such that*

- *for all $i \in \mathbb{N}$, $B^i \supseteq B^{i+1}$,*
- *and $\bigcap_{i \in \mathbb{N}} B^i = \{d\}$, where the degree of truth of ϕ at the variable assignment d is negative,*

there is a k, such that for all $l \geq k$, the free-variable bound of $A\!\uparrow (\phi, B^l)$ is empty.

Note that narrowing operators that implement (our abstract version of) box consistency or hull consistency always fulfill this property. However, it is in general impossible to fulfill for any narrowing operator based on fixed-precision floating-point arithmetic. However, the application of the implementation [22] of an older method [25] to real-life problems has shown that floating-point arithmetic almost always suffices in practice.

Lemma 2. *Let $A{\uparrow}$ be a converging up-narrowing operator and let the sequence $(\phi^1, B^1), (\phi^2, B^2), \ldots$ be such that*

- *for all $i \in \mathbb{N}$, $B^i \supseteq B^{i+1}$, and*
- *$\bigcap_{i \in \mathbb{N}} B^i = \{d\}$ such that the degree of truth of ϕ at the variable assignment d is negative,*
- *for all $i \in N$, ϕ^{i+1} results from ϕ^i by branching, and*
- *for all $\varepsilon > 0$ there is a k such that for all $l \geq k$, the volume of all quantification sets in ϕ^l is less or equal ε.[2]*

Then there is a k such that for all $l \geq k$, the free-variable bound of $N{\uparrow}_A (\phi^l, B^l)$ is empty.

Proof. We proceed by induction over the structure of the constraint ϕ^1. For atomic constraints the lemma holds because $A{\uparrow}$ is converging. Now consider the following cases:

- For constraints of the form $\phi_1 \wedge \phi_2$, $\llbracket \phi_1 \wedge \phi_2 \rrbracket^\circ(d) = \min\{\llbracket \phi_1 \rrbracket^\circ(d), \llbracket \phi_2 \rrbracket^\circ(d)\}$ being negative implies that either $\llbracket \phi_1 \rrbracket^\circ(d)$ is negative or $\llbracket \phi_2 \rrbracket^\circ(d)$ is negative. Therefore at least one of the sequences $(\phi_1^1, B^1), (\phi_1^2, B^2), \ldots$ and $(\phi_2^1, B^1), (\phi_2^2, B^2), \ldots$ where ϕ_1^i is the sub-constraint of ϕ^i corresponding to ϕ_1 and ϕ_2^i is the sub-constraint of ϕ^i corresponding to ϕ_2, fulfills the preconditions of the induction hypothesis. Let $r \in \{1, 2\}$ be the number of this sequence. Then there is a k, such for all $k \geq l$, the free-variable bound of $N{\uparrow}_A (\phi_r^l, B^l)$ is empty. Thus, by definition of $N{\uparrow}_A$, also the corresponding free-variable bound in the original sequence is empty.
- For constraints of the form $\phi_1 \vee \phi_2$, $\llbracket \phi_1 \vee \phi_2 \rrbracket^\circ(d) = \max\{\llbracket \phi_1 \rrbracket^\circ(d), \llbracket \phi_2 \rrbracket^\circ(d)\}$ being negative implies that both $\llbracket \phi_1 \rrbracket^\circ(d)$ and $\llbracket \phi_2 \rrbracket^\circ(d)$ are negative. Therefore both sequences $(\phi_1^1, B^1), (\phi_1^2, B^2), \ldots$ and $(\phi_2^1, B^1), (\phi_2^2, B^2), \ldots$ where ϕ_1^i is the sub-constraint of ϕ^i corresponding to ϕ_1 and ϕ_2^i is the sub-constraint of ϕ^i corresponding to ϕ_2, fulfill the preconditions of the induction hypothesis. As a consequence there is a k_1, such for all $l \geq k_1$, the free-variable bound of $N{\uparrow}_A (\phi_1^l, B^l)$ is empty, and there is a k_2, such for all $l \geq k_2$, the free-variable bound of $N{\uparrow}_A (\phi_1^l, B^l)$ is empty,
  Thus, by definition of $N{\uparrow}_A$, for all $l \geq \max\{k_1, k_2\}$ the free-variable bound of the l-th element in the original sequence is empty.
- Constraints of the form $\forall x \in I \; \phi'$, are replaced by branching into the form $\forall x \in I_1 \; \phi' \; \wedge \; \ldots \; \wedge \; \forall x \in I_k \phi'$. Since the degree of truth of $\forall x \in I \; \phi'$ at d is negative, by definition of infimum, there is a $b \in I$ for which the degree of truth of ϕ' at $d \times b$ is negative. Consider the sequence for which the i-th element consists of the branch of ϕ^i that contains $d \times b$, and of B^i. This sequence fulfills the preconditions of the induction hypothesis, and as a consequence there is a k, such for all $k \geq l$, the k-th free-variable bound in

---

[2] This item formalizes the notion of a fair branching strategy.

this sequence is empty. Thus, by definition of $N\!\uparrow_A$, also the corresponding free-variable bound in the original sequence is empty.

- Constraints of the form $\exists x \in I \; \phi'$, are replaced by branching into the form $\exists x \in I_1 \phi' \lor \ldots \lor \exists x \in I_k \phi'$. Since the degree of truth of $\exists x \in I \; \phi'$ at d is negative, by definition of supremum, for all $b \in I$ the degree of truth of ϕ' at $d \times b$ is negative. This means that each sequence for which the i-th element consists of a branch of ϕ^i and of B^i fulfills the preconditions of the induction hypothesis, and as a consequence there is a k, such for all $k \geq l$, the k-th free-variable bound in this sequence is empty. Thus, by definition of $N\!\uparrow_A$, also the corresponding free-variable bound in the original sequence is empty.   □

From Lemma 2 and its dual version we get:

Theorem 5. *Algorithm 1 terminates for well-posed inputs.*

7   Relation to Classical Algorithms

A. Tarski [31] showed that real first-order constraints with equality and inequality predicates, multiplication and addition are decidable. Adding additional function symbols (e.g., sin, tan), usually removes this property [28,32,17]. Using the method in this paper one can still compute useful information for these cases, provided that the input is numerically well-posed.

The complexity bound supplied by Tarski's method has been improved several times [8,27,1]—but the problem is inherently doubly exponential [10,33] in the number of variables.

The only general algorithm for which a practically useful implementation exists, is the method of *quantifier elimination by cylindrical algebraic decomposition* [8]. This algorithm employs similar branching as the algorithm presented in this paper. However, its branching operation is much more complicated because it branches into a finite set of *truth-invariant cells*, that is, into pieces whose value can be computed by evaluation on a single *sample point*. For being able to do this, its quantifier bounds can depend on the free variables, and branching is done based on information from *projection polynomials*. For implementing these operations one needs expensive real algebraic number computations.

Instead of branching, quantifier elimination by *partial cylindrical algebraic decomposition* [9] employs pruning in a similar sense as described in this paper. However it still decomposes into truth-invariant cells, which again needs expensive computation of projection polynomials, and real algebraic numbers.

In contrast to this, the narrowing operator provided in this paper is cheap, and can do pruning in polynomial time. As a result, we have a clear separation between polynomial time pruning, and exponential branching. So we have a way of working around the high worst-case complexity of the problem, whenever a small amount of branching is necessary.

8 Conclusion

In this paper we have extended constraint satisfaction techniques to first-order constraints over the reals. The result has several advantages over earlier approaches: Compared to symbolic approaches [8,7] it is not restricted to polynomials, and avoids complicated and inefficient computation with real algebraic numbers. Furthermore it decreases the necessity for expensive space decomposition by extracting information using fast consistency techniques. Compared to earlier interval approaches that could deal with quantifiers of some form, it can either handle a more general case [15,18,3,30], or provides a much cleaner, more elegant, and efficient framework [25].

In future work we will provide an implementation of an instantiation of our framework with box-consistency [5], explore different branching strategies, exploit continuity information for efficiently dealing with equalities, and study analogous algorithms for discrete domains.

This work has been supported by a Marie Curie fellowship of the European Union under contract number HPMF-CT-2001-01255.

References

1. S. Basu, R. Pollack, and M.-F. Roy. On the combinatorial and algebraic complexity of quantifier elimination. In S. Goldwasser, editor, *Proceedings fo the 35th Annual Symposium on Foundations of Computer Science*, pages 632–641, Los Alamitos, CA, USA, 1994. IEEE Computer Society Press.
2. F. Benhamou. Interval constraint logic programming. In Podelski [21].
3. F. Benhamou and F. Goualard. Universally quantified interval constraints. In *Proc. of the Sixth Intl. Conf. on Principles and Practice of Constraint Programming (CP'2000)*, number 1894 in LNCS, Singapore, 2000. Springer Verlag.
4. F. Benhamou, F. Goualard, L. Granvilliers, and J. F. Puget. Revising hull and box consistency. In *Int. Conf. on Logic Programming*, pages 230–244, 1999.
5. F. Benhamou, D. McAllester, and P. V. Hentenryck. CLP(Intervals) Revisited. In *International Symposium on Logic Programming*, pages 124–138, Ithaca, NY, USA, 1994. MIT Press.
6. F. Benhamou and W. J. Older. Applying interval arithmetic to real, integer and Boolean constraints. *Journal of Logic Programming*, 32(1):1–24, 1997.
7. B. F. Caviness and J. R. Johnson, editors. *Quantifier Elimination and Cylindrical Algebraic Decomposition*. Springer, 1998.
8. G. E. Collins. Quantifier elimination for the elementary theory of real closed fields by cylindrical algebraic decomposition. In *Second GI Conf. Automata Theory and Formal Languages*, volume 33 of *LNCS*, pages 134–183. Springer Verlag, 1975. Also in [7].
9. G. E. Collins and H. Hong. Partial cylindrical algebraic decomposition for quantifier elimination. *Journal of Symbolic Computation*, 12:299–328, 1991. Also in [7].
10. J. H. Davenport and J. Heintz. Real quantifier elimination is doubly exponential. *Journal of Symbolic Computation*, 5:29–35, 1988.
11. L. Granvilliers. A symbolic-numerical branch and prune algorithm for solving nonlinear polynomial systems. *Journal of Universal Computer Science*, 4(2):125–146, 1998.

12. P. V. Hentenryck, D. McAllester, and D. Kapur. Solving polynomial systems using a branch and prune approach. *SIAM Journal on Numerical Analysis*, 34(2), 1997.
13. P. V. Hentenryck, V. Saraswat, and Y. Deville. The design, implementation, and evaluation of the constraint language cc(FD). In Podelski [21].
14. H. Hong and V. Stahl. Safe starting regions by fixed points and tightening. *Computing*, 53:323–335, 1994.
15. L. Jaulin and É. Walter. Guaranteed tuning, with application to robust control and motion planning. *Automatica*, 32(8):1217–1221, 1996.
16. J. Jourdan and T. Sola. The versatility of handling disjunctions as constraints. In M. Bruynooghe and J. Penjam, editors, *Proc. of the 5th Intl. Symp. on Programming Language Implementation and Logic Programming, PLILP'93*, number 714 in LNCS, pages 60–74. Springer Verlag, 1993.
17. A. Macintyre and A. Wilkie. On the decidability of the real exponential field. In P. Odifreddi, editor, *Kreiseliana—About and Around Georg Kreisel*, pages 441–467. A K Peters, 1996.
18. S. Malan, M. Milanese, and M. Taragna. Robust analysis and design of control systems using interval arithmetic. *Automatica*, 33(7):1363–1372, 1997.
19. K. Marriott, P. Moulder, P. J. Stuckey, and A. Borning. Solving disjunctive constraints for interactive graphical applications. In *Seventh Intl. Conf. on Principles and Practice of Constraint Programming - CP2001*, number 2239 in LNCS, pages 361–376. Springer, 2001.
20. A. Neumaier. *Interval Methods for Systems of Equations*. Cambridge Univ. Press, Cambridge, 1990.
21. A. Podelski, editor. *Constraint Programming: Basics and Trends*, volume 910 of *LNCS*. Springer Verlag, 1995.
22. S. Ratschan. Approximate quantified constraint solving (AQCS). http://www.risc.uni-linz.ac.at/research/software/AQCS, 2000. Software package.
23. S. Ratschan. Applications of real first-order constraint solving — bibliography. http://www.risc.uni-linz.ac.at/people/sratscha/appFOC.html, 2001.
24. S. Ratschan. Real first-order constraints are stable with probability one. http://www.risc.uni-linz.ac.at/people/sratscha, 2001. Draft.
25. S. Ratschan. Approximate quantified constraint solving by cylindrical box decomposition. *Reliable Computing*, 8(1):21–42, 2002.
26. S. Ratschan. Quantified constraints under perturbations. *Journal of Symbolic Computation*, 33(4):493–505, 2002.
27. J. Renegar. On the computational complexity and geometry of the first-order theory of the reals. *Journal of Symbolic Computation*, 13(3):255–352, March 1992.
28. D. Richardson. Some undecidable problems involving elementary functions of a real variable. *Journal of Symbolic Logic*, 33:514–520, 1968.
29. D. Sam-Haroud and B. Faltings. Consistency techniques for continuous constraints. *Constraints*, 1(1/2):85–118, September 1996.
30. S. P. Shary. Outer estimation of generalized solution sets to interval linear systems. *Reliable Computing*, 5:323–335, 1999.
31. A. Tarski. *A Decision Method for Elementary Algebra and Geometry*. Univ. of California Press, Berkeley, 1951. Also in [7].
32. L. van den Dries. Alfred Tarski's elimination theory for real closed fields. *Journal of Symbolic Logic*, 53(1):7–19, 1988.
33. V. Weispfenning. The complexity of linear problems in fields. *Journal of Symbolic Computation*, 5(1–2):3–27, 1988.

Coloring Algorithms for Tolerance Graphs: Reasoning and Scheduling with Interval Constraints

Martin Charles Golumbic[1] and Assaf Siani[2]

[1] Dept. of Computer Science, University of Haifa, Haifa, Israel
golumbic@cs.haifa.ac.il
[2] Dept. of Computer Science, Bar-Ilan University, Ramat-Gan, Israel
siani@cs.biu.ac.il

Abstract. Interval relations play a significant role in constraint-based temporal reasoning, resource allocation and scheduling problems. For example, the intervals may represent events in time which may conflict or may be compatible, or they may represent tasks to be performed according to a timetable which must be assigned distinct resources like processors or people. In previous work [G93, GS93, G98], we explored the interaction between the interval algebras studied in artificial intelligence and the interval graphs and orders studied in combinatorial mathematics, extending results in both disciplines.
In this paper, we investigate algorithmic problems on tolerance graphs, a family which generalizes interval graphs, and which therefore have broader application. Tolerance graph models can represent qualitative and quantitative relations in situations where the intervals can tolerate a certain degree of overlap without being in conflict. We present a coloring algorithm for a tolerance graph on n vertices whose running time is $O(n^2)$, given the tolerance representation, thus improving previously known results. The coloring problem on intervals has direct application to resource allocation and scheduling temporal processes.

Keywords and Topics: AI, OR applications, reasoning, coloring tolerance graphs

1 Introduction

Graph $G=(V,E)$ is a *tolerance graph* if each vertex $v \in V$ can be assigned an interval on the real line that represents it, denoted I_v, and a real number $t_v > 0$ referred to as its tolerance, such that for each pair of adjacent vertices, $uv \in E$ if and only if $|I_u \cap I_v| \geq \min\{t_u, t_v\}$. The intervals represented by the vertices and the tolerances assigned to the intervals form the *tolerance representation* of graph G [see Figure 1]. If the graph has a tolerance representation for which the tolerance t_v of the interval I_v representing each vertex $v \in V$ is smaller than or equal to the interval's length, i.e., $t_v \leq |I_v|$, then that graph is called a *bounded tolerance graph* and its representation is a *bounded representation*. The family of tolerance graphs was first introduced by

J. Calmet et al. (Eds.): AISC-Calculemus 2002, LNAI 2385, pp. 196–207, 2002.

Golumbic and Monma [GM82]. Their motivation was the need to solve scheduling problems in which resources that normally we would use exclusively, like rooms, vehicles, etc. can tolerate some sharing. Since then, properties of this model and quite a number of variations of it have appeared, and are the topic of a forthcoming monograph [GT02]. The tolerance graphs are a generalization of *probe graphs*, a graph family used for Genome research, which is in itself a generalization of the well-known family of *interval graphs*. Tolerance graphs are a sub-family of the Perfect Graphs [GMT84], and shares the latter's variety of mathematical and algorithmic properties [G80]. We are particularly interested in finding algorithms to calculate specific problems concerning these graphs, like graph coloring or maximum clique cover, because of their application to constraint-based temporal reasoning, resource allocation and scheduling problems.

As part of our algorithmic research, we present a coloring algorithm for a tolerance graph on n vertices whose running time is $O(n^2)$, given the interval tolerance representation. This is an improvement over previously known results of [NM92]. Another problem we deal with in this research is finding all the maximal cliques in a tolerance graph. Let k be the number of a maximal cliques of a graph G. Since it is known that k is not polynomially bounded by n for tolerance graphs, we present an algorithm for iteratively generating all maximal cliques in a graph, which uses as a subroutine an efficient algorithm for finding all the maximal cliques in a bounded tolerance graph.

2 Basic Definitions and Background

Let $G = (V, E)$ be an undirected graph. We call $N(v)=\{w \mid (v,w)\in E\}$ the *(open) neighborhood* of vertex v, and we call $N[v] = N(v)\cup\{v\}$ the *closed neighborhood* of v. The pair $(v,w)\in E$ is an *edge* and we say that w is adjacent to v, and v and w are *endpoints* of the edge (v,w). When it is clear, we will often drop the parenthesis and the comma when denoting an edge. Thus $xy\in E$ and $(x,y)\in E$ will have the same meaning.

We define the *complement* of G to be the graph $\overline{G} = (V, \overline{E})$, where $\overline{E}=\{(x,y)\in V \times V \mid x \neq y$ and $xy\notin E\}$. Given a subset $A\subseteq V$ of the vertices, we define the *subgraph induced* by A to be $G_A=(A,E_A)$, where $E_A = \{xy\in E \mid x\in A$ and $y\in A\}$.

A graph in which every pair of distinct vertices are adjacent is called a *complete graph*. K_n denotes the complete graph on n vertices. A subset $A\subseteq V$ of r vertices is an *r-clique* if it induces a complete subgraph, i.e., if G_A isomorphic to K_r. A single vertex is a 1-clique. A clique A is *maximal* if there is no clique of G, which properly contains A as a subset. A clique is *maximum* if there is no clique of G of larger cardinality. We denote by $\omega(G)$ the number of vertices in a maximum clique of G; it is called the *clique number* of G.

A *clique cover* of size k is a partition of the vertices $V=A_1+A_2+...+A_k$ such that each A_i is a clique. We denote by $k(G)$ the size of the smallest possible clique cover of G and is called the *clique cover number* of G.

A *stable set* (or *independent set*) is a subset X of vertices no two of which are adjacent. We denote $\alpha(G)$ to be the number of vertices in a stable set of maximum cardinality; it is called the *stability number* of G.

A *proper c-coloring* is a partition of the vertices $V=X_1+X_2+...+X_c$ such that each X_i is a stable set. In such a case, the members of X_i are "painted" with the color i and adjacent vertices will receive different colors. We denote by $\chi(G)$ the smallest possible c for which there exists a proper c-coloring of G; it is called the *chromatic number* of G.

A *graph G* is a *perfect graph* if G has the property that for every induced subgraph G_A of G, its chromatic number equals its clique number (i.e., $\chi(G_A) = \alpha(G_A)$). Due to a theorem of Lovász [L72] perfect graphs may be defined alternatively, G has the property that for every induced subgraph G_A of G, its stability number equals its clique cover number (i.e., $\alpha(G_A) = k(G_A)$). Perfect graphs are important for their applications and because certain decision problems that are NP-complete in general have polynomial-time algorithms when the graphs under consideration are perfect. Tolerance graphs were shown to be perfect in [GMT84].

Let F be a family of nonempty sets. The *intersection* graph of F is obtained by representing each set in F by a vertex and connecting two vertices by an edge if and only if their corresponding sets intersect. The intersection graph of a family of intervals on a linearly ordered set is called an *interval graph*, see Figure 2. The interval graph's real world applications vary from classic computer science problems such as scheduling or storage, to a chemical, biological and even archeological problems (refer to [G80] for details).

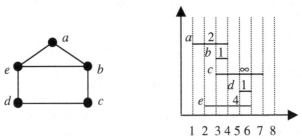

Fig. 1. A tolerance graph and a tolerance representation for it.

It can be easily shown that an interval graph is the special case of a tolerance graph where the tolerance t_v of interval I_v for all $v \in V$ equals some very small constant $\varepsilon > 0$.

Being an interval graph is a hereditary property, i.e., an induced subgraph of an interval graph is an interval graph. Another hereditary property of interval graph is that every simple cycle of length strictly greater than 3 possesses a chord. Graphs that satisfy this property are called *chordal graphs*. The graph in Figure 2 is chordal, but the graph in Figure 1 is not chordal because it contains a chordless 4-cycle. Therefore, it is not an interval graph. It is, however, a tolerance graph, which can be seen by the tolerance representation in Figure 1.

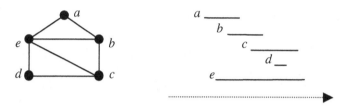

Fig. 2. An interval graph and an interval representation for it.

A graph said to be *weakly chordal* if it contains no induced subgraph isomorphic to C_n or to the complement of C_n for $n \geq 5$. Golumbic, Monma and Trotter [GMT84] showed that tolerance graphs are weakly chordal. Hayward [H85] showed that weakly chordal graphs are perfect, and gave some polynomial time algorithms for problems relating with the chromatic number and the stability number for weakly chordal graphs. Raghunathan [R89] improve these results and provide $O(|E||V|^2)$ algorithms to find a maximum clique and a minimum coloring of a weakly chordal graph. He also obtained an $O(|V|^4)$ algorithm to find a maximum independent set and a minimum clique cover of such a graph, and $O(|V|^5)$ algorithms for weighted versions of these problems

A graph G is a *comparability graph* if each edge in E can be assigned a direction so that the resulting oriented graph (V, F) satisfies the following condition: $ij \in F$ and $jk \in F$ implies that $ik \in F$ for all $i, j, k \in V(G)$. Such an orientation is called a *transitive orientation* of G and when such an orientation exists we say that G is *transitively orientable*. A *co-comparability graph* is a graph whose complement is a comparability graph.

Theorem 2.1 [GM82]. Bounded tolerance graphs are co-comparability graphs.

One proof of this theorem is given in [BFIL95] and uses parallelogram graphs, which provide another useful way to think about bounded tolerance graphs. A graph G is a *parallelogram graph* if we can fix two parallel lines L_1 and L_2 with L_1

above L_2, and for each vertex $i \in V(G)$ we can assign a parallelogram P_i with parallel sides along L_1 and L_2 so that G is the intersection graph of $\{P_i \mid i \in V(G)\}$.

***Lemma 2.2* [BFIL95].** A graph is a bounded tolerance graph if and only if it is a parallelogram graph.

Finally, it is sometimes convenient to assume, without loss of generality as shown in [GMT84], that any tolerance graph has a *regular* tolerance representation which satisfies the following addition properties:

1. Any tolerance larger than the length of its corresponding interval is set to infinity.
2. All tolerances are distinct (except for those set to infinity).
3. No two different intervals share an endpoint.

3   Coloring Tolerance Graphs

The problem of coloring a graph has many applications. Basically, a graph can represent a collection of objects: the vertices, which relate to each other through the edges. It is very common to use this structure to model objects that consume resources, whereby the edges symbolize some restriction on two objects that cannot consume the same resource simultaneously. The act of coloring a graph, i.e., labeling the vertices of the graph such that each of any two adjacent vertices receives a different label, is actually the act of allocating a resource (the label) to an object (the vertex). Minimal coloring in that sense is therefore utilizing minimal resources without violating any restriction. For example, consider a model in which lectures are represented by vertices, any two lectures which take place simultaneously cause an edge. A minimal coloring of the graph may signify a minimal allocation of classrooms for this set of lectures. The graph coloring problem is NP-complete for general graphs, however, efficient polynomial time algorithms are known for many classes of perfect graphs.

Narasimhan and Manber [NM92] suggested a simple method for computing the chromatic number $\chi(G)$ of a tolerance graph. This method is based on the facts that (1) the bounded intervals in the tolerance model form a co-comparability graph (Theorem 2.1) and that there exists a known algorithm for computing the chromatic number of co-comparability graphs. Their algorithm results in $O(qn^3)$ time, where q is the number of unbounded intervals in the tolerance model, n is the number of vertices and $O(n^3)$ is the time required for computing the chromatic number of a co-comparability graph. However, as was shown subsequently, bounded tolerance graphs are parallelogram graphs (Lemma 2.2), and Felsner et al. [FMW97] suggested an algorithm for computing the chromatic number of trapezoid graphs (which contain parallelogram graphs) in optimal $O(n\log n)$ time. Hence, we can use this algorithm for computing the chromatic number of the bounded intervals, and using the same idea of [NM92] obtain the chromatic number $\chi(G)$ of a tolerance graph in $O(qn\log n)$ time. However, their method does not give a coloring for the graph.

We took this problem a step forward to produce a coloring of the graph. Our algorithm has complexity $O(n^2)$ for general tolerance. In all these algorithms, it is assumed that a tolerance representation for the graph is given as input, which is typically the case in most applications. The problem of recognizing tolerance graphs is a longstanding open question.

3.1 Definitions, Terminology, and Lemmas

Let $G=(V,E)$ be a tolerance graph with a regular tolerance representation $<I,t>$. We define the subset of bounded vertices of V as $V_B=\{v\in V \mid t_v \leq |I_v|\}$, and the unbounded vertex subset of V as $V_U=\{v\in V \mid t_v>|I_v|\}$. Similarly, the induced bounded subgraph is $G_B=(V_B,E)$, and the induced unbounded subgraph is $G_U=(V_U,E)$.

Definition 3.1. An unbounded tolerance vertex $v\in V$ is labeled an *inevitable unbounded* vertex in G (for a certain tolerance representation), if the following holds:
(1) v is not an isolated vertex.
(2) $t_v > |I_v|$.
(3) Setting v's tolerance to its length (i.e. $t_v=|I_v|$) creates a new edge in the representation (i.e., the representation is no longer a tolerance representation for G).

Definition 3.2. An *inevitable unbounded* tolerance representation for G is a tolerance representation, where every unbounded tolerance vertex is an inevitable unbounded vertex.

Lemma 3.3. Every tolerance representation can be transformed into an inevitable unbounded representation in $O(n^2)$ time.

Proof. We will scan the representation from left to right. At any point during the algorithm we are aware of the *active* intervals, i.e., the intervals whose left endpoint has been scanned but whose right endpoint hasn't. Whenever a left endpoint of unbounded interval I is scanned, we check whether there is, among the active intervals, an interval which contains the interval I (i.e., its right endpoint is larger than $r(I)$ and the interval's tolerance is larger than $|I|$). If such an interval exists, then reducing I's tolerance to its length will cause a new edge in the representation between I and that interval. If no such interval exists, reduce I to its length. Otherwise, check whether there is some interval that forms an edge with I. If no such interval exists, then I is a isolated interval. In this case, make it the rightmost interval and make its tolerance equal to its length. If there is an interval adjacent to I, then I is left with unbounded tolerance. The scan is continued with the modified representation. $\square$

Definition 3.4. Let $v\in V$ be an inevitable unbounded vertex in G (for some representation $<I,t>$), then there is (at least) one bounded vertex $u\in V$ for which $uv\notin E$ and $I_v\subset I_u$ ($t_u > |I_v|$). The vertex u is called a *hovering vertex* for v and I_u is called a *hovering interval* for I_v.

Definition 3.5. We define the *hovering vertex set* (abbreviated *HVS*) for some inevitable unbounded vertex $v \in V$ to be the set of all hovering vertices of v, i.e., $HVS(v) = \{u \mid u$ is a hovering vertex for $v)$.

Lemma 3.6. Let $v \in V$ be an inevitable unbounded vertex in G (for some representation $<I,t>$). The set *HVS* always contains (at least) one bounded interval.

Proof. Since v is an inevitable unbounded vertex, hence there is some vertex $w_1 \in V$ where $I_v \subset I_{w_1}$. Supposing $t_{w_1} = \infty$, then w_1 is an inevitable unbounded vertex and there is some vertex $w_2 \in V$ where $I_{w_1} \subset I_{w_2}$. By induction this continues until we get $I_v \subset I_{w_1} \subset I_{w_2} \subset ... \subset I_{w_k}$ where w_k is a bounded vertex and $\left| I_{w_k} \right| > \left| I_{w_{k-1}} \right| > ... > \left| I_{w_1} \right| > \left| I_v \right|$, hence $w_k v \notin E$ and therefore $I_{w_k} \in HVS(v)$. □

Lemma 3.7. Let $v \in V$ be an inevitable unbounded vertex in G (for a representation $<I,t>$). We can find some bounded interval I_w, $I_w \in HVS(v)$, to be a representative interval for the set $HVS(v)$ in $O(n^2)$ time.

Proof. We will scan the representation from left to right. At any point during the algorithm we are aware of the *active* intervals, i.e., the intervals whose left endpoint has been scanned but their right endpoint hasn't. We will search amongst the active intervals for some bounded interval I_w that contains I_v, which is not adjacent to I_v (we know that such an interval exists from Lemma 3.6). We choose this interval to be a representative for $HVS(v)$. □

3.2 Algorithm for Coloring Tolerance Graphs

On one hand, if every maximum clique in G contains an unbounded vertex, then we can color G_B as a co-comparability graph (or parallelogram graph), give the set of vertices V_U a different color, and we are done. If, on the other hand, there exists a maximum clique in G which does not contain an unbounded vertex, then we should do the following: (1) color G_B as a co-comparability graph or parallelogram graph, (2) for every vertex $v \in V_U$, insert v into the color-set of a representative vertex $w \in HVS(v)$. This is justified since $vw \notin E$ and so $N(v) \subseteq N(w)$, hence no neighbor of v is present in the color-set of w, and thus this is a proper coloring of G.

Thus, a preliminary algorithm could require that we find all maximum cliques in the graph and check whether they contain an unbounded tolerance vertex or not. However, for every inevitable unbounded tolerance vertex $v \in V$, $HVS(v)$ contains some bounded vertex which would be colored when coloring G_B, and thus can determine a color for v. It follows that we do not need the maximum cliques after the transformation of Lemma 3.3 has been applied. Knowing the above, we conclude the following algorithm:

Algorithm 3.1.

Color G_B as a co-comparability graph (or parallelogram graph).
For every inevitable unbounded vertex $v \in V_U$,

insert v into the color-set of some representative vertex $w \in HVS(v)$.

Correctness

Clearly, the coloring for G_B is proper. For every unbounded interval I_i we have some bounded interval I_j such that $I_i \subset I_j$ but $v_i v_j \notin E$, which implies that $t_j > |I_i|$. Clearly, every interval I_k which forms an edge with I_i has to be bounded, and thus we get $|I_j \cap I_k| \geq |I_k \cap I_i| \geq \min\{t_k, t_i\} = t_k$, implying that $v_j v_k \in E$, and that v_k is not in the color set of v_j. Therefore, if we insert v_i into the color set of v_j, there will be no vertex in that color set to form an edge with v_i, and thus the coloring is proper.

Complexity

The algorithm consists of three stages: the first stage is a transformation of the tolerance representation into an inevitable unbounded one. The transformation takes $O(n^2)$ time (because we need to check each intersecting interval of every unbounded tolerance interval). The second stage is finding a representative interval for the HVS set of every unbounded tolerance interval which also takes $O(n^2)$ time, for the reasons mentioned above. Finally, the third stage is the coloring itself, which is also divided into three parts: the first part is a transformation of the tolerance representation (of the bounded intervals) into a parallelogram representation, which takes linear time. The second part is the coloring of the parallelogram representation, which is done in $O(n \log n)$ time [FMW97]. The third part is the coloring of each unbounded tolerance interval; this takes linear time, for the procedure is simply labeling the unbounded intervals with the color of their HVS's set representative. The total time is therefore $O(n^2)$.

We conclude by noting that since tolerance graphs are perfect, $\chi(G) = \omega(G)$, and our coloring algorithm can easily find a maximum clique at the same time.

4 All Maximal Cliques of Tolerance Graphs

A maximal clique in a graph G is a complete set X of vertices such that no superset of X is a clique of G. Finding all maximal cliques is sometimes necessary to find a solution for other problems. For example, cut vertices, bridges and vertex-disjoint paths can all be determined easily once the maximal cliques are known. In this section, we will present a method for computing all maximal cliques of tolerance graphs, given the tolerance model of the graph. For solving the all-maximal cliques problem we used the same method as with the coloring problem in section 3, i.e., we solve the problem for the induced graph containing only the bounded vertices and later we deal with the unbounded vertices. Unfortunately, the number of maximal cliques for the induced bounded intervals may be exponential. However, for interval graphs, as with all chordal graphs, the number of maximal cliques is at most $|V|$. Our algorithm especially suits subfamilies of tolerance graphs where computing all maximal cliques for the induced subgraph containing only bounded vertices is done in polynomial time.

4.1 All Maximal Cliques of a Tolerance Graph

Let $G=(V,E)$ be a tolerance graph and let $<I,t>$ be a regular tolerance representation for G. Let V_B be the set of all bounded vertices in V in this representation, and let V_U be the set of all unbounded vertices in V. We will use K_X to denote the set of all maximal cliques of any set of vertices X, $X \subseteq V$. We define the set K_G to be the set of all maximal cliques in G, and let K_B denote the set of all maximal cliques in G_B. For every $u \in V_U$, let $K_{N(u)}$ denote the set of all maximal cliques in the graph $G_{N(u)}$, which is the subgraph induced from G by the neighbors of the vertex u. Clearly, for every $u \in V_U$ and $Y \in K_{N(u)}$, $Y \cup \{u\}$ is a maximal clique in G. Finally, let $K_{N[u]}$ denote the set of all maximal cliques in $G_{N[u]}$, i.e., $K_{N[u]} = \{ Y \cup \{u\} \mid Y \in K_{N(u)} \}$.

***Theorem* 4.1.** $K_G = \left\{ K_B \setminus \left\{ \bigcup_{u \in U} K_{N(u)} \right\} \right\} \cup \left\{ \bigcup_{u \in U} K_{N[u]} \right\}.$

Proof. ("$\subseteq$" part.) Let Y be a maximal clique in G, i.e. $Y \in K_G$. If Y does not contain any unbounded vertex, then $Y \in K_B$ and is not in any $K_{N[u]}$ where $u \in U$. Otherwise, Y contains an unbounded vertex $u \in V_U$ (Y may contain only one such vertex since Y is a clique). $Y \setminus \{u\}$ is a maximal clique in $G_{N(u)}$ and $Y \in K_{N[u]}$.

("$\supseteq$" part.) Assume $Y \in \left\{ K_B \setminus \left\{ \bigcup_{u \in U} K_{N(u)} \right\} \right\} \cup \left\{ \bigcup_{u \in U} K_{N[u]} \right\}.$

Suppose to the contrary that there is some $Y' \in K_G$ such that $Y \subset Y'$. If Y' does not contain an unbounded vertex, then $Y \in K_B$ - a contradiction. Hence, Y' contains an unbounded vertex $u \in U$. But in that case $Y' \setminus \{u\} \in K_{N(u)}$, and $Y' \in K_{N[u]}$, a contradiction. □

Now, if G_B belongs to a known hereditary graph family, for which we have an algorithm for computing all maximal cliques of that family, we can use that algorithm to compute K_B and $K_{N[u]}$ for every $u \in U$, since $G_{N(u)}$ is also a graph of the same family because all the vertices in $N(u)$ are bounded.

Recall from section 3, that we defined the terms *inevitable unbounded representation* and the *HVS* set of an inevitable unbounded interval/vertex. Also recall that each neighbor vertex of some unbounded vertex $u \in V_U$ is a neighbor vertex of every vertex $v \in HVS(u)$. Algorithm 4.1 is based on the following lemma.

Lemma 4.2. Let $v \in B$ be a hovering vertex of u, i.e., $v \in HVS(u)$. For every $Y \in K_{N(u)}$, there is some $Y' \in K_{N[v]}$ for which $Y \subseteq Y'$.

Proof. Clearly, for every vertex $w \in N(u)$, $w \in N(v)$. Hence, for every maximal clique $Y \in K_{N(u)}$, either Y is maximal clique of $K_{N[v]}$ or there is some other maximal clique $Y' \in K_{N[v]}$ such that $Y \subset Y'$. $\square$

Let $v \in B$ be a hovering vertex of u. By lemma 4.2, we can compute all the maximal cliques of $G_{N(u)}$ by taking all the maximal cliques in G containing v and deleting any vertex which is not adjacent to u, as we do next in algorithm 4.1.

Algorithm 4.1

Let $G=(V,E)$ be a tolerance graph and let $<I,t>$ be an inevitable unbounded representation for G.

(1) Compute K_B, let $K_B^{Computed} \leftarrow K_B$.

(2) For any unbounded vertex $u \in V_U$,

        (2.1) Let $w \in V_B$ be a hovering vertex of u, i.e., w is some representative of $HVS(u)$.

        (2.2) For any maximal clique $Y \in K_B$ containing w, compute the set $Y' = \{Y \backslash \{w\}\} \cap N(u)$.

        (2.3) If $Y' \neq \phi$ then $K_U^{Computed} \leftarrow K_U^{Computed} \cup \{Y' \cup \{u\}\}$.

Theorem 4.3. $K_G = K_B^{Computed} \cup K_U^{Computed}$

Proof. . ("$\subseteq$" part) Let $Y \in K_G$ be a maximal clique in G. If Y does not contain any unbounded vertex, then $Y \in K_B^{Computed}$. Otherwise, Y contains some unbounded vertex $u \in U$ (Y may contain only one such vertex since Y is a clique). For each $v \in HVS(u)$, there exists some maximal clique Y' such that $Y \backslash \{u\} \subset Y'$, hence by the algorithm $Y \in K_U^{Computed}$.

("$\supseteq$" part) Assume that $Y \in K_B^{Computed} \cup K_U^{Computed}$. Suppose to the contrary that there is some $Y' \in K_G$ where $Y \subset Y'$. If Y' contains only bounded tolerance vertices it is a contradiction for $K_B^{Computed}$ being the set of all maximal cliques of G_B. Hence, Y' must contain some $u \in U$. The set $Y \backslash \{u\}$ is a set of bounded vertices, hence there is a clique set $Y'' \in K_B^{Computed}$ such that $Y \backslash \{u\} \subset Y''$, and Y'' contains $v \in HVS(u)$ (v is the reason for the proper inclusion of $Y \backslash \{u\}$ in Y''). But this is a contradiction since $Y'' \backslash \{v\} \cap N(u) \in K_U^{Computed}$. $\square$

4.2 The Number of All Maximal Cliques in a Tolerance Graph

As previously mentioned, the number of all maximal cliques of tolerance graph may be exponential. Bounded tolerance graphs, a subfamily of tolerance graphs, contains the permutation graph family. Consider the permutation diagram in Figure 3, which has $2n$ lines, such that each line intersects with all the other lines except for one line. We can choose one line of any two parallel lines, and the result would be a maximal clique in the permutation graph. It is easy to see that we have 2^n maximal cliques in this graph. Hence, the number of maximal cliques of a tolerance graph may be exponential.

As a result, the algorithm presented for computing all maximal cliques of tolerance graphs is one, which iteratively generates them. In the case where the bounded vertices of the graph form a subgraph of a structured family whose maximal cliques may be computed in polynomial time using a known algorithm, as with the family of probe graphs, the method then becomes polynomially efficient.

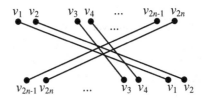

Fig. 3. Permutation diagram whose permutation graph has $2n$ vertices and exactly 2^n maximal cliques.

References

[BFIL95] K. Bogart, P. Fishburn, G. Isaak and P. Langley, Proper and unit tolerance graphs, *Discrete Applied Math.* 60 (1995) 37-51.

[F98] S. Felsner, Tolerance graphs and orders, *J. of Graph Theory* 28 (1998) 129-140.

[FMW97] S. Felsner, R. Müller, L. Wernisch, Trapezoid graphs and generalizations, geometry and algorithms, *Discrete Applied Math.* 74 (1997) 13-32.

[G80] M.C. Golumbic, *Algorithmic Graph Theory and Perfect Graphs*, Academic Press, New York, 1980.

[G93] M.C. Golumbic, Reasoning about time, Invited talk, AISMC-1 Karlsruhe, Germany, August 3-6, 1992, abstract in *LNCS* 737 (1993) p. 276.

[G98] M.C. Golumbic, Reasoning about time, in *"Mathematical Aspects of Artificial Intelligence"*, F. Hoffman, ed., American Math. Society, *Proc. Symposia in Applied Math.*, vol. 55, 1998, pp. 19-53.

[GM82] M.C. Golumbic and C.L. Monma, A generalization of interval graphs with tolerances, in: Proceedings 13th Southeastern Conference on Combinatorics, Graph Theory and Computing, *Congressus Numerantium* 35 (1982) 321-331.

[GMT84] M.C. Golumbic, C.L. Monma and W.T. Trotter Jr., Tolerance graphs, *Discrete Applied Math.* 9 (1984) 157-170.

[GS93]   M.C. Golumbic and R. Shamir, Complexity and algorithms for reasoning about time: a graph-theoretic approach, *J. Assoc. Comput. Mach.* 40 (1993), 1108-1133.

[GT02]   M.C. Golumbic and Ann N. Trenk, *Tolerance Graphs*, monograph in preparation.

[H85]   R. Hayward, Weakly triangulated graphs, *J. Combin. Theo. Ser. B* 39 (1985) 200-209.

[HHM90]   R. Hayward, C. Hoàng, and F. Maffray, Optimizing weakly triangulated graphs, *Graphs and Combinatorics* 6 (1990) 33-35. Erratum to *ibid*, 5:339-349.

[L72]   L. Lovász, Normal hypergraphs and the perfect graph conjecture, *Discrete Math.* 2 (1972), 253-267.

[NM92]   G. Narasimhan and R. Manber, Stability number and chromatic number of tolerance graphs. *Discrete Applied Math.* 36 (1992) 47-56.

[R89]   A. Raghunathan, Algorithms for weakly triangulated graphs, Univ. of Calif. Berkeley, Technical Report CSD-89-503 (April 1989).

A Genetic-Based Approach for Satisfiability Problems

Mohamed Tounsi

Computer Science Department,
Ecole des Mines de Nantes
4, Rue Alfred Kastler, 44307 Nantes, France
`Mohamed.Tounsi@emn.fr`

Abstract. We present a genetic-based approach to solve SAT problem and NP-complete problems. The main idea of the approach presented here is to exploit the fact that, although all NP-complete problems are equally difficult in a general computational sense, some have much better genetic representations than others, leading to much more successful use of genetic-based algorithm on some NP-complete problems than on others. Since any NP-complete problem can be mapped into any other one in polynomial time by a transformation, the approach described here consists of identifying and finding a canonical or generic NP-complete problem on which genetic algorithm work well, and solving other NP-complete problems indirectly by translating them onto the canonical problem. We presented some initial results where we have the Boolean Satisfiability Problem (SAT) as a canonical problem, and results on Hamiltonian Circuit problem which represent a family of NP-complete problems, it can be solved efficiently by mapping them first onto SAT problems.

Keywords: SAT, Genetic Algorithm, NP-complete Problem, Hamiltonian Circuit.

1 Introduction

A strong progress on solving SAT problems, as local search methods [LSM92] and many of its variants were made last decade. These methods were applied on strong instances of SAT and given a good results. The canonical example of a problem in NP is the boolean satisfiability problem (SAT): given an arbitrary boolean expression of n variables, does exist an assignment to those variables such that the expression is true? Other examples include job shop scheduling and traveling salesman problems. The concept of NP-completeness comes from the observation that, although every problem L in NP can be transformed into an equivalent SAT problem in polynomial time[1], the reverse polynomial time transformation may not exist. Those problems in NP which do have 2 way transformations form an equivalence class of "equally hard" problems and have been

---

[1] Cookes theorem

J. Calmet et al. (Eds.): AISC-Calculemus 2002, LNAI 2385, pp. 208–216, 2002.
© Springer-Verlag Berlin Heidelberg 2002

called NP-complete problems [GJ79]. Although NP-complete problems are computationally equivalent in this complexity theoretic sense, they do not appear to be equivalent at all with respect to how well they map onto genetic representations (GAs). For example, in the case of GAs, the SAT problem has a very natural representation while finding effective representations for bin packing problem, job shop scheduling problem, and travel salesman problem seems to be quite difficult [Gol89] [GL85]. Those observations suggest the following idea: suppose we are able to identify an NP-complete problem which has an effective representation in the methodology of interest Genetic Algorithms (GAs) and develop an efficient problem solver for that particular case. Other NP-complete problems which don't have effective representations can then be solved by transforming them into the canonical problem, solving it, and transforming the solution back to the original one. We have explored this strategy in detail for genetic algorithms using SAT as the canonical NP-complete problem. The (GAs) can achieve power and generality by demanding that problems be mapped into their own particular representation in order to be solved. If a natural mapping exists, good performance results. On the other hand, if the mapping is difficult, our approach behave much like the more traditional weak methods yielding mediocre, unsatisfying results. These observations suggest two general issues which deserve further study. First, we need to understand how severe the mapping problem is. Are there large classes of problems for which effective mappings exist? Clearly, if we have to spend large amounts of time and effort in constructing a mapping for each new problem, then it is not a better way than the more traditional complete methods. It focuses on GAs and how they can be applied to a large, well known class of combinatorially explosive problems: NP-complete problems. We introduce in this paper a genetic algorithm to SAT and recall what is genetic and SAT problem and the components of our algorithm: population, mutation and crossover operators, then followed by a new valuation function to be used in our algorithm. Some preliminary results on two families of SAT problem are presented with results on a NP-complete problem : the Hamiltonien Circuit problem (HC). We conclude with a discussion and future works.

2   Genetic Algorithms for SAT

The field of Genetic Algorithms has grown into a huge area over the last few years [Raw91]. GAs are adaptvie methods [EvdH97] [Koz92], which can be used to solve search and optimization problems over a period of generations, based upon the genetic processes of biological organisms over principles of natural selection and survival of fittest. In order to apply GAs to a particular problem, we need to select an internal string representation for the solution space and define an external evaluation function which assigns utility to candidate solutions. Both components are critical to the success or failure of the GAs on the problem of interest [Rud94]. We have selected SAT as the choice for our canonical NP-complete problem because it appears to have a good string representation, namely, binary strings of length N in which the ith bit represents the truth value of the ith boolean variable of the N boolean variables present in the expression. It is hard to imagine a representation much better suited for use with GAs [ARS93]: it is fixed length, binary, and context independent in the sense that the meaning of one bit is unaffected by changing the value of other bits [H79]. Thus, the use of genetic algorithm to solve an optimization problem requires a data encoding to build gene strings [Lan98].

Our basic genetic algorithm fellows the following steps :

1. **Encoding data problems on binary variables:** The encoding function is easy, so population is composed of individuals, an individual is a suite of N bits, each bit represent one variable.
2. **Valuation function:** The fitness function or valuation function is computed during the search, it gives the adequation of the bits sequence. Takes value in $[0, 1]$ (cf. next section), it is function to maximize.
3. **Crossover:** the crossover operator is easily done, for two individuals of N bits randomly chosen, they are crossed by randomly choosing a position, cutting each individuals in this position and glue the first part of each with the second part of the others and inversely. In our experiments, we have taken the crossover rate equal 60%.
4. **Mutation:** from each individual of N bits, we inverse one bit randomly, the rate of mutation taken is 1%.
5. **Back to step 2**

2.1   A New Valuation Function

In order to compare between two individuals, we must have an utility function called "valuation function" or "fitness function". The simplest and most natural function assigns a valuation of 1 to a candidate solution (string) if the values specified by that string result in the boolean expression evaluating to TRUE, and 0 otherwise. However, for problems of interest the valuation function would be 0 almost every where and would not support the formation of useful intermediate building blocks. Even though in the real problem domain, partial solutions to SAT are not of much interest, they are critical components of a genetic approach [Gat98]. One approach to providing intermediate feedback would be to transform a given boolean expression into conjunctive normal form (CNF) and define the valuation to be the total number of conjuncts which evaluate to true. While this makes some intuitive sense, one cannot in general perform such transformations in polynomial time without introducing a large number of additional boolean variables which, in turn, combinatorially increase the size of the search space [BEV98]. An alternative would be to assign valuation to individual clauses in the original expression and combine them in some way to generate a total valuation value. In this context the most natural approach is to define the value of TRUE to be 1, the value of FALSE to be 0, and to define the value of simple expressions as follows:

$$valuation(NOT \quad x) = 1 - valuation(x)$$
$$valuation(AND \quad x_1...x_n) = Min(valuation(x_1), ..., valuation(x_n))$$
$$valuation(OR \quad x_1...x_n) = Max(valuation(x_1), ..., valuation(x_n))$$

Since any boolean expression can be writen into these basic elements, one has a systematic mechanism for assigning valuation. Unfortunately, this mechanism is no better than the original one since it still only assigns valuation values of 0 and 1 to both individual clauses and the entire expression. However, a minor change to this mechanism can generate differential valuations:

$$valuation(AND \quad x_1...x_n) = average(valuation(x_1), ..., valuation(x_n))$$

This suggestion was made first by Smith [H79] and justified by the expression can be "more nearly true" [EvdH97], for example the (table 1) gives the results of valuation: $X_1 \quad AND \quad (X_1 \quad OR \quad \overline{X_2})$.

Table 1. Valuation Function

X_1	X_2	VALUATION
0	0	(average 0 (Max(0(1 − 0))))= 0.5
0	1	(average 0 (Max(0(1 − 1))))= 0.0
1	0	(average 1 (Max(1(1 − 0))))= 1.0
1	1	(average 1 (Max(1(1 − 1))))= 1.0

Notice that both of the correct solutions (lines 3 and 4) are assigned a valuation of 1 and, of the incorrect solutions (lines 1 and 2), line 1 gets higher valuation because it got half of the AND right. However, there were a number of features of this valuation function that must be improved. The first and obvious property of using *average* to evaluate AND clauses is that the valuation function is not invariant under standard boolean equivalence transformations. For example, it violates the associativity law:

$$valuation((X_1 \quad AND \quad X_2) \quad AND \quad X_3) \neq valuation(X_1 \quad AND \quad (X_2 \quad AND \quad X_3))$$
$$since : (average(average \quad X_1 \quad X_2) \quad X_3) \neq (average \quad X_1(average \quad X_2 \quad X_3))$$

However, one could argue that a weaker form of invariance could be adequate for use with genetic algorithms. By that we mean that the valuation function should assign the same value (1, but could even be a set of values) to all correct solutions of the given boolean expression, and should map all incorrect solutions into a set of values ($0 \leq value < 1$) which is distinct and lower than the correct ones. In general, it can be shown that, although the valuation does not assign the value of 1 to non-solutions, it frequently assigns values ≤ 1 to good solutions and can potentially give higher valuation to non-solutions.
A careful analysis, indicates that these problems only arise when DeMorgan laws are involved in introducing terms of the form $\overline{(AND...)}$. This suggests a simple preprocess: for each boolean expression apply DeMorgan laws to remove such

constructs. It also suggests another interesting opportunity. Constructs of the form $\overline{OR...}$ are computed correctly, but only take on value 0 or 1. By using DeMorgan laws to convert these to AND constructs, we introduce additional differential valuation. Fortunately, unlike the conversion to CNF, this process has only linear complexity and can be done quickly and efficiently.

In summary, with the addition of this preprocessing steps, we now have an effective valuation function for applying genetic algorithm to boolean satisfiability problems.

3   Experiments on SAT Problems

Our first set of experiments involves constructing several families of boolean expressions for which we can control the size and the difficulty of the problem. The first family selected consists of family of two solutions (SAT_{2S}) of the form:

$(AND \quad x_1...x_n) \quad OR \quad (AND \quad \overline{x_1}...\overline{x_n})$

which have exactly two solutions (all 0s and all 1s). By varying the number n of boolean variables, one can observe how the genetic algorithm perform as the size of the search space increases exponentially while the number of solutions remains fixed. Figure 1 presents the results of varying N between 10 and 90 (i.e. for search spaces ranging in size from 2^{10} to 2^{90}). It is clear that the differential valuation function is working as intended, and that our algorithm can locate solutions to (SAT_{2S}) problems without much difficulty (figures 1,2).

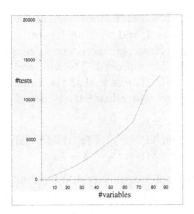

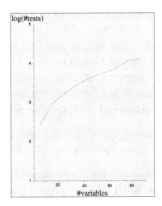

Fig. 1. Results on $SAT_{(2S)}$

Fig. 2. The log-log results on $SAT_{(2S)}$

To make things more difficult, we changed the problem by turning one of the solutions into a one solution problem (SAT_{1S}) as follows:

$(AND \quad x_1...x_n) \quad OR \quad (AND \quad x_1\overline{x_1}...\overline{x_n})$

so that the solution is all at 1.

Figure 3 presents the results of log-log of applying our algorithm to the SAT_{1S} family with n ranging from 10 to 90. As before, we see that the GAs have no difficulty in finding the correct solution.

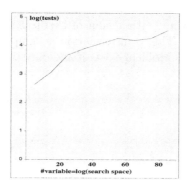

Fig. 3. The log-log results on $SAT_{(1S)}$

Since we are dealing with problems for which there are no known polynomial time algorithms, we have been particularly interested in the log-log graphs.

Notice that, for both the (2S) and (1S) problems, a sub-linear curve is generated, indicating an improvement over systematic search. The form that these sublinear curves take give some indication of the speedup (over systematic search) obtained by using genetic algorithm. If, for example, these curves are all logarithmic in form, we will have a good algorithm. Additional discussion of these curves will occur in a later section after more data has been presented. With these first results, we were eager to apply a genetic algorithm to more naturally arising boolean expressions. So, we have chosen to look at other NP-complete problems: Hamiltonian Circuit problems.

3.1 Solving Hamiltonian Circuit Problems

The Hamiltonian Circuit (HC) problem consists of finding a tour through a directed graph that touches all nodes exactly once. It is a particular case of travel salesman problem. Clearly, if a graph is fully connected, this is an easy task. However, as edges are removed the problem becomes much more difficult, and the general problem is known to be NP-Complete. Attempting to solve this problem directly with genetic algorithm raises many of the same representation issues as in the case of traveling salesman problems [Ga84]. However, it is not difficult to construct a polynomial time transformation from HC problems to SAT problems. The definition of the HC problem implies that, for any solution, each node must have exactly one input edge and one output edge. If any tour violates this constraint, it cannot be a solution. Therefore, an equivalent boolean expression is simply the conjunction of terms indicating valid edge combinations for each node. As an example, consider node X_5. Node X_5 has two output edges and one input edge. The output edge constraints are given by the exclusive OR, $((X_5X_1$ AND $\overline{X_5X_2})$ OR $(\overline{X_5X_1}$ AND $X_5X_2))$. The input edge is described simply by X_4X_5. The assignments to the edge variables indicate which edges make up a tour, with a value of 1 indicating an edge is included and a value of 0 if it is not. This transformation is computed in polynomial time, and a solution to the HC problem exists if and only if the boolean expression is satisfiable.

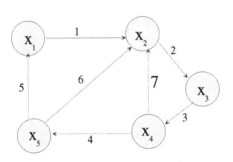

Fig. 4. An example of circuit Hamiltonian problem

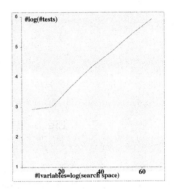

Fig. 5. The log-log results on HC problem

As before, we wish to systematically study the performance of our genetic algorithm on a series of HC problems. Clearly, the complexity in this case is a function of both the number of nodes and the number of directed edges. For a given number N of nodes, problems with only a small number of edges N or nearly fully connected (approximately N^2 edges) are not very interesting. We feel that problems with approximately $\frac{N^2}{2}$ edges would, in general, present the most difficult problems. In addition, we wanted the problems to have exactly one solution. So, we have defined the following family of HC problems for our experiments :

Consider a graph of N nodes, which are labeled using consecutive integers. Suppose the first node has directed edges to all nodes with larger labels (except for the last node). The next $N-2$ nodes have directed edges to all nodes with larger labels (including the last one). The last node has a directed edge back to the first node. A complete tour consists of following the node labels in increasing order, until you reach the last node. From the last node you travel back to the first. Because the edges are directed, it is clear that this is also the only legal tour. Intuitively, such instances of HC problems should be difficult because only one tour exists in each instance.

In summary, our experimental framework consists of varying the number N of nodes in the range $4 \leq N \leq 60$ and, for each value of N, generating a directed graph of the form described above containing approximately $\frac{N^2}{2}$ edges and exactly one solution.

Each of these HC problems is transformed into its equivalent SAT problem using the transformation described above. GAs are then used to solve each of the corresponding SAT problems which, in turn, describes a legal HC tour. Figures 5 presents the results of these experiments. Notice that we have succeeded in generating significantly more difficult SAT problems. However, even with these difficult problems, the log-log plot is still sub-linear.

One of the theoretical results in Holland [Hol75] analysis of the power of genetic algorithm is sets a lower bound of an N^3 speedup over systematic search [Hol75]. It suggests that, in the worst case, genetic algorithm should not have to search more than $\sqrt[3]{N}$ of the search space in order to find a solution. One of the our results presented here is to improvements of speedups. As noted earlier, the log-log curves appear to be sub-linear. To get a better idea for the form of these curves, we have tried to fit a linear and quadratic curves to the data. For each of the families of SAT problems, a quadratic form produces a better fit. So, we can calculate the observed speedup (table 2). Clearly, on the easier problems (1S and 2S) we are better than the predicted lower bound, however, for HC problems, we have the same lower bound as the theoretical one.

Table 2. Improvements on SAT

improve on $SAT_{(1s)}$	N^7
improve on $SAT_{(2S)}$	$N^{6.5}$
improve on HC	N^3

4   Conclusions and Future Works

This paper presents a series of initial results regarding a strategy for using genetic algorithm to solve SAT and NP-complete problems. This strategy avoids many of the genetic representation difficulties associated with various NP-complete problems by mapping them into SAT problems for which an effective GA representation exists. These initial results based on a basic genetic algorithm support the view that genetic algorithms are an effective, robust search procedure for NP-complete problems in the sense that, although they may not outperform highly problem specific algorithms, genetic algorithm can be easily applied to a broad range of NP-complete problems with performance characteristics no worse than the theoretical lower bound of an N^3 speedup. More expreriments on strong instance of SAT and comparaison with specific HC algorithms should valid our results, we feel also that more strong valuation function and very specific operators may give a better results.

References

[ARS93]   Rudolf F. Albrecht, Colin R. Reeves, and Nigel C. Steele, editors. *Artificial neural nets and genetic algorithms*. Springer, April 14-16 1993. ANNGA 93, International Conference on Artificial Neural Networks & Genetic Algorithms.

[BEV98]   Back, Eiben, and Vink. A superior evolutionary algorithm for 3-SAT. In *International Conference on Evolutionary Programming, in cooperation with IEEE Neural Networks Council*. LNCS, 1998.

[EvdH97]   Eiben and van der Hauw. Solving 3-SAT by (gas) adapting constraint weights. In *Proceedings of The IEEE Conference on Evolutionary Computation, World Congress on Computational Intelligence*, 1997.

[Ga84]   J. Grefenstette and al. Genetic algorithms fot the traveling salesman problem. In *Conference on Intelligent Systems and Machines*, 1984.

[Gat98]   Chris Gathercole. An investigation of supervised learning in genetic programming. In *Ph.D. thesis. Department of Electronics and Electrical Engineering. University of Edinburg, 1998*, 1998.

[GJ79]   M. Garey and D. Johnsond. Computers and intractability : A guide to the theory of np-completeness. In *W. H. Freeman and Company, CA.*, 1979.

[GL85]   D. Goldberg and Robert Lingle. Alleles, loci, and the traveling salesman problem. In *Conference on Intelligent Systems and Machines 161-165*, 1985.

[Gol89]   D. Goldberg. Genetic algorithms in search, optmization and machine learning. In *Addison Wesley Pubslishing, ISBN :0-201-15767-5*, 1989.

[H79]   Gerald Smith H. Adaptive genetic algorithms and the boolean satisfiability problem. In *Conference on genetic algorithm*, 1979.

[Hol75]   J. H. Holland. Adaptation in natural and artificial systems. In *The university of Michigan Press*, 1975.

[Koz92]   J.R. Koza. *Genetic Programming*. MIT Press, 1992.

[Lan98]   William B. Langdon. Data structures and genetic programming. In *Ph.D. thesis. University College, London*, 1998.

[LSM92]   H. Levesque, B. Selman, and D. Mitchell. A new method for solving hard satisfiability problems. In *Proceeding of AAAI, pages 440-446*, 1992.

[Raw91]   Gregory J. E. Rawlins, editor. *Foundations of Genetic Algorithms*, San Mateo, California, 1991. Morgan Kaufmann.

[Rud94]   Günter Rudolph. Convergence analysis of canonical genetic algorithms. *IEEE Trans. Neural Networks, Special Issue on Evolutionary Computation*, 5(1):96–101, 1994.

On Identifying Simple and Quantified Lattice Points in the 2SAT Polytope

K. Subramani

LDCSEE,
West Virginia University,
Morgantown, WV
ksmani@csee.wvu.edu

Abstract. This paper is concerned with techniques for identifying simple and quantified lattice points in 2SAT polytopes. 2SAT polytopes generalize the polyhedra corresponding to Boolean 2SAT formulae, Vertex-Packing (Covering, Partitioning) and Network flow problems; they find wide application in the domains of Program verification (Software Engineering) and State-Space search (Artificial Intelligence). Our techniques are based on the symbolic elimination strategy called the Fourier-Motzkin elimination procedure and thus have the advantages of being extremely simple (from an implementational perspective) and incremental. We also provide a characterization of the 2SAT polytope in terms of its extreme points and derive some interesting hardness results for associated optimization problems.

1 Introduction

Consider the standard integer feasibility (IP) program $\exists \vec{x}\ \mathbf{A}.\vec{x} \leq \vec{b}, \vec{x} \geq \vec{0}, x_i \in \{0,1\}$; this problem is also referred to as the problem of finding a simple lattice point in a polytope and is NP-complete [GJ79,CLR92]. A related problem is the Quantified Integer feasibility (QIP) problem, in which one or more of the problem variables are universally quantified. It is well known that the quantified integer programming problem is PSPACE-complete although there are polynomial time algorithms for certain special cases [Sub02a]. Both the IP and QIP models have been used to model problems within a range of domains, ranging from state-space search in AI applications [Nil98,Gen90] to Program Verification in Software Engineering [AO97,LS85].

In this paper we are concerned with a special structure called the 2SAT polytope; these polytopes are a generalization of the polytopes representing 2SAT formulae and those representing vertex packing and covering problems in an undirected graph.

Definition 1. *A polyhedral system* $\mathbf{A} \cdot \vec{x} \leq \vec{b}$ *is said to be a 2SAT polytope system if*

1. *Every entry* $a_{ij} \in \mathbf{A}$ *belongs to the set* $\{-1, 0, 1\}$,

J. Calmet et al. (Eds.): AISC-Calculemus 2002, LNAI 2385, pp. 217–230, 2002.
© Springer-Verlag Berlin Heidelberg 2002

2. *There are at most 2 non-zero entries per row,*
3. $\vec{b}$ *is an arbitrary integral vector.*

Our focus is on the development of algorithms for determining the existence of simple and quantified lattice points in a given 2SAT polytope. We show that 2SAT polytope problems admit polyhedral elimination techniques; in particular we show that the Fourier-Motzkin elimination procedure can be applied to these problems. The advantage of using elimination procedures are twofold: (a) they are inherently incremental, and (b) a number of symbolic computation programs have such procedures built-in [CH99]. To the best of our knowledge, our results are the first of their kind.

The rest of this paper is organized as follows: Section §2 provides a formal description of the problem(s) under study; the motivation for our work is detailed in Section §3, while Section §4 discusses related work in the literature. In Section §5, we present our algorithm for the problem of finding a simple lattice point within a 2SAT polytope; in Section §6, we extend these ideas to develop an algorithm for the problem of finding a quantified lattice point in a 2SAT polytope. Important properties of 2SAT polytopes are identified and proved in Section §7; these properties are important from the perspective of optimization problems. We conclude in Section §8, by summarizing our contributions and pointing out directions for future research.

2   Statement of Problem(s)

We consider the following two problems in this paper:

> **P$_1$** : *Does an arbitrary 2SAT polytope system $\mathbf{A} \cdot \vec{x} \leq \vec{b}$, as defined in Definition (1), have a lattice point solution, i.e. $\exists \vec{x}, \mathbf{A} \cdot \vec{x} \leq \vec{b}, \vec{x}$ integer?*

P$_1$ is referred to as the simple lattice point problem for 2SAT polytopes. Note that the only restriction on $\vec{x}$ is integrality, i.e. the solution points belong to $\mathcal{Z}_+^n$ and not $\{0,1\}^n$.

Let Q be an arbitrary quantifier string over the n variables $x_1, x_2, \ldots, x_n$; i.e. $Q = Q_1 x_1 Q_2 x_2 \ldots Q_n x_n$, where each Q_i is either an existential quantifier ($\exists$) or a discrete-range universal quantifier ($\forall x_i \in \{a_i - b_i\}, a_i, b_i$ integer)[1].

> **P$_2$** : *Does a given 2SAT polytope system $\mathbf{A} \cdot \vec{x} \leq \vec{b}$, as defined in Definition (1), have a quantified lattice point solution, i.e. $Q_1 x_1 Q_2 x_2 \ldots Q_n x_n \mathbf{A} \cdot \vec{x} \leq \vec{b}, \vec{x}$ integer, for arbitrary Q_i?*

---

[1] Note that $\forall x_i \in \{a_i - b_i\}$ means that the variable x_i can assume any integral value in the discrete range $\{a_i, a_i + 1, \ldots, b_i\}$

P₂ is referred to as the quantified lattice point problem for 2SAT polytopes. In order to simplify the analysis, we distinguish between the existentially and universally quantified variables as follows:

1. The ith existentially quantified variable is represented by x_i,
2. The ith universally quantified variable is represented by y_i

Without loss of generality, we can assume that the variables are strictly alternating, so that problem **P₂** can be represented as:

$$\exists x_1 \forall y_1 \in \{a_1 - b_1\} \exists x_2 \forall y_2 \in \{a_2 - b_2\} \ldots \exists x_n \forall y_n \in \{a_n - b_n\}$$

$$\mathbf{A} \cdot [\vec{x} \ \vec{y}]^{\mathbf{T}} \leq \vec{b}? \tag{1}$$

where $\mathbf{A} \cdot [\vec{x} \ \vec{y}]^{\mathbf{T}} \leq \vec{b}$ is a 2SAT polytope system and $\vec{x}$ is integral.

In the rest of this paper, we shall discuss algorithms for deciding problems **P₁** and **P₂**.

3   Motivation

The 2SAT polytope generalizes three types of polytopes that are widely studied, viz.

1. The network polytope - In this case, every row of $\mathbf{A}$ has precisely 2 entries, one of them being 1 and the other being -1. For formulations of network polytopes, see [NW99];
2. Vertex Packing, Partitioning and Covering Polytopes of an undirected graph - Consider the vertex covering problem; in this case, every row of $\mathbf{A}$ has precisely 2 entries, with both entries being -1; in case of the vertex packing problem, both entries are 1, while for the partitioning problem, the entries are 1 but the polytope system has the form: $\mathbf{A} \cdot \vec{x} = \vec{b}$, as opposed to the form $\mathbf{A} \cdot \vec{x} \leq \vec{b}$ in Defintion (1). In all these cases, the values assumed by the variables belong to the set $\{0, 1\}$; further $\vec{b} = [1, 1, \ldots, 1]$. For formulations of Vertex Packing, Partitioning and Covering problems, see [Hoc96,Sch87].
3. The clausal 2SAT polytope - In this case, the (at most) 2 entries of each row may have the same sign or the opposite sign; however the solution vectors are in $\{0, 1\}^n$; further each $b_i \in \{0, 1, -1\}$. There is an additional restriction, viz. $b_i = 0$, if and only if the entries in the ith row are of opposite sign, $b_i = 1$, if and only if both entries in the ith row are 1 and $b_i = -1$, if and only of both entries in the ith row are -1.

We now show that the Clausal Quantified 2SAT problem can be expressed as problem **P₂**. Let

$$F = Q_1 x_1 Q_2 x_2 \ldots Q_n x_n \ C \tag{2}$$

where C is a conjunction of m clauses in which each clause has at most 2 literals. The quantified satisfiability problem is concerned with answering the question: Is F true ?

We first note that F has the following Integer Programming formulation,

$$F = Q_1 x_1 \in \{0,1\} Q_2 x_2 \in \{0,1\} \ldots Q_n x_n \in \{0,1\} \quad \mathbf{A}.\vec{x} \leq \vec{b} \qquad (3)$$

where,

1. **A** has n columns corresponding to the n variables and m rows corresponding to the m constraints,
2. A clause (x_i, x_j) is replaced by the integer constraint $x_i + x_j \geq 1$; a clause $(\bar{x}_i, x_j)$ is replaced by $1 - x_i + x_j \geq 1 \Rightarrow -x_i + x_j \geq 0$; a clause of the form $(x_i, \bar{x}_j)$ is replaced by $x_i + 1 - x_j \geq 1 \Rightarrow x_i - x_j \geq 0$, and a clause of the form $(\bar{x}_i, \bar{x}_j)$ is replaced by $1 - x_i + 1 - x_j \geq 1 \Rightarrow -x_i - x_j \geq -1$.
3. Each Q_i is one of $\exists$ or $\forall$.

The equivalence of of the clausal system and the integer programming formulation is obvious and has been argued in [Pap94].

From the above discussion, it is clear that the study of 2SAT polytopes unifies the study of problems in a wide variety of domains including Network Optimization, Operations Research and Logic. In this paper, we show that a very generic symbolic computation technique called the Fourier-Motzkin elimination procedure can be used for deciding 2SAT polytope problems.

4   Related Work

In this section, we briefly review the work on finding feasible linear and integer solutions over polyhedral systems composed of 2-support constraints. A constraint is said to be 2-support, if at most 2 entries are non-zero. Note that a 2-support constraint generalizes a 2SAT polytope constraint since the non-zero entries are not restricted to $\{1, -1\}$.

LP(2) i.e. the linear programming feasibility problem when each constraint has at most 2 non-zero variables has been studied extensively in [Sho81,AS79] and [HN94]. [HN94] provides the fastest algorithm to date, for this problem $(O(mn^2 \log m)$, assuming m constraints over n variables). [HN94] also studied the IP(2) problem, i.e the problem of checking whether a 2−support polyhedron contains a lattice point. This problem is `NP-complete` and they provide a pseudo-polynomial time algorithm for the same.

Packing and covering problems for graphs as well as for set systems have been studied extensively in the literature; [NT74] showed that packing and covering polytopes were half-integral, while [NT75] discussed algorithms for optimization problems over these polytopes.

[Sch78] argued that the clausal Q2SAT problem could be solved in polynomial time, but provided no algorithm for the same; [APT79] gave the first algorithm (linear time) for the clausal SAT and QSAT problems. The parallel complexity of this problem has been explored in [Che92], [CM88] and [Gav93] and it has been demonstrated that Q2SAT is in the parallel complexity class `NC`$_2$. Although our algorithms take $O(n^3)$ time, it must be noted that we are solving a far more general class of problems than those studied in [APT79] and [CM88]; it is not clear how their techniques (verifying reachability conditions in the constraint graph) extend to finding arbitrary integral solutions over 2SAT polytopes.

Fourier-Motzkin elimination as a technique for resolving feasibility in linear systems was proposed in [Fou24] and elaborated in [DE73]. Extending the technique to resolving integer programs was the thrust of [Wil76] and [Wil83]. A direct application of Fourier's theorem to integer programs results in congruences and modulo arithmetic. We show that for 2SAT polytopes, we can avoid modulo arithmetic altogether.

5 The Simple Lattice Point Problem

Algorithm (1) outlines our strategy for problem $\mathbf{P_1}$; details of the Fourier-Motzkin elimination procedure are available in Section §A.

Algorithm 1 An elimination strategy for checking whether a 2SAT polytope has a simple lattice point

---

Function SIMPLE-LATTICE-POINT $(\mathbf{A}, \vec{\mathbf{b}})$
1: **for** ($i = n$ **down to** 2) **do**
2:    Eliminate x_i using Fourier-Motzkin
3:    PRUNE-CONSTRAINTS()
4:    **if** (CHECK-INCONSISTENCY()) **then**
5:       **return (false)**
6:    **end if**
7: **end for**
8: **if** $(a \leq x_1 \leq b, a \leq b)$ **then**
9:    **return (true)**
10: **else**
11:    **return (false)**
12: **end if**

When a variable x_i is eliminated from 2 constraints containing the same variable x_j it is possible to derive a constraint of the form $x_j \leq \frac{1}{2}c_1$, c_1 *odd*; since x_j must be integral, this constraint is replaced by the constraint $x_j \leq \frac{c_1-1}{2}$. (Note that $\frac{c_1-1}{2}$ is always integral, when c_1 is odd.) This is carried out in PRUNE-CONSTRAINTS(). Note that there can be at most 4 non-redundant constraints between any 2 variables. When a variable is eliminated, it gives rise to new constraints between other variables. These constraints could be the redundant, in which case they are eliminated (through PRUNE-CONSTRAINTS()) or they could create an inconsistency, in which case the 2SAT polytope is declared infeasible. Identification of inconsistencies is carried out in CHECK-INCONSISTENCY().

During the elimination of variable x_n, each constraint involving x_n is put into the set $\mathbf{L_{x_n}} = \{x_n \leq ()\}$ or the set $\mathbf{U_{x_n}} = \{x_n \geq ()\}$. Thus, we have

$\max \mathbf{U_{x_n}} \le x_n \le \mathbf{L_{x_n}}$. The sets $\mathbf{L_{x_i}}, \mathbf{U_{x_i}}, i = 1, 2, \ldots, n$ represent a model for the 2SAT polytope; in that the value of x_i can be computed from $x_1, x_2, \ldots, x_{i-1}$ and the sets $\mathbf{L_{x_j}}, \mathbf{U_{x_j}}, j = 1, 2, \ldots, (i-1)$.

5.1   Correctness

The correctness of the Fourier-Motzkin procedure for linear programs has been argued in [DE73] and [Sch87]; we need to argue that it works over 2SAT polytopes as well. Assume that we are eliminating variable x_n of an n-dimensional 2SAT polytope $\mathbf{P}$, by pairing off constraints which appear as $x_n \le ()$ and $-x_n \le ()$ as described in Section §A. Let $\mathbf{P}'$ be the $(n-1)$-dimensional polytope that results after the elimination. Clearly, if $\mathbf{P}'$ is infeasible, i.e. it does not have a lattice point, then $\mathbf{P}$ cannot have a lattice point solution. Consider the case when $\mathbf{P}'$ has a lattice point solution $\vec{\mathbf{x}'} = [x_1', x_2', \ldots, x_{n-1}']^T$. We show that this lattice point can be complented to obtain a lattice point in $\mathbf{P}$. Substitute the components of $\vec{\mathbf{x}'}$ in the linear system defining $\mathbf{P}$; we get a system of inequalities of the form $x_n \le c_i$, c_i *integer*. This is because the co-efficient of x_n is either 1 or -1 in all the constraints and $\vec{\mathbf{b}}$ is integral. This system defines a single integer point or an interval with integral end-points. We cannot get an infeasible interval, since that would imply that the point $\vec{\mathbf{x}'}$ does not belong to the polyhedron $\mathbf{P}'$! To see this, note that $\max \mathbf{U_{x_n}} \le x_n \le \min \mathbf{L_{x_n}}$. Further both $\max \mathbf{U_{x_n}}$ and $\min \mathbf{L_{x_n}}$ are integers. Thus, in either case, we can pick an integral value of x_n and obtain a lattice point solution for the 2SAT polytope system.

We must point out that the technique breaks down, even in 2 dimensions, if we allow arbitrary constraints between 2 variables [Wil76].

5.2   Complexity

Let us store the 2SAT polytope as a graph structure; the variables x_i represent the vertices of the graph, while a constraint between x_i and x_j is represented as an edge of the graph. This graph structure can have at most 4 edges between 2 vertices. Every edge has 2 labels describing its direction with respect to its end-points. The edge $x_i + x_j \le 7$ is labeled $(+, +)$ to indicate that it is positive for both x_i and x_j; likewise $x_i - x_j \le 3$ is labeled $(+, -)$ and so on. We use the special variable x_0 to represent constraints involving only one variable; for instance the constraint $x_i \le 7$ will be represented by the edge $x_i + x_0 \le 7$ between x_i and x_0. When variable x_i is to be eliminated, all the edges that are positive for x_i are combined with all the edges that are negative for x_i to derive new edges in the graph structure. This step takes time at most $O(n^2)$, since the number of positive and negative edges could each be at most $O(n)$ in number. It follows that Algorithm (1) takes time at most $O(n^3)$.

6   The Quantified Lattice Point Problem

In this section, we discuss an elimination procedure to decide problem $\mathbf{P_2}$; Algorithm (2) represents our strategy.

Algorithm 2 A Quantifier Elimination Algorithm for deciding whether a 2SAT polytope has a quantified lattice point

---

Function QUANTIFIED-LATTICE-POINT $(\mathbf{A}, \vec{b})$
1: **for** ($i = n$ **down to** 1) **do**
2:    **if** ($Q_i = \exists$) **then**
3:      Eliminate x_i using Fourier-Motzkin
4:      PRUNE-CONSTRAINTS()
5:      **if** (CHECK-INCONSISTENCY()) **then**
6:        **return** (**false**)
7:      **end if**
8:    **end if**
9:    **if** ($Q_i = \forall$) **then**
10:      ELIM-UNIV-VARIABLE(y_i)
11:      PRUNE-CONSTRAINTS()
12:      **if** (CHECK-INCONSISTENCY()) **then**
13:        **return** (**false**)
14:      **end if**
15:    **end if**
16: **end for**
17: **return** (**true**)

Algorithm 3 Eliminating Universally Quantified variable $y_i \in \{a_i - b_i\}$

Function ELIM-UNIV-VARIABLE $(\mathbf{A}, \vec{b}, y_i)$
1: Substitute $y_i = a_i$ in each constraint that can be written in the form $y_i \geq ()$
2: Substitute $y_i = b_i$ in each constraint that can be written in the form $y_i \leq ()$

The elimination of a variable (existential or universal) results in one or more of the following consequences:

1. Some redundant constraints result, which can be pruned out. This is the function of the subroutine PRUNE-CONSTRAINTS().
2. An inconsistency results and we return **false** through the function CHECK-INCONSISTENCY().
3. We get a smaller subset of variables to work with. Observe that the total number of constraints is always bounded by 4 times the square of the number of variables.

6.1   Correctness

We need to argue the correctness of the procedure for eliminating Universally quantified variables. We have already established that the procedure for eliminating existentially quantified variables is solution-preserving (Section §5.1).

    Lemma (1) establishes that if the *last quantifier* is universal, the interval of the corresponding interval can be relaxed to a continuous interval. This permits the use of polyhedral techniques, in particular Algorithm (3).

Lemma 1. *Let*

$$\mathbf{L} : \exists x_1 \in \{a^1 - b^1\} \; \forall y_1 \in \{c^i - d^i\} \; \exists x_2 \in \{a^2 - b^2\} \; \forall y_2 \in \{c^2 - d^2\}$$
$$\ldots \exists x_n \in \{a^n - b^n\} \; \forall y_n \in \{c^n - d^n\} \; \mathbf{A} \cdot [\vec{\mathbf{x}} \; \vec{\mathbf{y}}]^{\mathbf{T}} \leq \vec{\mathbf{b}}$$

and

$$\mathbf{R} :; \exists x_1 \in \{a^1 - b^1\} \; \forall y_1 \in \{c^i - d^i\} \; \exists x_2 \in \{a^2 - b^2\} \; \forall y_2 \in \{c^2 - d^2\}$$
$$\ldots \exists x_n \in \{a^n - b^n\} \; \forall y_n \in [c^n, d^n]$$
$$\mathbf{A} \cdot [\vec{\mathbf{x}} \; \vec{\mathbf{y}}]^{\mathbf{T}} \leq \vec{\mathbf{b}}?$$

Then $\mathbf{L} \Leftrightarrow \mathbf{R}$. *i.e. if the last quantifier of a Quantified Integer Program (QIP) is universal, then the discrete range can be relaxed into a continuous range, while preserving the solution space.*

Proof: *The detailed proof is available in [Sub02a].* □

Since the interval of the outermost universally quantified variable is continuous, we can use the strategy outlined in [Sub02b] to eliminate it, which is precisely Algorithm (3).

6.2   Complexity

Once again, we assume that the 2SAT polytope is stored as a graph structure, as described in Section §5.2. Note that the elimination of a universally quantified variable through substitution, could result in constraints between a program variable and the special variable x_0. In $O(1)$ time, each such constraint can be identified as required, redundant or inconsistency-creating. Thus the total time taken for eliminating a universally quantified variable is at most $O(n)$ and hence the total time taken to eliminate all the universally quantified variables is at most $O(n^2)$.

The total time taken for the elimination of all the existentially quantified variables is at most $O(n^3)$ as argued in Section §5.2. It follows that the total time taken by Algorithm (2) is at most $O(n^3)$.

7   Properties of the 2SAT Polytope

In this section, we identify and prove some properties of 2SAT polytopes, that are useful from the perspective of optimization.

Theorem 1. *The 2SAT polytope is half-integral, i.e. every component of an extreme point of the 2SAT polyhedron is either an integer or an integral multiple of $\frac{1}{2}$.*

Proof: *Our proof is based on the correctness of the Fourier-Motzkin elimination procedure and is much simpler than the proof in [NT74] (which analyzed covering and packing polytopes only). We use mathematical induction on the number of variables, i.e. the dimension of the 2SAT polytope $\mathbf{P} = \mathbf{A} \cdot \vec{x} \leq \vec{b}$. We assume without loss of generality that $\mathbf{P}$ is non-empty. Clearly, the theorem is true for $n = 1$; in this case, every constraint is of the form $x_1 \leq \pm c_i$, where the c_i are integral. Hence the end-points of the interval describing x_1 are integral and Theorem (1) holds. Let us assume that the theorem holds for all 2SAT polytopes having dimension at most $n - 1$. Now consider a 2SAT polytope having dimension n. When we eliminate the variable x_n we are in fact projecting the 2SAT polytope onto $n - 1$ dimensions to get a 2SAT polytope $\mathbf{P}' = \mathbf{A}' \cdot \vec{x}' \leq \vec{b}'^2$, where $\vec{x}' = [x_1 \ x_2 \ \ldots x_{n-1}]$. From the induction hypothesis, we know that all the extreme points of $\mathbf{P}'$ are half-integral. Each extreme point of $\mathbf{P}$ can be generated as follows:*

1. *Let $\vec{x}'$ be an extreme point of $\mathbf{P}'$; substitute the components of $\vec{x}'$ in the polyhedral system $\mathbf{P}$;*
2. *The resultant system of linear inequalities in one variable results in either a single half-integral point or an interval with half-integral end-points. This is because, every constraint involving x_n is of the form $b_i - x'_j$, for some b_i and x'_j representing the jth component of $\vec{x}'$. Since b_i is integral and x'_j is half-integral (by the inductive hypothesis), the claim follows.*

□

Theorem 2. *Every square submatrix of a 2SAT polytope has determinant $\{0, \pm 1, \pm 2\}$*

Proof: *Consider a 2SAT polytope $\mathbf{A} \cdot \vec{x} \leq \vec{b}$ ($\mathbf{A}$ is an $m \times n$ matrix); by Theorem (1) every basic solution to the system $[\mathbf{A}, \mathbf{I}] \cdot [\vec{x} \ \vec{x_I}]^T = \vec{b}$ is half-integral (Not all basic solutions are feasible). Pick some arbitrary $k \times k$ square submatrix in $\mathbf{A}$ (say $\mathbf{A}'$). Without loss of generality, we can assume that $\mathbf{A}'$ is formed using the first k rows and first k columns of $\mathbf{A}$ (otherwise perform the appropriate permutations!). Consider the $m \times m$ square matrix $\mathbf{C}$ formed by combining the first k columns of $\mathbf{A}$ with the last $m - k$ columns of $\mathbf{I}$. Figure (1) describes the composition.*
We thus have

$$\mathbf{C} = \begin{bmatrix} \mathbf{A}' & \mathbf{0} \\ \mathbf{B} & \mathbf{I} \end{bmatrix} \qquad (4)$$

It is not hard to see that $|det(\mathbf{C})| = |det(\mathbf{A}')|$; we now show that $det(\mathbf{C}) \in \{0, \pm 1, \pm 2\}$. If $det(\mathbf{C}) = 0$, we are done; let us consider the case, where $det(\mathbf{C}) \neq 0$. Consider the sequence of linear equality systems $\mathbf{C} \cdot \vec{x_i} = \vec{e_i}, i = 1, 2, \ldots, m$, where $\vec{e_i}$ is the unit-vector in the ith dimension. Since $\vec{x_i}$ is half-integral, by Theorem (1) ($\vec{x_i}$ is a basic solution), it follows that $\mathbf{C}^{-1} \cdot \vec{e_i}$ is half-integral.

---

[2] We argued in Section §5 that the elimination procedure preserves the 2SAT structure

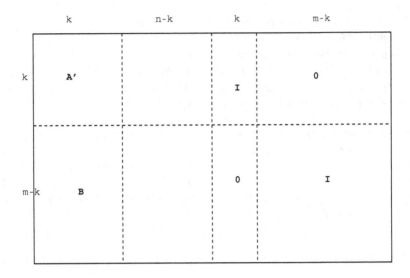

Fig. 1. Creating a basis

This is possible, only if all the entries of the ith column of $\mathbf{C}^{-1}$ are half-integral. We can thus conclude that all the entries of $\mathbf{C}^{-1}$ must be half-integral. We write $\mathbf{C}^{-1} = \frac{1}{2}\mathbf{C}'^{-1}$, i.e. the factor $\frac{1}{2}$ can be taken out from every element in $\mathbf{C}^{-1}$ to give an integral matrix $\mathbf{C}'^{-1}$. Thus, $det(\mathbf{C}'^{-1})$ is integral and hence $det(\mathbf{C}^{-1})$ must be half-integral. Observe that $det(\mathbf{C})$ is integral and $det(\mathbf{C}).det(\mathbf{C}^{-1}) = 1$. Thus if, $det(\mathbf{C}^{-1})$ is integral, then $|det(\mathbf{C})| = 1$; likewise if $det(\mathbf{C}^{-1})$ is an odd multiple of $\frac{1}{2}$, then $|det(\mathbf{C})| = 2$, since that is the only way for $det(\mathbf{C}).det(\mathbf{C}^{-1}) = 1$ to hold. We have thus shown that for any square sub-matrix $\mathbf{A}'$ of a 2SAT polytope $det(\mathbf{A}') = \pm 1$ or ± 2.

Our proof follows the lines of [VD68] in [KV00] for totally unimodular matrices. □

Theorem 3. *If $\vec{c}$ is an integral vector, $\max \vec{c}.\vec{x}$ is over a 2SAT polytope is reached at a half-integral point.*

Proof: *Follows immediately from Theorem (1), since the maximum of a linear function is reached at an extreme point of the polyhedron.* □

We say that a cost vector $\vec{c}$ is positive, if all its components are positive. Observe that the problems of finding an integer minimum or the integer maximum of a positive cost vector over a network polyhedron can be carried out in polynomial time, since network polyhedra are totally unimodular. Likewise, maximizing (integer maximum) a positive cost vector over a covering polytope is trivial, whereas minimizing (integer minimum) a positive cost vector over the same is NP-complete. In case of packing polytopes, finding the integral minimum of a positive cost vector is trivial, whereas the integral maximization problem is NP-complete.

Theorem 4. *Finding the integral maximum or integral minimum of a positive cost vector over a 2SAT polytope is* NP-complete.

Proof: *We will argue that the problems of maximizing and minimizing the number of* **true** *variables in a clausal 2SAT system are* NP-complete; *the theorem follows. Clearly the problem of minimizing the number of* **true** *variables in a clausal 2SAT system (called* $\mathbf{P_{min}}$*) is* NP-complete, *since the standard vertex cover problem is a special case of this problem.*

Now consider the problem of maximizing the number of **true** *variables in a clausal 2SAT system (called* $\mathbf{P_{max}}$*). We show that this problem is also* NP-complete *by reducing* $\mathbf{P_{min}}$ *to it. Given an instance of* $\mathbf{P_{min}}$ *we construct an instance of* $\mathbf{P_{max}}$ *as follows:*

1. *Replace each positive literal* x_i *with a negative literal* $\bar{y}_i$ *and each negative literal* $\bar{x}_i$ *with a positive literal* y_i.
2. *Thus minimizing the number of* **true** *variables in the original system corresponds to minimizing the number of* **false** *variables in the new system, which is equivalent to the problem of maximizing the number of* **true** *variables in the new system.*

□

Theorem 5. *Finding the integer optimum of any linear function over an arbitrary 2SAT polytope is* NP-complete.

Proof: *We can use the substitution technique of Theorem (4) to reduce either* $\mathbf{P_{min}}$ *or* $\mathbf{P_{max}}$ *to this problem.* □

8   Conclusions

In this paper, we demonstrated the existence of polynomial time algorithms for the problems of finding simple and quantified lattice points in a 2SAT polytope. To the best of our knowledge, polynomial time algorithms for these problems do not exist in the literature, although various special cases have been well studied. Our techniques are useful because of their simplicity, generality and incrementality.

An important property of polytopes from the property of Operations Research, is the *fixing variables property*. Let $\vec{x_1}$ be the (unique) solution to the polyhedral system: $\max \vec{c}.\vec{x} \ \mathbf{A} \cdot \vec{x} \le \vec{b}$. The polyhedral system is said to have the fixing variables property, if there exists an integer optimum of a linear function over that system which is equal to the linear optimum ($\vec{x_1}$) in all its integer components. While it is known that the vertex covering and packing polytopes have the fixing variables property, the question of whether 2SAT polytopes have the same property is open. This property is useful from the perspective of deriving approximation algorithms for integer optimization problems over the polytope [Hoc96].

We can define Horn-SAT polytopes in a fashion, similar to the definition of 2SAT polytopes; however, the techniques discussed in this paper do not yield polynomial time strategies. The interesting question then is whether there are larger classes of polytopes for which our techniques do provide polynomial time strategies.

Finally, this paper focussed on deterministic approaches only, which we believe was the reason for the rather pessimistic bound of $O(n^3)$; we are currently exploring the existence of randomized approaches with a view to improving the running time of our algorithms.

Algorithm 4 The Fourier-Motzkin Elimination Procedure

Function FOURIER-MOTZKIN ELIMINATION $(\mathbf{A}, \vec{\mathbf{b}})$

1: **for** ($i = n$ **down to** 2) **do**
2:    Let $\mathbf{I}^+ = \{$ set of constraints that can be written in the form $x_i \geq ()\}$
3:    Let $\mathbf{I}^- = \{$ set of constraints that can be written in the form $x_i \leq ()\}$
4:    **for** (each constraint $k \in \mathbf{I}^+$) **do**
5:       **for** (each constraint $l \in \mathbf{I}^-$) **do**
6:          Add $k \leq l$ to the original constraints
7:       **end for**
8:    **end for**
9:    Delete all constraints containing x_i
10: **end for**
11: **if** ($a \leq x_1 \leq b, \quad a, b \geq 0$) **then**
12:    Linear program is consistent
13:    **return**
14: **else**
15:    Linear program is inconsistent
16:    **return**
17: **end if**

A   Fourier-Motzkin Elimination

The Fourier-Motzkin elimination procedure is an elegant, syntactic, variable elimination scheme to solve constraint systems that are comprised of linear inequalities. It was discovered initially by Fourier [Fou24] and later by Motzkin [DE73], who used it to solve general purpose Linear programs.

The key idea in the elimination procedure is that a constraint system in n variables (i.e. $\Re^n$), can be projected onto a space of $n-1$ variables (i.e. $\Re^{n-1}$), without altering the solution space. In other words, polyhedral projection of a constraint set is solution preserving. This idea is applied recursively, till we are left with a single variable (say x_1). If we have $a \leq x_1 \leq b, a \leq b$, then the system is consistent, for any value of x_1 in the interval $[a, b]$. Working backwards, we can deduce the values of all the variables $x_2, \ldots, x_n$. If $a > b$, we conclude that the system is infeasible.

Algorithm (4) is a formal description of the above procedure.

Though elegant, this syntactic procedure suffers from an exponential growth in the constraint set, as it progresses. This growth has been observed both in theory [Sch87] and in practice [HLL90,LM91]. By appropriately choosing the constraint matrix $\mathbf{A}$, it can be shown that eliminating k variables causes the size of the constraint set to increase from m to $O(m^{2^k})$ [Sch87]. Algorithm (4) remains useful though as a tool for proving theorems on polyhedral spaces [VR99]. [Sch87] gives a detailed exposition of this procedure.

References

[AO97]   Krzystof R. Apt and Ernst R. Olderog. *Verification of Sequential and Concurrent Programs*. Springer-Verlag, 1997.

[APT79] Bengt Aspvall, Michael F. Plass, and Robert Tarjan. A linear-time algorithm for testing the truth of certain quantified boolean formulas. *Information Processing Letters*, (3), 1979.

[AS79]   Bengt Aspvall and Yossi Shiloach. A polynomial time algorithm for solving systems of linear inequalities with two variables per inequality. In *20th Annual Symposium on Foundations of Computer Science*, pages 205–217, San Juan, Puerto Rico, 29–31 October 1979. IEEE.

[CH99]   V. Chandru and J. N. Hooker. *Optimization Methods for Logical Inference*. Series in Discrete Mathematics and Optimization. John Wiley & Sons Inc., 1999.

[Che92] Z. Chen. A fast and efficient parallel algorithm for finding a satisfying truth assignment to a 2-cnf formula. *Information Processing Letters*, pages 191–193, 1992.

[CLR92] T. H. Cormen, C. E. Leiserson, and R. L. Rivest. *Introduction to Algorithms*. MIT Press and McGraw-Hill Book Company, 6th edition, 1992.

[CM88]   S.A. Cook and M.Luby. A simple parallel algorithm for finding a satisfying truth assignment to a 2-cnf formula. *Information Processing Letters*, pages 141–145, 1988.

[DE73]   G. B. Dantzig and B. C. Eaves. Fourier-Motzkin Elimination and its Dual. *Journal of Combinatorial Theory (A)*, 14:288–297, 1973.

[Fou24] J. B. J. Fourier. *Reported in: Analyse de travaux de l'Academie Royale des Sciences, pendant l'annee 1824, Partie Mathematique, Historyde l'Academie Royale de Sciences de l'Institue de France 7 (1827) xlvii-lv. (Partial English translation in: D.A. Kohler, Translation of a Report by Fourier on his work on Linear Inequalities. Opsearch 10 (1973) 38-42.)*. Academic Press, 1824.

[Gav93] F. Gavril. An efficiently solvable graph partition, problem to which many problems are reducible. *Information Processing Letters*, pages 285–290, 1993.

[Gen90] Michael Genesereth. *Logical Foundations of Artificial Intelligence*. Morgan Kaufmann, 1990.

[GJ79]   M. R. Garey and D. S. Johnson. *Computers and Intractability: A Guide to the Theory of NP-Completeness*. W. H. Freeman Company, San Francisco, 1979.

[HLL90]   Tien Huynh, Catherine Lassez, and Jean-Louis Lassez. Fourier Algorithm Revisited. In Hélène Kirchner and W. Wechler, editors, *Proceedings Second International Conference on Algebraic and Logic Programming*, volume 463 of *Lecture Notes in Computer Science*, pages 117–131, Nancy, France, October 1990. Springer-Verlag.

[HN94]   Dorit S. Hochbaum and Joseph (Seffi) Naor. Simple and fast algorithms for linear and integer programs with two variables per inequality. *SIAM Journal on Computing*, 23(6):1179–1192, December 1994.

[Hoc96]   Hochbaum, editor. *Approximation Algorithms for NP-Hard Problems*. PWS Publishing Company, 1996.

[KV00]   B. Korte and J. Vygen. *Combinatorial Optimization*. Number 21. Springer-Verlag, New York, 2000.

[LM91]   Jean-Louis Lassez and Michael Maher. On fourier's algorithm for linear constraints. *Journal of Automated Reasoning, to appear*, 1991.

[LS85]   Jacques Loeckx and Kurt Sieber. *The Foundations of Program Verification*. John Wiley and Sons, 1985.

[Nil98]   Nils J. Nilsson. *Artificial Intelligence: A New Synthesis*. Morgan Kaufmann, 1998.

[NT74]   G. L. Nemhauser and L. E. Trotter Jr. Properties of vertex packing and independence system polyhedra. *mathprog*, 6:48–61, 1974.

[NT75]   G. L. Nemhauser and J. L. E. Trotter. Vertex packing: structural properties and algorithms. *Mathematical Programming*, 8:232–248, 1975.

[NW99]   G. L. Nemhauser and L. A. Wolsey. *Integer and Combinatorial Optimization*. John Wiley & Sons, New York, 1999.

[Pap94]   Christos H. Papadimitriou. *Computational Complexity*. Addison-Wesley, New York, 1994.

[Sch78]   T.J. Schaefer. The complexity of satisfiability problems. In Alfred Aho, editor, *Proceedings of the 10th Annual ACM Symposium on Theory of Computing*, pages 216–226, New York City, NY, 1978. ACM Press.

[Sch87]   Alexander Schrijver. *Theory of Linear and Integer Programming*. John Wiley and Sons, New York, 1987.

[Sho81]   Robert Shostak. Deciding linear inequalities by computing loop residues. *Journal of the ACM*, 28(4):769–779, October 1981.

[Sub02a] K. Subramani. An analysis os selected qips, 2002. Submitted to Journal of Logic and Computation.

[Sub02b] K. Subramani. A polynomial time algorithm for a class of quantified linear programs, 2002. Submitted to Mathematical Programming.

[VD68]   A.F. Veinott and G.B. Dantzig. Integral extreme points. *SIAM Review*, 10:371–372, 1968.

[VR99]   V.Chandru and M.R. Rao. Linear programming. In *Algorithms and Theory of Computation Handbook, CRC Press, 1999*. CRC Press, 1999.

[Wil76]   H.P. Williams. Fourier-motzkin elimination extension to integer programming. *J. Combinatorial Theory*, (21):118–123, 1976.

[Wil83]   H.P. Williams. A characterisation of all feasible solutions to an integer program. *Discrete Appl. Math.*, (5):147–155, 1983.

Integrating Boolean and Mathematical Solving: Foundations, Basic Algorithms, and Requirements*

Gilles Audemard[1,2], Piergiorgio Bertoli[1], Alessandro Cimatti[1], Artur Korniłowicz[1,3], and Roberto Sebastiani[1,4]

[1] ITC-IRST, Povo, Trento, Italy
{audemard,bertoli,cimatti,kornilow}@itc.it
[2] LSIS, University of Provence, Marseille, France
[3] Institute of Computer Science, University of Białystok, Poland
[4] DIT, Università di Trento, Povo, Trento, Italy
roberto.sebastiani@dit.unitn.it

Abstract. In the last years we have witnessed an impressive advance in the efficiency of boolean solving techniques, which has brought large previously intractable problems at the reach of state-of-the-art solvers. Unfortunately, simple boolean expressions are not expressive enough for representing many real-world problems, which require handling also integer or real values and operators. On the other hand, mathematical solvers, like computer-algebra systems or constraint solvers, cannot handle efficiently problems involving heavy boolean search, or do not handle them at all. In this paper we present the foundations and the basic algorithms for a new class of procedures for solving boolean combinations of mathematical propositions, which combine boolean and mathematical solvers, and we highlight the main requirements that boolean and mathematical solvers must fulfill in order to achieve the maximum benefits from their integration. Finally we show how existing systems are captured by our framework.

1 Motivation and Goals

In the last years we have witnessed an impressive advance in the efficiency of boolean solving techniques (SAT), which has brought large previously intractable problems at the reach of state-of-the-art solvers.[1] As a consequence, some hard real-world problems have been successfully solved by encoding them into SAT. Propositional planning [KMS96] and boolean model-checking [BCCZ99] are among the best achievements.

---

* This work is sponsored by the CALCULEMUS! IHP-RTN EC project, contract code HPRN-CT-2000-00102, and has thus benefited of the financial contribution of the Commission through the IHP programme.

[1] SAT procedures are commonly called *solvers* in the SAT community, although the distinction between *solving*, *proving* and *computing* services may suggest to call them *provers*.

J. Calmet et al. (Eds.): AISC-Calculemus 2002, LNAI 2385, pp. 231–245, 2002.
© Springer-Verlag Berlin Heidelberg 2002

Unfortunately, simple boolean expressions are not expressive enough for representing many real-world problems. For example, problem domains like temporal reasoning, resource planning, verification of systems with numerical data or of timed systems, require handling also constraints on integer or real quantities (see, e.g., [ACG99,WW99,CABN97,MLAH01]). Moreover, some problem domains like model checking often require an explicit representation of integers and arithmetic operators, which cannot be represented efficiently by simple boolean expressions (see, e.g., [CABN97,MLAH01]). On the other hand, mathematical solvers, like computer-algebra systems or constraint solvers, cannot handle efficiently problems involving heavy boolean search, or do not handle them at all.

In 1996 we have proposed a new general approach to build domain-specific decision procedures on top of SAT solvers [GS96,GS00]. The basic idea was to decompose the search into two orthogonal components, one purely propositional component and one "boolean-free" domain-specific component and to use a (modified) Davis Putnam Longemann Loveland (DPLL) SAT solver [DLL62] for the former and a pure domain-specific procedure for the latter. So far the SAT based approach proved very effective in various problem domains like, e.g., modal and description logics [GS00], temporal reasoning [ACG99], resource planning [WW99].

In this paper we present the foundations and the basic algorithms for a new class of procedures for solving boolean combinations of mathematical propositions, which integrate SAT and mathematical solvers, and we highlight the main requirement SAT and mathematical solvers must fulfill in order to achieve the maximum benefits from their integration. The ultimate goal is to develop solvers able to handle complex problems like those hinted above.

The paper is structured as follows. In Section 2 we describe formally the problem we are addressing. In Section 3 we present the logic framework on which the procedures are based. In Section 4 we present a generalized search procedure which combine boolean and mathematical solvers and introduce some efficiency issues. In Section 5 we highlight the main requirements that boolean and mathematical solvers must fulfill in order to achieve the maximum benefits from their integration. In Section 6 we briefly describe some existing systems which are captured by our framework,and our own implemented procedure.

For lack of space, in this paper we omit the proofs of all the theoretical results presented, which can be found in [Seb01].

2   The Problem

We address the problem of checking the satisfiability of boolean combinations of primitive and mathematical propositions. Let $\mathcal{D}$ be the domain of either integer numbers $\mathbb{Z}$ or real numbers $\mathbb{R}$, with the respective set $\mathcal{OP}_{\mathcal{D}}$ of arithmetical operators $\{+, -, \cdot, /, mod\}$ or $\{+, -, \cdot, /\}$ respectively. Let $\{\bot, \top\}$ denote the *false* and *true* boolean values. Given the standard boolean connectives $\{\neg, \wedge\}$ and math operators $\{=, \neq, >, <, \geq, \leq\}$, let $\mathcal{A} = \{A_1, A_2, \ldots\}$ be a set of primitive propositions, let $\mathcal{C} = \{c_1, c_2, \ldots\}$ and $\mathcal{V} = \{v_1, v_2, \ldots\}$ respectively be a set of numerical constants in $\mathcal{D}$ and variables over the $\mathcal{D}$.

We call *Math-terms* the mathematical expressions built up from constants, variables and arithmetical operators over $\mathcal{D}$:

- a constant $c_i \in \mathcal{C}$ is a Math-term;
- a variable $v_i \in \mathcal{V}$ is a Math-term;
- if t_1 is a Math-term, then $-t_1$ is a Math-term;
- if t_1, t_2 are Math-terms, then $(t_1 \otimes t_2)$ is a Math-term, $\otimes \in \mathcal{OP}_\mathcal{D}$.

We call *Math-formulas* the mathematical formulas built on primitive propositions, Math-terms, operators and boolean connectives:

- a primitive proposition $A_i \in \mathcal{A}$ is a Math-formula;
- if t_1, t_2 are Math-terms, then $(t_1 \bowtie t_2)$ is a Math-formula, $\bowtie \in \{=, \neq, >, < , \geq, \leq\}$;
- if φ_1 is a Math-formula, then $\neg\varphi_1$ is a Math-formula;
- if φ_1, φ_2 are Math-formulas, then $(\varphi_1 \wedge \varphi_2)$ is a Math-formula.

For instance, $A_1 \wedge ((v_1 + 5.0) \leq (2.0 \cdot v_3))$ and $A_2 \wedge \neg(((2 \cdot v_1) \bmod v_2) > 5)$ are Math-formulas.[2]

Notationally, we use the lower case letters $t, t_1, \ldots$ to denote Math-terms, and the Greek letters $\alpha, \beta, \varphi, \psi$ to denote Math-formulas. We use the standard abbreviations, that is: "$\varphi_1 \vee \varphi_2$" for "$\neg(\neg\varphi_1 \wedge \neg\varphi_2)$", "$\varphi_1 \to \varphi_2$" for "$\neg(\varphi_1 \wedge \neg\varphi_2)$", "$\varphi_1 \leftrightarrow \varphi_2$" for "$\neg(\varphi_1 \wedge \neg\varphi_2) \wedge \neg(\varphi_2 \wedge \neg\varphi_1)$", "$\top$" for any valid formula, and "$\bot$" for "$\neg\top$". When this does not cause ambiguities, we use the associativity and precedence rules of arithmetical operators to simplify the appearance of Math-terms; e.g., we write "$(c_1(v_2 - v_1) - c_1v_3 + c_3v_4)$" instead of "$(((c_1 \cdot (v_2 - v_1)) - (c_1 \cdot v_3)) + (c_3 \cdot v_4))$".

We call *interpretation* a map $\mathcal{I}$ which assigns $\mathcal{D}$ values and boolean values to Math-terms and Math-formulas respectively and preserves constants and arithmetical operators:[3]

- $\mathcal{I}(A_i) \in \{\top, \bot\}$, for every $A_i \in \mathcal{A}$;
- $\mathcal{I}(c_i) = c_i$, for every $c_i \in \mathcal{C}$;
- $\mathcal{I}(v_i) \in \mathcal{D}$, for every $v_i \in \mathcal{V}$;
- $\mathcal{I}(t_1 \otimes t_2) = (\mathcal{I}(t_1) \otimes \mathcal{I}_\mathcal{D}(t_2))$, for all Math-terms t_1, t_2 and $\otimes \in \mathcal{OP}_\mathcal{D}$.

---

[2] The assumption that the domain is the whole $\mathbb{Z}$ or $\mathbb{R}$ is not restrictive, as we can restrict the domain of any variable v_i at will by adding to the formula some constraints like, e.g., $(v_1 \neq 0.0)$, $(v_1 \leq 5.0)$, etc.

[3] Here we make a little abuse of notation with the constants and the operators in $\mathcal{OP}_\mathcal{D}$. In fact, e.g., we denote by the same symbol "$+$" both the language symbol in $\mathcal{I}_\mathcal{D}(t_1 + t_2)$ and the arithmetic operator in $(\mathcal{I}_\mathcal{D}(t_1) + \mathcal{I}_\mathcal{D}(t_2))$. The same discourse holds for the constants $c_i \in \mathcal{C}$ and also for the operators $\{=, \neq, >, <, \geq, \leq\}$.

The binary relation $\models$ between a interpretation $\mathcal{I}$ and a Math-formula φ, written "$\mathcal{I} \models \varphi$" ("$\mathcal{I}$ *satisfies* φ" or "$\mathcal{I}$ satisfies φ") is defined as follows:

$$
\begin{aligned}
\mathcal{I} &\models A_i, \ A_i \in \mathcal{A} & &\Longleftrightarrow \mathcal{I}(A_i) = \top; \\
\mathcal{I} &\models (t_1 \bowtie t_2), \bowtie \ \in \{=, \neq, >, <, \geq, \leq\} & &\Longleftrightarrow \mathcal{I}(t_1) \bowtie \mathcal{I}(t_2); \\
\mathcal{I} &\models \neg\varphi_1 & &\Longleftrightarrow \mathcal{I} \not\models \varphi_1; \\
\mathcal{I} &\models (\varphi_1 \wedge \varphi_2) & &\Longleftrightarrow \mathcal{I} \models \varphi_1 \ and \ \mathcal{I} \models \varphi_2.
\end{aligned}
$$

We say that a Math-formula φ is *satisfiable* if and only if there exists an interpretation $\mathcal{I}$ such that $\mathcal{I} \models \varphi$. E.g., if $\mathcal{D} = \mathbb{R}$, then $A_1 \rightarrow ((v_1 + 2v_2) \leq 4.5)$ is satisfied by an interpretation $\mathcal{I}$ such that $\mathcal{I}(A_1) = \top$, $\mathcal{I}(v_1) = 1.1$, and $\mathcal{I}(v_2) = 0.6$. For every φ_1 and φ_2, we say that $\varphi_1 \models \varphi_2$ if and only if $\mathcal{I} \models \varphi_2$ for every $\mathcal{I}$ such that $\mathcal{I} \models \varphi_1$. We also say that $\models \varphi$ (φ is *valid*) if and only if $\mathcal{I} \models \varphi$ for every $\mathcal{I}$. It is easy to verify that $\varphi_1 \models \varphi_2$ if and only if $\models \varphi_1 \rightarrow \varphi_2$, and that $\models \varphi$ if and only if $\neg\varphi$ is unsatisfiable.

3   The Formal Framework

3.1   Basic Definitions and Results

Definition 1. *We call* **atom** *any Math-formula that cannot be decomposed propositionally, that is, any Math-formula whose main connective is not a boolean operator. A* **literal** *is either an atom (a* **positive** *literal) or its negation (a* **negative** *literal).*

Examples of literals are, A_1, $\neg A_2$, $(v_1 + 5.0 \leq 2.0 v_3)$, $\neg((2v_1 \ mod \ v_2) > 5)$. If l is a negative literal $\neg\psi$, then by "$\neg l$" we conventionally mean ψ rather than $\neg\neg\psi$. We denote by $Atoms(\varphi)$ the set of atoms in φ.

Definition 2. *We call a* **total truth assignment** μ *for a Math-formula* φ *a set*

$$
\mu = \{\alpha_1, \ldots, \alpha_N, \neg\beta_1, \ldots, \neg\beta_M, A_1, \ldots, A_R, \neg A_{R+1}, \ldots, \neg A_S\}, \tag{1}
$$

such that every atom in $Atoms(\varphi)$ *occurs as either a positive or a negative literal in* μ. *A* **partial truth assignment** μ *for* φ *is a subset of a total truth assignment for* φ. *If* $\mu_2 \subseteq \mu_1$, *then we say that* μ_1 **extends** μ_2 *and that* μ_2 **subsumes** μ_1.

A total truth assignment μ like (1) is interpreted as a truth value assignment to all the atoms of φ: $\alpha_i \in \mu$ means that α_i is assigned to $\top$, $\neg\beta_i \in \mu$ means that β_i is assigned to $\bot$. Syntactically identical instances of the same atom are always assigned identical truth values; syntactically different atoms, e.g., $(t_1 \geq t_2)$ and $(t_2 \leq t_1)$, are treated differently and may thus be assigned different truth values.

Notationally, we use the Greek letters μ, η to represent truth assignments. We often write a truth assignment μ as the conjunction of its elements. To this extent, we say that μ is satisfiable if the conjunction of its elements is satisfiable.

Definition 3. *We say that a total truth assignment μ for φ* **propositionally satisfies** *φ, written $\mu \models_p \varphi$, if and only if it makes φ evaluate to $\top$, that is, for all sub-formulas φ_1, φ_2 of φ:*

$$\mu \models_p \varphi_1, \ \varphi_1 \in Atoms(\varphi) \iff \varphi_1 \in \mu;$$
$$\mu \models_p \neg\varphi_1 \iff \mu \not\models_p \varphi_1;$$
$$\mu \models_p \varphi_1 \wedge \varphi_2 \iff \mu \models_p \varphi_1 \ and \ \mu \models_p \varphi_2.$$

We say that a partial truth assignment μ **propositionally satisfies** *φ if and only if all the total truth assignments for φ which extend μ propositionally satisfy φ.*

From now on, if not specified, when dealing with propositional satisfiability we do not distinguish between total and partial assignments.

We say that φ is *propositionally satisfiable* if and only if there exist an assignment μ such that $\mu \models_p \varphi$. Intuitively, if we consider a Math-formula φ as a propositional formula in its atoms, then $\models_p$ is the standard satisfiability in propositional logic. Thus, for every φ_1 and φ_2, we say that $\varphi_1 \models_p \varphi_2$ if and only if $\mu \models_p \varphi_2$ for every μ such that $\mu \models_p \varphi_1$. We also say that $\models_p \varphi$ (φ is *propositionally valid*) if and only if $\mu \models_p \varphi$ for every assignment μ for φ. It is easy to verify that $\varphi_1 \models_p \varphi_2$ if and only if $\models_p \varphi_1 \rightarrow \varphi_2$, and that $\models_p \varphi$ if and only if $\neg\varphi$ is propositionally unsatisfiable.

Notice that $\models_p$ is stronger than $\models$, that is, if $\varphi_1 \models_p \varphi_2$, then $\varphi_1 \models \varphi_2$, but not vice versa. E.g., $(v_1 \leq v_2) \wedge (v_2 \leq v_3) \models (v_1 \leq v_3)$, but $(v_1 \leq v_2) \wedge (v_2 \leq v_3) \not\models_p (v_1 \leq v_3)$.

Example 1. Consider the following math-formula φ:

$$\varphi = \{\neg\underline{(2v_2 - v_3 > 2)} \vee A_1\} \wedge$$
$$\{\neg A_2 \vee (2v_1 - 4v_5 > 3)\} \wedge$$
$$\{\underline{(3v_1 - 2v_2 \leq 3)} \vee A_2\} \wedge$$
$$\{\neg(2v_3 + v_4 \geq 5) \vee \neg\underline{(3v_1 - v_3 \leq 6)} \vee \neg A_1\} \wedge$$
$$\{A_1 \vee (3v_1 - 2v_2 \leq 3)\} \wedge$$
$$\{\underline{(v_1 - v_5 \leq 1)} \vee (v_5 = 5 - 3v_4) \vee \neg A_1\} \wedge$$
$$\{A_1 \vee \underline{(v_3 = 3v_5 + 4)} \vee A_2\}.$$

The truth assignment given by the underlined literals above is:

$$\mu = \{\neg(2v_2 - v_3 > 2), \neg A_2, (3v_1 - 2v_2 \leq 3), (v_1 - v_5 \leq 1), \neg(3v_1 - v_3 \leq 6), (v_3 = 3v_5 + 4)\}.$$

Notice that the two occurrences of $(3v_1 - 2v_2 \leq 3)$ in rows 3 and 5 of φ are both assigned $\top$. μ is an assignment which propositionally satisfies φ, as it sets to true one literal of every disjunction in φ. Notice that μ is not satisfiable, as both the following sub-assignments of μ

$$\{(3v_1 - 2v_2 \leq 3), \neg(2v_2 - v_3 > 2), \neg(3v_1 - v_3 \leq 6)\} \qquad (2)$$
$$\{(v_1 - v_5 \leq 1), (v_3 = 3v_5 + 4), \neg(3v_1 - v_3 \leq 6)\} \qquad (3)$$

do not have any satisfying interpretation. $\diamond$

Definition 4. *We say that a collection* $\mathcal{M} = \{\mu_1, \ldots, \mu_n\}$ *of (possibly partial) assignments propositionally satisfying* φ *is* **complete** *if and only if*

$$\models_p \varphi \leftrightarrow \bigvee_j \mu_j. \tag{4}$$

where each assignment μ_j *is written as a conjunction of its elements.*

$\mathcal{M}$ is complete in the sense that, for every total assignment η such that $\eta \models_p \varphi$, there exists $\mu_j \in \mathcal{M}$ such that $\mu_j \subseteq \eta$. Therefore $\mathcal{M}$ is a compact representation of the whole set of total assignments propositionally satisfying φ. Notice however that $\|\mathcal{M}\|$ is worst-case exponential in the size of φ, though typically much smaller than the set of all total assignments satisfying φ.

Definition 5. *We say that a complete collection* $\mathcal{M} = \{\mu_1, \ldots, \mu_n\}$ *of assignments propositionally satisfying* φ *is* **non-redundant** *if for every* $\mu_j \in \mathcal{M}$, $\mathcal{M} \setminus \{\mu_j\}$ *is no more complete, it is* **redundant** *otherwise.* $\mathcal{M}$ *is* **strongly non-redundant** *if, for every* $\mu_i, \mu_j \in \mathcal{M}$, $(\mu_1 \wedge \mu_2)$ *is propositionally unsatisfiable.*

It is easy to verify that, if $\mathcal{M}$ is redundant, then $\mu_i \subseteq \mu_j$ for some i, j, and that, if $\mathcal{M}$ is strongly non-redundant, then it is non-redundant too, but the vice versa does not hold.

Example 2. Let $\varphi := (\alpha \vee \beta \vee \gamma) \wedge (\alpha \vee \beta \vee \neg\gamma)$, α, β and γ being atoms. Then

1. $\{\{\alpha, \beta, \gamma\}, \{\alpha, \beta, \neg\gamma\}, \{\alpha, \neg\beta, \gamma\}, \{\alpha, \neg\beta, \neg\gamma\}, \{\neg\alpha, \beta, \gamma\}, \{\neg\alpha, \beta, \neg\gamma\}\}$ is the set of all total assignments propositionally satisfying φ;
2. $\{\{\alpha\}, \{\alpha, \beta\}, \{\alpha, \neg\gamma\}, \{\alpha, \beta\}, \{\beta\}, \{\beta, \neg\gamma\}, \{\alpha, \gamma\}, \{\beta, \gamma\}\}$ is complete but redundant;
3. $\{\{\alpha\}, \{\beta\}\}$ is complete, non redundant but not strongly non-redundant;
4. $\{\{\alpha\}, \{\neg\alpha, \beta\}\}$ is complete and strongly non-redundant. $\quad\diamond$

Theorem 1. *Let* φ *be a Math-formula and let* $\mathcal{M} = \{\mu_1, \ldots, \mu_n\}$ *be a complete collection of truth assignments propositionally satisfying* φ. *Then* φ *is satisfiable if and only if* μ_j *is satisfiable for some* $\mu_j \in \mathcal{M}$.

3.2 Decidability and Complexity

Having a Math-formula φ, it is always possible to find a complete collection of satisfying assignments for φ (see later). Thus from Theorem 1 we have trivially the following fact.

Proposition 1. *The satisfiability problem for a Math-formula over atoms of a given class is decidable if and only if the satisfiability of sets of literals of the same class is decidable.*

For instance, the satisfiability of a set of linear constraints on $\mathbb{R}$ or on $\mathbb{Z}$, or a set of non-linear constraints on $\mathbb{R}$ is decidable, whilst a set of non-linear (polynomial) constraints on $\mathbb{Z}$ is not decidable (see, e.g., [RV01]). Consequently, the satisfiability of Math-formulas over linear constraints on $\mathbb{R}$ or on $\mathbb{Z}$, or over non-linear constraints on $\mathbb{R}$ is decidable, whilst the satisfiability of Math-formulas over non-linear constraints over $\mathbb{Z}$ is undecidable.

For the decidable cases, as standard boolean formulas are a strict subcase of Math-formulas, it follows trivially that deciding the satisfiability of Math-formulas is "at least as hard" as boolean satisfiability.

Proposition 2. *The problem of deciding the satisfiability of a Math-formula φ is NP-hard.*

Thus, deciding satisfiability is computationally very expensive. The complexity upper bound may depend on the kind of mathematical problems we are dealing. For instance, if we are dealing with arithmetical expressions over bounded integers, then for every $\mathcal{I}$ we can verify $\mathcal{I} \models \varphi$ in a polynomial amount of time, and thus the problem is also NP-complete.

4   A Generalized Search Procedure

Theorem 1 allows us to split the notion of satisfiability of a Math-formula φ into two orthogonal components:

- a *purely boolean* component, consisting of the existence of a propositional model for φ;
- a *purely mathematical* component, consisting of the existence of an interpretation for a set of atomic (possibly negated) mathematical propositions.

These two aspects are handled, respectively, by a *truth assignment enumerator* and by a *mathematical solver*.

Definition 6. *We call a* **truth assignment enumerator** *a total function* ASSIGN_ENUMERATOR *which takes as input a Math-formula φ and returns a complete collection $\{\mu_1, \ldots, \mu_n\}$ of assignments satisfying φ.*

Notice the difference between a truth assignment enumerator and a standard boolean solver: a boolean solver has to find *only one* satisfying assignment —or to decide there is none— while an enumerator has to find a *complete collection* of satisfying assignments. (We will show later how some boolean solvers can be modified to be used as enumerators.)

We say that ASSIGN_ENUMERATOR is

- *strongly non-redundant* if ASSIGN_ENUMERATOR(φ) is strongly non-redundant, for every φ,
- *non-redundant* if ASSIGN_ENUMERATOR(φ) is non-redundant for every φ,
- *redundant* otherwise.

boolean MATH-SAT*(formula φ, assignment & μ, interpretation & $\mathcal{I}$)*
   do
       $\mu := Next(\text{ASSIGN\_ENUMERATOR}(\varphi))$ /* next in $\{\mu_1, ..., \mu_n\}$ */
       if $(\mu \neq Null)$
          $\mathcal{I} :=$ MATHSOLVER(μ);
   while $((\mathcal{I} = Null)$ **and** $(\mu \neq Null))$
   if $(\mathcal{I} \neq Null)$
   then return *True*;   /* a $\mathcal{D}$-satisfiable assignment found */
   else return *False*;   /* no $\mathcal{D}$-satisfiable assignment found */

Fig. 1. Schema of the generalized search procedure for $\mathcal{D}$-satisfiability.

Definition 7. *We call a* **mathematical solver** *a total function* MATHSOLVER *which takes as input a set of (possibly negated) atomic Math-formulas μ and returns an interpretation satisfying μ, or $Null$ if there is none.*

The general schema of a search procedure for satisfiability is reported in Figure 1. MATH-SAT takes as input a formula φ and (by reference) an initially empty assignment μ and an initially null interpretation $\mathcal{I}$. For every assignment μ in the collection $\{\mu_1, .., \mu_n\}$ generated by ASSIGN_ENUMERATOR(φ), MATH-SAT invokes MATHSOLVER over μ, which either returns a interpretation satisfying μ, or $Null$ if there is none. This is done until either one satisfiable assignment is found, or no more assignments are available in $\{\mu_1, ..., \mu_n\}$. In the former case φ is satisfiable, in the later case it is not.

MATH-SAT performs at most $||\mathcal{M}||$ loops. Thus, if every call to MATH-SOLVER terminates, then MATH-SAT terminates. Moreover, it follows from Theorem 1 that MATH-SAT is correct and complete if MATHSOLVER is correct and complete. Notice that, it is not necessary to check the whole set of total truth assignments satisfying φ, rather it is sufficient to check an arbitrary complete collection $\mathcal{M}$ of partial assignments propositionally satisfying φ, which is typically much smaller.

It is very important to notice that the search procedure schema of Figure 1 is completely independent on the kind of mathematical domain we are addressing, once we have a mathematical solver for it. This means that the expressivity of MATH-SAT, that is, the kind of math-formulas MATH-SAT can handle, depends only on the kind of sets of mathematical atomic propositions MATHSOLVER can handle.

4.1   Suitable ASSIGN_ENUMERATORs

The following are the most significant boolean reasoning techniques that we can adapt to be used as assignment enumerators.

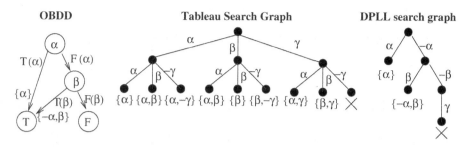

Fig. 2. OBDD, Tableau search graph and DPLL search graph for the formula $(\alpha \lor \beta \lor \gamma) \land (\alpha \lor \beta \lor \neg\gamma)$.

DNF. The simplest technique we can use as an enumerator is the *Disjunctive Normal Form (DNF)* conversion. A propositional formula φ can be converted into a formula $DNF(\varphi)$ by (i) recursively applying DeMorgan's rewriting rules to φ until the result is a disjunction of conjunction of literals, and (ii) removing all duplicated and subsumed disjuncts. The resulting formula is normal, in the sense that $DNF(\varphi)$ is propositionally equivalent to φ, and that propositionally equivalent formulas generate the same DNF modulo reordering.

By Definition 4, we can see (the set of disjuncts of) $DNF(\varphi)$ as a complete and non-redundant —but not strongly non-redundant— collection of assignments propositionally satisfying φ. For instance, in Example 2, the set of assignments at point 2. and 3. are respectively the results of step (i) and (ii) above.

OBDD. A more effective normal form for representing a boolean formula if given by the *Ordered Binary Decision Diagrams (OBDDs)* [Bry86], which are extensively used in hardware verification and model checking. Given a total ordering $v_1, ..., v_n$ on the atoms of φ, the OBDD representing φ $(OBDD(\varphi))$ is a directed acyclic graph such that (i) each node is either one of the two terminal nodes T, F, or an internal node labeled by an atom v of φ, with two outcoming edges $T(v)$ ("v is true") and $F(v)$ ("v is false"), (ii) each arc $v_i \to v_j$ is such that $v_i < v_j$ in the total order. If a node n labeled with v is the root of $OBDD(\phi)$ and n_1, n_2 are the two son nodes of n through the edges $T(v)$ and $F(v)$ respectively, then n_1, n_2 are the roots of $OBDD(\phi[v = \top])$ and $OBDD(\phi[v = \bot])$ respectively. A path from the root of $OBDD(\varphi)$ to T [resp. F] is a propositional model [resp. counter-model] of φ, and the disjunction of such paths is propositionally equivalent to φ [resp. $\neg\varphi$].

Thus, we can see $OBDD(\varphi)$ as a complete collection of assignments propositionally satisfying φ. As every pair of paths differ for the truth value of at least one variable, $OBDD(\varphi)$ is also strongly non-redundant. For instance, in Figure 2 (left) the OBDD of the formula in Example 2 is represented. The paths to T are those given by the set of assignments at point 4. of Example 2.

Semantic tableaux. A standard boolean solving technique is that of semantic tableaux [Smu68]. Given an input formula φ, in each branch of the search tree the set $\{\varphi\}$ is decomposed into a set of literals μ by the recursive application of the rules:

$$\frac{\mu' \cup \{\varphi_1, ..., \varphi_n\}}{\mu' \cup \{\bigwedge_{i=1}^{n} \varphi_i\}} (\wedge) \qquad \frac{\mu' \cup \{\varphi_1\} \quad .... \quad \mu' \cup \{\varphi_n\}}{\mu' \cup \{\bigvee_{i=1}^{n} \varphi_i\}} (\vee),$$

plus analogous rules for $(\rightarrow)$, $(\leftrightarrow)$, $(\neg\wedge)$, $(\neg\vee)$, $(\neg \rightarrow)$, $(\neg \leftrightarrow)$. The main steps are:

- (closed branch) if μ contains both φ_i and φ_i for some subformula φ_i of φ, then μ is said to be *closed* and cannot be decomposed any further;
- (solution branch) if μ contains only literals, then it is an assignment such that $\mu \models_p \varphi$;
- ($\wedge$-rule) if μ contains a conjunction, then the latter is unrolled into the set of its conjuncts;
- ($\vee$-rule) if μ contains a disjunction, then the search branches on one of the disjuncts.

The search tree resulting from the decomposition is such that all its solution branches are assignments in a collection $Tableau(\varphi)$, whose disjunction is propositionally equivalent to φ. Thus $Tableau(\varphi)$ is complete, but it may be redundant. For instance, in Figure 2 (center) the search tree of a semantic tableau applied on the formula in Example 2 is represented. The solutions branches give rise to the redundant collection of assignments at point 2. of Example 2.

DPLL. The most commonly used boolean solving procedure is DPLL [DLL62]. Given φ in input, DPLL tries to build recursively one assignment μ satisfying φ, at each step adding a new literal to μ and simplifying φ, according to the following steps:

- (base) If $\varphi = \top$, then μ propositionally satisfies the original formula, so that μ is returned;
- (backtrack) if $\varphi = \bot$, μ propositionally falsifies the original formula, so that DPLL backtracks;
- (unit propagation) if a literal l occurs in φ as a unit clause, then DPLL is invoked recursively on $\varphi_{l=\top}$ and $\mu \cup \{l\}$, $\varphi_{l=\top}$ being the result of substituting $\top$ for l in φ and simplifying;
- (split) otherwise, a literal l is selected, and DPLL is invoked recursively on $\varphi_{l=\top}$ and $\mu \cup \{l\}$. If this call succeeds, then the result is returned, otherwise DPLL is invoked recursively on $\varphi_{l=\bot}$ and $\mu \cup \{\neg l\}$.

Standard DPLL can be adapted to work as a boolean enumerator by simply modifying in the base case "μ is returned" with "μ is added to the collection, and backtrack" [GS96,Seb01].

The resulting set of assignments $DPLL(\varphi)$ is complete and strongly non-redundant [Seb01]. For instance, in Figure 2 (right) the search tree of DPLL applied on the formula in Example 2 is represented. The non closed branches give rise to the set of assignments at point 4. of Example 2.

Notice the difference between an OBDD and (the search tree of) DPLL: first, the former is a direct acyclic graph whilst the second is a tree; second, in OBDDs the order of branching variables if fixed a priori, while DPLL can choose each time the best variable to split.

4.2 Non-suitable ASSIGN_ENUMERATORS

It is very important to notice that, in general, not every boolean solver can be adapted to work as a boolean enumerator. For instance, many implementations of DPLL include also the following step between unit propagation and split:

– (pure literal) if an atom ψ occurs only positively [resp. negatively] in φ, then DPLL is invoked recursively on $\varphi_{\psi=\top}$ and $\mu \cup \{\psi\}$ [resp. $\varphi_{\psi=\bot}$ and $\mu \cup \{\neg\psi\}$];

(we call this variant DPLL+PL). DPLL+PL is complete as a boolean solver, but does not generate a complete collection of assignments, so that it cannot be used as an enumerator.

Example 3. If we used DPLL+PL as ASSIGN_ENUMERATOR in MATH-SAT and gave in input the formula in Example 2, DPLL+PL might return the one-element collection $\{\{\alpha\}\}$, which is not complete. If α is $(x^2 + 1 \leq 0)$ and β is $(y \leq x)$, $x, y \in \mathbb{R}$, then $\{\alpha\}$ is not satisfiable, so that MATH-SAT would return unsatisfiable. On the other hand, the formula φ is satisfiable because, e.g., the assignment $\{\neg\alpha, \beta\}$ is satisfied by $\mathcal{I}(x) = 1.0$ and $\mathcal{I}(y) = 0.0$.

5 Requirements for ASSIGN_ENUMERATOR and MATHSOLVER

Apart from the efficiency of MATHSOLVER —which varies with the kind of problem addressed and with the technique adopted, and will not be discussed here— and that of the SAT solver used —which does not necessarily imply its efficiency as an enumerator— many other factors influence the efficiency of MATH-SAT.

5.1 Efficiency Requirements for ASSIGN_ENUMERATOR

Polynomial vs. exponential space in ASSIGN_ENUMERATOR. We would rather MATH-SAT require polynomial space. As $\mathcal{M}$ can be exponentially big with respect to the size of φ, we would rather adopt a generate-check-and-drop paradigm: at each step, generate the next assignment $\mu_i \in \mathcal{M}$, check its satisfiability, and then drop it —or drop the part of it which is not common to the next assignment— before passing to the $i + 1$-step. This means that AS-SIGN_ENUMERATOR must be able to generate the assignments one at a time.

To this extent, both DNF and OBDD are not suitable, as they force generating the whole assignment collection $\mathcal{M}$ one-shot. Instead, both semantic tableaux and DPLL are a good choice, as their depth-first search strategy allows for generating and checking one assignment at a time.

Non-redundancy of ASSIGN_ENUMERATOR. We want to reduce as much as possible the number of assignments generated and checked. To do this, a key issue is avoiding MATHSOLVER being invoked on an assignment which either is identical to an already-checked one or extends one which has been already found unsatisfiable. This is obtained by using a non-redundant enumerator. To this extent, semantic tableaux are not a good choice.

Non-redundant enumerators avoid generating partial assignments whose unsatisfiability is a propositional consequence of those already generated. If $\mathcal{M}$ is strongly non-redundant, however, each total assignment η propositionally satisfying φ is represented by one and only one $\mu_j \in \mathcal{M}$, and every $\mu_j \in \mathcal{M}$ represents univocally $2^{|Atoms(\varphi)|-|\mu_j|}$ total assignments. Thus strongly non-redundant enumerators also avoid generating partial assignments covering areas of the search space which are covered by already-generated ones.

For enumerators that are not strongly non-redundant, there is a tradeoff between redundancy and polynomial memory. In fact, when adopting a generate-check-and-drop paradigm, the algorithm has no way to remember if it has already checked a given assignment or not, unless it explicitly keeps track of it, which requires up to exponential memory. Strong non-redundancy instead provides a *logical* warrant that an already checked assignment will never be checked again.

5.2 Exploiting the Interaction between ASSIGN_ENUMERATOR and MATHSOLVER

Intermediate assignment checking. If an assignment μ' is unsatisfiable, then all its extensions are unsatisfiable. Thus, when the unsatisfiability of μ' is detected during its recursive construction, this prevents checking the satisfiability of all the up to $2^{|Atoms(\varphi)|-|\mu'|}$ truth assignments which extend μ'. Thus, another key issue for efficiency is the possibility of modifying ASSIGN_ENUMERATOR so that it can perform intermediate calls to MATHSOLVER and it can take advantage of the (un)satisfiability information returned to prune the search space.

With semantic tableaux and DPLL, this can be easily obtained by introducing an intermediate test, immediately before the ($\vee$-rule) and the (split) step respectively, in which MATHSOLVER is invoked on an intermediate assignment μ' and, if it is inconsistent, the whole branch is cut [GS96,ABC+02]. With OBDDs, it is possible to reduce an existing OBDD by traversing it depth-first and redirecting to the F node the paths representing inconsistent assignments [CABN97]. However, this requires generating the non-reduced OBDD anyway.

Generating and handling conflict sets. Given an unsatisfiable assignment μ, we call a *conflict set* any unsatisfiable sub-assignment $\mu' \subset \mu$. (E.g., in Example 1 (2) and (3) are conflict sets for the assignment μ.) A key efficiency issue for MATH-SAT is the capability of MATHSOLVER to return the conflict set which has caused the inconsistency of an input assignment, and the capability of ASSIGN_ENUMERATOR to use this information to prune search.

For instance, both Belman-Ford algorithm and Simplex LP procedures can produce conflict sets [ABC+02,WW99]. Semantic tableaux and DPLL can be enhanced by a technique called *mathematical backjumping* [HPS98, WW99,ABC⁺02]: when MATHSOLVER(μ) returns a conflict set η, AS-SIGN_ENUMERATOR can jump back in its search to the deepest branching point in which a literal $l \in \eta$ is assigned a truth value, pruning the search tree below. DPLL can be enhanced also with *learning* [WW99,ABC⁺02]: the negation of the conflict set $\neg\eta$ is added in conjunction to the input formula, so that DPLL will never again generate an assignment containing the conflict set η.

Generating and handling derived assignments. Another efficiency issue for MATH-SAT is the capability of MATHSOLVER to produce an extra assignment η derived deterministically from a satisfiable input assignment μ, and the capability of ASSIGN_ENUMERATOR to use this information to narrow the search.

For instance, in the procedure presented in [ABC⁺02,ACKS02], MATH-SOLVER computes equivalence classes of real variables and performs substitutions which can produce further assignments. E.g., if $(v_1 = v_2), (v_2 = v_3) \in \mu$, $(v_1 - v_3 > 2) \notin \mu$ and μ is satisfiable, then MATHSOLVER(μ) finds that v_1 and v_3 belong to the same equivalence class and returns an extra assignment η containing $\neg(v_1 - v_3 > 2)$, which is unit-propagated away by DPLL.

Incrementality of MATHSOLVER. Another efficiency issue of MATHSOLVER is that of being *incremental*, so that to avoid restarting computation from scratch whenever it is given in input an assignment μ' such that $\mu' \supset \mu$ and μ has already proved satisfiable. (This happens, e.g., at the intermediate assignments checking steps.) Thus, MATHSOLVER should "remember" the status of the computation from one call to the other, whilst ASSIGN_ENUMERATOR should be able to keep track of the computation status of MATHSOLVER.

For instance, it is possible to modify a Simplex LP procedure so that to make it incremental, and to make DPLL call it incrementally after every unit propagation [WW99].

6   Implemented Systems

In order to provide evidence of the generality of our approach, in this section we briefly present some examples. First we enumerate some existing procedures which are captured by our framework. Then we present a brief description of our own solver MATH-SAT.

6.1   Examples

Our framework captures a significant amount of existing procedure used in various application domains. We briefly recall some of them.

Omega [Pug92] is a procedure used for dependence analysis of software. It is an integer programming algorithm based on an extension of Fourier-Motzkin variable elimination method. It handles boolean combinations of linear constraints by simply pre-computing the DNF of the input formula.

TSAT [ACG99] is an optimized procedure for temporal reasoning able to handle sets of disjunctive temporal constraints. It integrates DPLL with a simplex LP tool, adding some form of forward checking and static learning.

LPSAT [WW99] is an optimized procedure for Math-formulas over linear real constraints, used to solve problems in the domain of resource planning. It accept only formulas with positive mathematical constraints. LPSAT integrates DPLL with an incremental simplex LP tool, and performs backjumping and learning.

SMV+QUAD-CLP [CABN97] integrates OBDDs with a quadratic constraint solver to verify transition systems with integer data values. It performs a form of intermediate assignment checking.

DDD [MLAH01] are OBDD-like data structures handling boolean combinations of temporal constraints in the form $(x - z \leq 3)$, which are used to verify timed systems. They combine OBDDs with an incremental version of Belman-Ford minimal path and cycle detection algorithm.

Unfortunately, the last two approaches inherit from OBDDs the drawback of requiring exponential space in worst case.

6.2   A DPLL-Based Implementation of MATH-SAT

In [ABC$^+$02,ACKS02] we presented MATH-SAT, a decision procedure for Math-formulas over boolean and linear mathematical propositions over the reals. MATH-SAT uses as ASSIGN_ENUMERATOR an implementation of DPLL, and as MATHSOLVER a hierarchical set of mathematical procedures for linear constraints on real variables able to handle theories of increasing expressive power. The latter include a procedure for computing and exploiting equivalence classes from equality constraints like $(x = y)$, a Bellman-Ford minimal path algorithm with cycle detection for handling differences like $(x - y \leq 4)$, and a Simplex LP procedure for handling the remaining linear constraints. MATH-SAT implements and uses most of the tricks and optimizations described in Section 5. Technical details can be found in [ABC$^+$02]. MATH-SAT is available at http://www.dit.unitn.it/~rseba/Mathsat.html.

In [ABC$^+$02,ACKS02] preliminary experimental evaluations were carried out on tests arising from temporal reasoning [ACG99] and formal verification of timed systems [ACKS02]. In the first class of problems, we have compared our results with the results of the specialized procedure TSAT; although MATH-SAT is able to tackle a wider class of problems, it runs faster that the TSAT solver, which is specialized to the problem class. In the second class, we have encoded bounded model checking problems for timed systems into the satisfiability of Math-formulas, and run MATH-SAT on them. It turned out that our approach was comparable in efficiency with two well-established model checkers for timed systems, and significantly more expressive [ACKS02].

References

[ABC⁺02]   G. Audemard, P. Bertoli, A. Cimatti, A. Kornilowicz, and R. Sebastiani. A SAT Based Approach for Solving Formulas over Boolean and Linear Mathematical Propositions. In *Proc. CADE'2002.*, 2002. To appear. Available at http://www.dit.unitn.it/~rseba/publist.html.

[ACG99]   A. Armando, C. Castellini, and E. Giunchiglia. SAT-based procedures for temporal reasoning. In *Proc. European Conference on Planning, CP-99*, 1999.

[ACKS02]   G. Audemard, A. Cimatti, A. Kornilowicz, and R. Sebastiani. SAT-Based Bounded Model Checking for Timed Systems. 2002. Available at http://www.dit.unitn.it/~rseba/publist.html.

[BCCZ99]   A. Biere, A. Cimatti, E. Clarke, and Y. Zhu. Symbolic model checking without BDDs. In *Proc. CAV'99*, 1999.

[Bry86]   R. E. Bryant. Graph-Based Algorithms for Boolean Function Manipulation. *IEEE Transactions on Computers*, C-35(8):677–691, August 1986.

[CABN97]   W. Chan, R. J. Anderson, P. Beame, and D. Notkin. Combining constraint solving and symbolic model checking for a class of systems with non-linear constraints. In *Proc. CAV'97*, volume 1254 of *LNCS*, pages 316–327, Haifa, Israel, June 1997. Springer-Verlag.

[DLL62]   M. Davis, G. Longemann, and D. Loveland. A machine program for theorem proving. *Journal of the ACM*, 5(7), 1962.

[GS96]   F. Giunchiglia and R. Sebastiani. Building decision procedures for modal logics from propositional decision procedures - the case study of modal K. In *Proc. of the 13th Conference on Automated Deduction*, LNAI, New Brunswick, NJ, USA, August 1996. Springer Verlag.

[GS00]   F. Giunchiglia and R. Sebastiani. Building decision procedures for modal logics from propositional decision procedures - the case study of modal K(m). *Information and Computation*, 162(1/2), October/November 2000.

[HPS98]   I. Horrocks and P. F. Patel-Schneider. FaCT and DLP. In *Procs. Tableaux'98*, number 1397 in LNAI, pages 27–30. Springer-Verlag, 1998.

[KMS96]   H. Kautz, D. McAllester, and Bart Selman. Encoding Plans in Propositional Logic. In *Proc. KR'96*, 1996.

[MLAH01]   J. Moeller, J. Lichtenberg, H. Andersen, and H. Hulgaard. Fully Symbolic Model Checking of Timed Systems using Difference Decision Diagrams. In *Electronic Notes in Theoretical Computer Science*, volume 23. Elsevier Science, 2001.

[Pug92]   W. Pugh. The Omega Test: a fast and practical integer programming algoprithm for dependence analysis. *Communication of the ACM*, August 1992.

[RV01]   A. Robinson and A. Voronkov, editors. *Handbook of Automated Reasoning.* Elsevier Science Publishers, 2001.

[Seb01]   R. Sebastiani. Integrating SAT Solvers with Math Reasoners: Foundations and Basic Algorithms. Technical Report 0111-22, ITC-IRST, November 2001. Available at http://www.dit.unitn.it/~rseba/publist.html.

[Smu68]   R. M. Smullyan. *First-Order Logic.* Springer-Verlag, NY, 1968.

[WW99]   S. Wolfman and D. Weld. The LPSAT Engine & its Application to Resource Planning. In *Proc. IJCAI*, 1999.

The Meaning of Infinity in Calculus and Computer Algebra Systems

Michael Beeson[1] and Freek Wiedijk[2]

[1] San José State University
[2] University of Nijmegen

Abstract. We use filters of open sets to provide a semantics justifying the use of infinity in informal limit calculations in calculus, and in the same kind of calculations in computer algebra. We compare the behavior of these filters to the way Mathematica behaves when calculating with infinity.
We stress the need to have a proper semantics for computer algebra expressions, especially if one wants to use results and methods from computer algebra in theorem provers. The computer algebra method under discussion in this paper is the use of rewrite rules to evaluate limits involving infinity.

1 Introduction

1.1 Problem

In calculus, when calculating limits, one often first uses the heuristic of 'calculating with infinity' before trying to evaluate the limit in a more formal way. For instance one 'calculates':

$$\lim_{x \to \infty} \frac{1}{x+1} = \frac{1}{\infty+1} = \frac{1}{\infty} = 0$$

which indeed gives the correct answer. However, it is not clear what the *meaning* of this use of the symbol '∞' is, and why this method works. This problem arises in calculus textbooks, which usually avoid examples of such calculations for fear of 'lack of rigor', although students are taught these methods at the blackboard. It arose in the design of the first author's software, MathXpert [1,2, 3]. This software, which is designed to assist a student in producing step-by-step solutions to calculus problems, had to be able to produce 'ideal' step-by-step solutions of limit problems. Are such 'ideal solutions' allowed to use calculations involving infinity? Or are those calculations just private preliminary considerations intended to guide a rigorous proof? MathXpert does allow calculations involving infinity, but not the full system justified in this paper, since that goes beyond what one finds in calculus textbooks.

In the Mathematica system [9] the approach of calculating with infinity is used. Since Mathematica gives answers, rather than step-by-step solutions, one will not notice the calculations with infinity, in cases where the limit turns out to

J. Calmet et al. (Eds.): AISC-Calculemus 2002, LNAI 2385, pp. 246–258, 2002.

exist (and be a finite number). But in fact, in Mathematica there is a complete 'calculus of infinity' (and some related symbols):

```
In[1] := 1/(Infinity + 1)
Out[1]= 0
In[2] := Sqrt[Infinity]
Out[2]= Infinity
In[3] := Infinity - Infinity
Infinity::indet:
   Indeterminate expression (-Infinity) + (Infinity) encountered.
Out[3]= Indeterminate
In[4] := Indeterminate + Infinity
Out[4]= Indeterminate
In[5] := Sin[Infinity]
Out[5]= Interval[{-1, 1}]
In[6] := 1/Interval[{-1, 1}]
Out[6]= Interval[{-Infinity, -1}, {1, Infinity}]
In[7] := Interval[{-1, 1}]*Interval[{-1, 1}]
Out[7]= Interval[{-1, 1}]
In[8] := Interval[{-1, 1}]^2
Out[8]= Interval[{0, 1}]
In[9] := 0*Sin[Infinity]
Out[9]= Interval[{0, 0}]
In[10] := Infinity/Sin[Infinity]
Out[10]= Interval[{-Infinity, -Infinity}, {Infinity, Infinity}]
In[11] := Infinity/Sin[Infinity]^2
Out[11]= Interval[{Infinity, Infinity}]
```

Other computer algebra systems implement similar calculi. For instance, the Maple system [6] uses the symbols infinity and undefined in answers to limit problems.[1]

It is well known that many computer algebra packages make errors. One of the reasons for that is that they fail to check the pre-conditions or 'side conditions' that must be satisfied for a simplification rule to be applicable. For example, before applying $\sqrt{x^2} = x$ we need to check that $x \geq 0$. Systematically keeping track of such assumptions is difficult. The errors in computer algebra systems sometimes give the impression that those systems place a higher priority

---

[1] There is also some notion of interval in Maple, written as 1 .. 2, but our attempts to calculate with these terms led only to error messages. These terms seem primarily to be used for generating integer sequences, although the answer to $\lim_{x \to \infty} \sin x$ comes out as -1 .. 1.

on performing as many simplifications as possible than on ensuring that only correct computations are performed. Generally, 'evaluation errors' which users complain about are taken care of on an ad hoc basis only, to get rid of the most embarrassing ones.

Related to these errors is the fact that these systems have no unified semantics for their expression language. In this paper we focus on the apparatus for limits and offer a solution: a semantics explaining and supporting the use of infinities in limit calculations. We will present a formal semantics of limits, which not only explains the calculations usually performed with infinities, but offers some extensions by introducing some other symbols for common ways in which a function can fail to have a limit. Thus, we will be able to get an answer by calculation for such a limit as $\lim_{x \to \infty} 1/(2 + \sin x)$ which will be 'oscillations through the interval $[\frac{1}{3}, 1]$'. We then compare the resulting semantics to the behavior of Mathematica as illustrated above. There is a rough general correspondence, and our semantics agrees with some of the examples above, but in some instances Mathematica does give incorrect answers, and in some cases we are able to distinguish between identical Mathematica expressions which are different in our semantics.

1.2   Approach

We will represent ∞ and its cousins indeterminate and interval by *filters* over some underlying topological space (which in calculus textbooks and Mathematica is the space of real numbers, but could also be the complex numbers or more general spaces). For each point of the space there will be a filter associated with it, which is called the principal filter of the point. For each function on the space there will be a *lifted* version that works on the filters instead of on the points.

Furthermore we will define classes of filters called the *interval* filters and the *connected* filters. It will turn out that those two classes coincide and that connectedness of filters is preserved under continuous mappings. Also we will define the *join* and the *meet* of two filters.

It turns out that the calculus used in Mathematica corresponds directly to the set of finite joins of interval filters.

1.3   Related Work

First, in topology, the two standard approaches for defining limits in topological spaces make use of *nets* or *filters*. There is therefore nothing original in the use of filters to analyze the notion of limits. However, our focus to use them in an applied setting, and identify specific filters associated with 'extra-mathematical' symbols such as ∞, seems to be new.

Second, the interval filters are directly related to the active field of *interval arithmetic*. We throw a new light on the calculations with intervals by looking at them as filters.

Third, justifying 'calculations with infinite objects' rigorously, is close to doing the same with 'calculations with infinitesimal objects', which is the domain of *nonstandard analysis*. In nonstandard analysis one also has infinity as a first class citizen. This relation is even more manifest when noting that the simplest way to get non-standard objects is as *ultrafilters*, a special kind of filter. The difference between filters and ultrafilters indicates what the difference between our approach and the non-standard one is. In nonstandard analysis there is not *one*, designated, infinity; instead there are many infinite nonstandard numbers, without a 'canonical' one. In our case there *is* a canonical infinity. To illustrate this difference concretely, let ω be the infinity of nonstandard analysis. Then we have $\omega + 1 \neq \omega$, but $\infty + 1 = \infty$. Another difference between the ultrafilters of nonstandard analysis and the filters that we study here is that our filters only contain open sets instead of arbitrary sets. Nonstandard analysis has been used in [4] to help in the computation of limits in a computer algebra system.

2 Filters, Lifting, Refinement, and Limits

Definition 1. *Let X be a topological space. Denote the open sets of X by $\mathcal{O}(X)$. A filter on X is a set $A \subseteq \mathcal{O}(X)$ that satisfies:*

$$\forall U \in A. \forall V \in \mathcal{O}(X). U \subseteq V \Rightarrow V \in A$$
$$\forall U \in A. \forall V \in A. U \cap V \in A$$

In words: a filter is a set of open sets that is closed under supersets and finite intersections. The collection of filters on X is written $\bar{X}$.

A filter that does not contain the empty set is called proper. *A filter that does not contain any set at all is called* empty.

Often the property of being proper is made part of the definition of a filter. However we did not do this, because otherwise we would be unable to define the notion of *meet* on page 254 below. Sometimes the property of being non-empty is made part of the definition of a filter too. However the empty filter, which is called domain-error below, is essential to our application. We found variants of the definition of filter in the literature, both allowing for improper [5] and for empty [7] filters. Therefore we feel free to define the concept of filter to suit our purposes.

In the topological literature a filter is generally not defined on a topological space but on an arbitrary set. In that case the restriction to open sets is not present. However, for our application it is more natural to restrict ourselves to filters of open sets.

Definition 2. *Here are some common filters on the real numbers, where $a \in \mathbb{R}$ is an arbitrary real number:*

$$\text{improper} \equiv \dagger \equiv \mathcal{O}(\mathbb{R})$$
$$\text{domain-error} \equiv \bot \equiv \emptyset$$
$$\text{indeterminate} \equiv \leftrightarrow \equiv \{\mathbb{R}\}$$
$$\text{principal}(a) \equiv \bar{a} \equiv \{U \in \mathcal{O}(\mathbb{R}) \mid a \in U\}$$

$$= \{U \in \mathcal{O}(\mathbb{R}) \mid \exists \varepsilon \in \mathbb{R}_{>0}. (a - \varepsilon, a + \varepsilon) \subseteq U\}$$

$$\mathsf{left}(a) \equiv a^- \equiv \{U \in \mathcal{O}(\mathbb{R}) \mid \exists \varepsilon \in \mathbb{R}_{>0}. (a - \varepsilon, a) \subseteq U\}$$

$$\mathsf{right}(a) \equiv a^+ \equiv \{U \in \mathcal{O}(\mathbb{R}) \mid \exists \varepsilon \in \mathbb{R}_{>0}. (a, a + \varepsilon) \subseteq U\}$$

$$\mathsf{punctured}(a) \equiv a^\pm \equiv \{U \in \mathcal{O}(\mathbb{R}) \mid \exists \varepsilon \in \mathbb{R}_{>0}. (a - \varepsilon, a) \cup (a, a + \varepsilon) \subseteq U\}$$

$$\mathsf{infinity} \equiv \infty \equiv \{U \in \mathcal{O}(\mathbb{R}) \mid \exists \varepsilon \in \mathbb{R}_{>0}. (1/\varepsilon, \infty) \subseteq U\}$$

$$\mathsf{minus\text{-}infinity} \equiv -\infty \equiv \{U \in \mathcal{O}(\mathbb{R}) \mid \exists \varepsilon \in \mathbb{R}_{>0}. (-\infty, -1/\varepsilon) \subseteq U\}$$

$$\mathsf{bi\text{-}infinity} \equiv \pm\infty \equiv \{U \in \mathcal{O}(\mathbb{R}) \mid \exists \varepsilon \in \mathbb{R}_{>0}. (-\infty, -1/\varepsilon) \cup (1/\varepsilon, \infty) \subseteq U\}$$

$$\mathsf{positive} \equiv \to \equiv \{U \in \mathcal{O}(\mathbb{R}) \mid (0, \infty) \subseteq U\}$$

$$\mathsf{negative} \equiv \leftarrow \equiv \{U \in \mathcal{O}(\mathbb{R}) \mid (-\infty, 0) \subseteq U\}$$

$$\mathsf{non\text{-}zero} \equiv \pm\to \equiv \{U \in \mathcal{O}(\mathbb{R}) \mid (-\infty, 0) \cup (0, \infty) \subseteq U\}$$

For each of these filters we have a 'long' and a 'short' notation. The first four filters can be defined for any topological space. The other filters have analogues in any order topology.

Definition 3. *Let again X be a topological space. Let A be a collection of subsets of X (not necessarily open) that satisfies:*

$$\forall U \in A. \forall V \in A. \exists W \in A. W \subseteq U \cap V \tag{$*$}$$

The filter generated by A *is defined to be:*

$$\mathsf{generated\text{-}by}(A) \equiv \{U \in \mathcal{O}(X) \mid \exists V \in A. V \subseteq U\}$$

The collection of sets A is called the basis *of the filter* $\mathsf{generated\text{-}by}(A)$.

Being closed under finite intersections implies $(*)$. If all elements of A are open sets, the filter generated by A is the intersection of all filters that contain A as a subset.

The filters given in Definition 2 can be defined more naturally using the notion of a generated filter. For instance, we have:

$$\mathsf{improper} = \mathsf{generated\text{-}by}(\{\emptyset\})$$
$$\mathsf{principal}(a) = \mathsf{generated\text{-}by}(\{\{a\}\})$$
$$\mathsf{right}(a) = \mathsf{generated\text{-}by}(\{(a, a + \varepsilon) \mid \varepsilon \in \mathbb{R}_{>0}\})$$
$$\mathsf{infinity} = \mathsf{generated\text{-}by}(\{(1/\varepsilon, \infty) \mid \varepsilon \in \mathbb{R}_{>0}\})$$

All other filters from Definition 2 can be defined in a similar way.

Definition 4. *Let $f : X \to X$ be some (possibly partial) function with domain $\mathrm{dom}(f)$. The* lift *of f is a function $\bar{f} : \bar{X} \to \bar{X}$, defined by:*

$$\bar{f}(A) \equiv \mathsf{generated\text{-}by}(\{f[U] \mid U \subseteq \mathrm{dom}(f) \wedge U \in A\})$$

This definition can be generalized to arbitrary arities. The function $\bar{f} : \bar{X} \times \bar{X} \times \ldots \times \bar{X} \to \bar{X}$ is defined by:

$$\bar{f}(A_1, A_2, \ldots, A_n) \equiv$$
$$\mathsf{generated\text{-}by}(\{f[U] \mid U \subseteq \mathrm{dom}(f) \wedge U = U_1 \times U_2 \times \ldots \times U_n \wedge$$
$$U_1 \in A_1 \wedge U_2 \in A_2 \wedge \ldots \wedge U_n \in A_n\})$$

Although f can be a partial function, the lift of f is always total. One can get rid of the problems of partial functions in calculus by lifting the whole theory to filters. In some sense by going to filters we are adding a 'bottom element' $\perp$ to the values of the theory. Looked at in this way, we have a *strict* partial logic, because a function applied to $\perp$ will always give $\perp$ again.

Note also that the definitions of $\bar{a}$ as a principal filter and as lift of a 0-ary constant function coincide. This justifies using one notation for both.

From now on we will often write f instead of $\bar{f}$ when one or more of the arguments of f are filters. So we will write $\sin(A)$ instead of $\overline{\sin}(A)$. This will allow us to write things like $\sqrt{A}$, and mean the lift of the square root function.

To state this convention more precisely: if $t[x_1, \ldots, x_n]$ is a term that does not involve filters (so $x_1, \ldots, x_n$ are variables ranging over the ordinary reals) then $t[A_1, \ldots, A_n]$ will mean the lift of the function $\lambda x_1 \cdots x_n . t[x_1, \ldots, x_n]$ applied to the filters $A_1, \ldots, A_n$. Note that with this convention $1/A$ means something different from $\bar{1}/A$. The first is the lift of the unary function $\lambda x . 1/x$ applied to A. The second is the lift of the binary function $\lambda x\, y . x/y$ applied to $\bar{1}$ and A. Those are not necessarily equal: $1/1^+ = 1^-$ but $\bar{1}/1^+ = \bar{1}$.

Definition 5. *The* filter limit *of the function* $f : X \to X$ *when taking the limit to the filter* A *is defined to be:*

$$\operatorname*{Lim}_{x \to A} f(x) \equiv \bar{f}(A)$$

We distinguish a filter limit from an ordinary limit by writing 'Lim' with a capital letter L. Note that the filter limit is always defined, even for non-continuous f. It might seem silly to introduce a new notation for this when we already have defined lifting, as it is the same operation. However, now we can write:

$$\operatorname*{Lim}_{x \to 0^+} x/x$$

which is something different from

$$0^+/0^+$$

The first is the lift of the unary function $\lambda x . x/x$ applied to 0^+ and has as value $\bar{1}$. The second is the lift of the binary function $\lambda x\, y . x/y$ applied to the pair $(0^+, 0^+)$ and has as value $\to$.

Definition 6. *A filter* A refines *a filter* B, *notation* $A \sqsubseteq B$ *when* $A \supseteq B$ *as collections of open sets. When the two filters* A *and* B *differ we write* $A \sqsubset B$.

Here are some refinement relations between the filters defined in Definition 2. For any proper and non-empty filter A we have:

$$\dagger \sqsubset A \sqsubseteq \leftrightarrow \sqsubset \perp$$

At any real number $a \in \mathbb{R}$ we have:

$$a^-, a^+ \sqsubset a^\pm \sqsubset \bar{a}$$

and the 'infinite' filters are related by:

$$-\infty, \infty \sqsubset \pm\infty \sqsubset \leftrightarrow, \quad -\infty \sqsubset \leftarrow \sqsubset \pm\rightarrow \sqsubset \leftrightarrow, \quad \infty \sqsubset \rightarrow \sqsubset \pm\rightarrow \sqsubset \leftrightarrow$$

Note that the filters from Definition 2 are not the only ones. There are many 'wild' filters refining $\bar{a}$ and ∞. For instance the filter generated by the sets $\{2\pi n \mid n > N\}$ is a filter which refines ∞. It has the property that the filter limit of sin to this filter is $\bar{0}$.

We can now state the first theorem[2], which lists some of the many calculation rules that one needs for arithmetic on filters:

Theorem 1. *Let* $a \in \mathbb{R}_{>0}$ *be some positive real number. Then:*

$\infty + \bar{a} = \infty$	$\bar{a}/\bar{0} = \bot$	$\bar{a}/\infty = 0^+$	$\bar{0}\,\infty = \leftrightarrow$
$\infty - \bar{a} = \infty$	$\bar{a}/0^+ = \infty$	$\bar{a}/\pm\infty = 0^\pm$	$0^+\infty = \rightarrow$
$\infty + \infty = \infty$	$\bar{a}/0^\pm = \pm\infty$	$\bar{a}/\rightarrow = \rightarrow$	$0^\pm\infty = \pm\rightarrow$
$\infty - \infty = \leftrightarrow$	$0^+/0^+ = \rightarrow$	$\bar{a}/\leftrightarrow = \bot$	$\infty\,\infty = \infty$

Note that, although the lift of division is a total function, 'division by zero' is still not allowed in a sense, because the result of $\bar{a}/\bar{0}$ is domain-error. This is essentially different from the way that Mathematica behaves. We will come back to this in Section 4

The next theorem tells us how to evaluate the lift of a continuous function in a point:

Theorem 2. *Let* f *be a function that is continuous at* a *and monotonically increasing in a neighborhood of* a. *Then:*

$$\bar{f}(\bar{a}) = \overline{f(a)}, \quad \bar{f}(a^\pm) = f(a)^\pm, \quad \bar{f}(a^-) = f(a)^-, \quad \bar{f}(a^+) = f(a)^+$$

Similar theorems hold for decreasing functions and functions at a local maximum or minimum.

The next theorems show how to evaluate filter limits:

Theorem 3. *Bringing filter limits inside expressions:*

$$\operatorname*{Lim}_{x \to A} f(g_1(x), g_2(x), \ldots, g_n(x)) \sqsubseteq \bar{f}(\operatorname*{Lim}_{x \to A} g_1(x), \operatorname*{Lim}_{x \to A} g_2(x), \ldots, \operatorname*{Lim}_{x \to A} g_n(x))$$

Note again that this theorem also holds for non-continuous f.

As an example of the fact that we do not always have equality here, not even when all functions are continuous, consider:

$$\operatorname{Lim}_{x \to \infty}(x - x) = \operatorname{Lim}_{x \to \infty} 0 = \bar{0}$$
$$(\operatorname{Lim}_{x \to \infty} x) - (\operatorname{Lim}_{x \to \infty} x) = \infty - \infty = \leftrightarrow$$

This agrees with the Theorem, since $\bar{0} \sqsubseteq \leftrightarrow$.

---

[2] We omit the straightforward proofs of the theorems in this paper. A paper containing a full development of the theory presented here, including all the proofs, can be found on the web pages of the authors.

Theorem 4. *Monotonicity with respect to refinement:*

$$A_1 \sqsubseteq B_1, A_2 \sqsubseteq B_2, \ldots, A_n \sqsubseteq B_n \Rightarrow \bar{f}(A_1, A_2, \ldots, A_n) \sqsubseteq \bar{f}(B_1, B_2, \ldots, B_n)$$

Together these two theorems allow one to evaluate a filter limit 'up to refinement' by substituting the filter inside the expression. Often this refinement does not hurt, because the right hand side will be a refinement of $\bar{a}$ or ∞ anyway, allowing us to apply the next theorem, which gives the relation between filter limits and the usual kind of limits:

Theorem 5. *Limit correspondence theorem:*

$$\lim_{x \to a} f(x) = b \Leftrightarrow \mathrm{Lim}_{x \to a^{\pm}} f(x) \sqsubseteq \bar{b}$$
$$\lim_{x \to a^+} f(x) = b \Leftrightarrow \mathrm{Lim}_{x \to a^+} f(x) \sqsubseteq \bar{b}$$
$$\lim_{x \to \infty} f(x) = b \Leftrightarrow \mathrm{Lim}_{x \to \infty} f(x) \sqsubseteq \bar{b}$$

Similar theorems hold at a^- and $-\infty$ and for limits to plus or minus infinity.

In Europe $\lim_{x \to a^+}$ is sometimes written as $\lim_{x \downarrow a}$. The ∞ and a^+ in the 'ordinary' limits on the left are not filters: those are just the customary notations for limits from the right and to infinity. The $a^{\pm}$, a^+ and ∞ on the right *are* filters.

Together these theorems now give us a method to rigorously evaluate ordinary limits using filters:

1. Replace the limit by the corresponding filter limit.
2. 'Evaluate' the filter limit using filter calculations, leading to a refinement.
3. If the right hand side of the refinement refines $\bar{a}$, $-\infty$ or ∞ then we have succeeded and can use Theorem 5 (or its analogue for infinite limits) to find the answer to the original question. If not, the method failed.

As an example, we use this method to evaluate $\lim_{x \to \infty} 1/(x+1)$:

$$\mathrm{Lim}_{x \to \infty} \frac{1}{x+1} \sqsubseteq \frac{\bar{1}}{\mathrm{Lim}_{x \to \infty}(x+1)} \sqsubseteq \frac{\bar{1}}{\infty + \bar{1}} = \frac{\bar{1}}{\infty} = 0^+$$

(The refinements here are really equalities but the theorems that we have do not give that, and in fact we do not need it.) Now $0^+ \sqsubseteq \bar{0}$ and so from Theorem 5 we find that:

$$\lim_{x \to \infty} \frac{1}{x+1} = 0$$

3 Interval Filters and Connected Filters

Definition 7. *We will define a class of filters on $\mathbb{R}$ called the* interval filters. *Consider the set:*

$$\mathcal{R} = \{-\infty^+\} \cup \{a^-, a^+ \mid a \in \mathbb{R}\} \cup \{\infty^-\}$$

For each pair of elements α and β from $\mathcal{R}$ for which $\alpha \leq \beta$ in the natural order on $\mathcal{R}$, we will define a filter interval(α, β). *We map the elements of $\mathcal{R}$ to formulas as:*

α	$\phi_l(x, \alpha, \varepsilon)$	$\phi_r(x, \alpha, \varepsilon)$
$-\infty^+$	$\top$	$x < -1/\varepsilon$
a^-	$a - \varepsilon < x$	$x < a$
a^+	$a < x$	$x < a + \varepsilon$
∞^-	$1/\varepsilon < x$	$\top$

and then we define:

interval$(\alpha, \beta) \equiv \{U \in \mathcal{O}(\mathbb{R}) \mid \exists \varepsilon \in \mathbb{R}_{>0}. \forall x \in \mathbb{R}. \phi_l(x, \alpha, \varepsilon) \wedge \phi_r(x, \beta, \varepsilon) \Rightarrow x \in U\}$

We will write interval filters using interval notation:

$$[a, b) \equiv \text{interval}(a^-, b^-) \qquad (a, b) \equiv \text{interval}(a^+, b^-)$$
$$[a, b] \equiv \text{interval}(a^-, b^+) \qquad (a, b] \equiv \text{interval}(a^+, b^+)$$

We suppose that it will be clear from the context when we mean an interval as a set of real numbers and when we mean an interval as an interval filter. Generally, for finite a and b they are related like:

$$(a, b] = \text{generated-by}(\{(a, b]\})$$

but not always. If $a = b$, then the left hand side is a^+ but the right hand side is improper because the set $(a, a]$ is empty.

When we analyze a two-sided limit into two one-sided limits, and then want to put the results back together, we need the concept of the 'join' of two filters, which we write $A \vee B$. For example, $0^- \vee 0^+ = 0^\pm$. This concept is defined as follows:

Definition 8. *The operations* join *and* meet *on filters are defined by:*

$$A \vee B = A \cap B$$
$$A \wedge B = \{U \cap V \mid U \in A \wedge V \in B\}$$

We can now write the filters from Definition 2 as interval filters or as joins of interval filters:

$\bar{a} = [a, a]$	$\infty = [\infty, \infty)$	$\rightarrow = (0, \infty)$
$a^- = [a, a)$	$-\infty = (-\infty, -\infty]$	$\leftarrow = (-\infty, 0)$
$a^+ = (a, a]$	$\pm\infty = (-\infty, -\infty] \vee [\infty, \infty)$	$\pm\rightarrow = (-\infty, 0) \vee (0, \infty)$
$a^\pm = [a, a) \vee (a, a]$	$\leftrightarrow = (-\infty, \infty)$	

Now that we have the class of interval filters, we will define the class of connected filters. This definition is much simpler:

Definition 9. *A filter A is called* connected *when:*

$$\forall U \in A. \exists V \in A. V \subseteq U \wedge V \text{ is a connected set}$$

Note that each of improper, domain-error and indeterminate is a connected filter.

The next three theorems give the relevant properties of the connected filters. Together they 'explain' why in practice one encounters only joins of interval filters: the filters one starts with are of that kind, and the operations that one applies to them conserve the property.

Theorem 6. *The interval filters are the proper non-empty connected filters.*

So all interval filters are connected, and the only connected filters which are not an interval filter are the 'error filters' improper and domain-error.

Theorem 7. *If f is a function that is continuous on its domain, and A is a connected filter, then $\bar{f}(A)$ is also connected.*

Theorem 8. $\bar{f}(A \vee B) = \bar{f}(A) \vee \bar{f}(B)$ *and* $\bar{f}(A \wedge B) \sqsubseteq \bar{f}(A) \wedge \bar{f}(B)$.

These last two theorems show that if one applies functions that are continuous on their domain to finite joins of interval filters, one always will end up with finite joins of interval filters again.

4 Mathematica Revisited

Now that we have given a calculus of filters that resembles the way Mathematica calculates with infinity, we will compare the behavior of our calculus and that of Mathematica in detail. This is what the calculations in the example Mathematica from Section 1.1 become when we redo them in our filter calculus:

$$
\begin{aligned}
1/(\infty + 1) &= 0^+ & 1/[-1,1] &= \bot \\
\sqrt{\infty} &= \infty & [-1,1] \cdot [-1,1] &= [-1,1] \\
\infty - \infty &= \leftrightarrow & [-1,1]^2 &= [0,1] \\
\leftrightarrow + \infty &= \leftrightarrow & \bar{0} \sin \infty &= \bar{0} \\
\sin \infty &= [-1,1] & \infty/\sin \infty &= \infty/(\sin \infty)^2 = \bot
\end{aligned}
$$

Here are some differences with Mathematica:

- Mathematica does not like to give 'no' for an answer. So it prefers not to complain about undefinedness of a function. According to Mathematica:

$$
\frac{1}{[-1,1]} = (-\infty, -1] \vee [1, \infty)
$$

instead of $\bot$. Our definitions have different behavior because we want the correspondence theorem about limits, Theorem 5, to hold. As an example of this difference in attitude consider the limit:

$$
\lim_{x \to 0^+} x \arctan(\tan \frac{1}{x})
$$

The graph of $x \arctan(\tan(1/x))$ looks like a 'saw tooth' converging to 0, and it is undefined infinitely often in each neighborhood of 0. Still Mathematica says[3]:

---

[3] In version 3.0. In version 4.1 it leaves the expression unevaluated.

```
In[12]:= Limit[x*ArcTan[Tan[1/x]], x->0, Direction->-1]
Out[12]= 0
```

If you ask MathXpert to evaluate this limit, you get the message: *This function is undefined for certain values arbitrarily close to the limit point, so the limit is undefined.*

- Mathematica does not identify as many expressions as it might. For instance, in the example session it might have simplified:

$$\text{Interval}[\{0, 0\}] = 0$$
$$\text{Interval}[\{\text{Infinity, Infinity}\}] = \text{Infinity}$$
$$\text{Interval}[\{-\text{Infinity, Infinity}\}] = \text{Indeterminate}$$

- Mathematica does not distinguish between open and closed intervals, nor does it have the concept of left and right filters to a point. In order to add this subtlety to its Interval calculus all that would be needed is to mark all the endpoints of the intervals with a $+$ or a $-$.

- We have *two* kinds of 'undefined' in our filter calculus: domain-error $= \bot$ and indeterminate $= \leftrightarrow$. (The third filter, improper $= \dagger$, only occurs as the meet of two disjoint interval filters, and never occurs in practice.) Mathematica only has Indeterminate, and does not distinguish between these two kinds of undefinedness.

- Mathematica issues a 'warning' message like:

Power::infy: Infinite expression $\dfrac{1}{0}$ encountered.

when it gets infinite or indeterminate results. This seems to imply that such results are errors. However, in our theory those results are not errors at all but the correct answers, and they should not generate such a message.

- We gave the details of the filter theory for the space of real numbers. However, the expression language of Mathematica is about the complex numbers. This is clear, for example, from the results of applying Sqrt and Log to negative numbers. It is therefore strange that Mathematica gives answers involving intervals to limit questions, since such answers are appropriate to *real* limits.

In any case, our filter theory can be adapted to the complex numbers. For example, complex infinity is represented by the filter generated by the exteriors of disks centered at 0 (i.e., 'neighborhoods of infinity'). The 'one-sided' filters a^+ and a^- are replaced by a wide variety of other filters representing different ways in which a complex number z can 'approach' a limit point a: for example, in complex analysis it is common to consider a limit restricted to an angular sector, such as $|\theta| < \pi/4$. It is easy to define a 'sector filter' generated by such a sector. Our theorems that do not involve interval filters carry over to the complex setting: pushing filter limits inside functions, the method of limit evaluation by refinement, etc. We have not given a characterization of the connected filters in the complex case. For example, there are more than just the sector filters: the filter generated by $|\theta| < r^2$ is not refined by any sector filter.

5   Conclusion and Future Directions

We have presented the filter approach to evaluating limits involving infinity. The usual way of calculating with infinities is not rigorous; indeed the central concept *infinity* is never defined in calculus textbooks. The issue is skirted by such statements as: 'The symbol ∞ does not represent a real number and we cannot use it in arithmetic in the usual way.' [8], p. 112.

Consider a student who says that $\lim_{x \to 0^+} 1/x$ is 'undefined', while the teacher says that ∞ is a better answer. 'But', says the student, 'you said ∞ *is* undefined.' Such dialogues do occur regularly in classrooms and teachers are unable to answer these questions on any rigorous basis. We have now, at least in principle, provided a remedy for this situation, since our theory of infinite limits is completely rigorous. Questions at the student level in our theory can usually be proved or refuted.

When computer algebra systems make use of a set of calculation rules, there should ideally be a semantics according to which these calculation rules are correct. Even for ordinary algebra, this is not usually the case. But it is usually the case that the rules are correct *except* that the system fails to check the preconditions. That is, the semantics of algebra is understood – but systems fail to implement the rules in a semantically correct way. Up until now, the semantics of limits has not been properly understood, and so the behavior of computer algebra systems did not even have a standard against which implementations could be measured. In using intervals as answers to limits, Mathematica has ventured into uncharted territory. We are now providing maps.

Our work, being completely rigorous, and based on simple set theory, is also completely formal.[4] Therefore computer-checking the theory from this paper is possible, and the resulting formalization would not only be an interesting exercise, but also probably could be used to make the prover automatically evaluate more limits. In another direction, this material is suitable for inclusion in an undergraduate real-analysis course, and the distinctions between different types of limits that it makes are suitable for inclusion in calculus books. In particular, calculus books need no longer steer away from calculations involving infinity. Simple rules for manipulating infinity can be given and the justifications omitted, as is usually the case now when the justifications involve epsilon-delta arguments.

References

1. Beeson, M., *Mathpert Calculus Assistant*. This software program (now known as MathXpert) was published in July, 1997 by Mathpert Systems, Santa Clara, CA, and is commercially available from <http://www.mathxpert.com/>.

---

[4] *Rigorous* implies that the concepts and theorems have a clear meaning and the theorems can be correctly proved. *Formal* implies that the concepts can be defined and the theorems proved in terms of set theory (or some other foundational theory of mathematics).

2. Beeson, M., Design Principles of Mathpert: Software to support education in algebra and calculus, in: Kajler, N. (ed.) *Computer-Human Interaction in Symbolic Computation*, Texts and Monographs in Symbolic Computation, Springer-Verlag, Berlin/Heidelberg/New York (1998), pp. 89-115.

3. Beeson, M. *MathXpert: un logiciel pour aider les élèves à apprendre les mathématiques par l'action*, to appear in *Sciences et Techniques Educatives*. An English translation of this article under the title *MathXpert: learning mathematics in the twenty-first century* is available at
   `<http://www.mathcs.sjsu.edu/faculty/beeson/Pubs/pubs.html>`.

4. Beeson, M., Using nonstandard analysis to verify the correctness of computations, *International Journal of Foundations of Computer Science*, **6** (3) (1995), pp. 299-338.

5. K. Kuratowski. *Topology*, volume 1. Academic Press, New York, London, 1966.

6. M. Monagan, K. Geddes, K. Heal, G. Labahn, and S. Vorkoetter. *Maple V Programming Guide for Release 5*. Springer-Verlag, Berlin/Heidelberg, 1997.

7. B. Sims. *Fundamentals of Topology*. MacMillan, New York, 1976.

8. Stewart. *Calculus, 3rd edition*, Brooks-Cole, Pacific Grove, CA 1995.

9. S. Wolfram. *The Mathematica book*. Cambridge University Press, Cambridge, 1996.

Making Conjectures about Maple Functions

Simon Colton

Division of Informatics
University of Edinburgh
United Kingdom
simonco@dai.ed.ac.uk
http://www.dai.ed.ac.uk/~simonco

Abstract. One of the main applications of computational techniques to pure mathematics has been the use of computer algebra systems to perform calculations which mathematicians cannot perform by hand. Because the data is produced within the computer algebra system, this becomes an environment for the exploration of new functions and the data produced is often analysed in order to make conjectures empirically. We add some automation to this by using the HR theory formation system to make conjectures about Maple functions supplied by the user. Experience has shown that HR produces too many conjectures which are easily proven from the definitions of the functions involved. Hence, we use the Otter theorem prover to discard any theorems which can be easily proven, leaving behind the more interesting ones which are empirically true but not trivially provable. By providing an application of HR's theory formation in number theory, we show that using Otter to prune HR's dull conjectures has much potential for producing interesting conjectures about standard computer algebra functions.

1 Introduction

There is an unfortunate dichotomy between the application of computer algebra systems (CASs) and automated theorem provers (ATPs) to pure mathematics: the concepts usually dealt with by computer algebra techniques are of too high a complexity to prove things about (at the moment) using automated techniques. There have been some attempts to bridge the gap in order to usefully apply automated theorem proving to computer algebra, including (i) the routine proving of fairly trivial theorems such as side conditions holding when calculating integrals and (ii) a less automated approach, where the user is actively involved in theory exploration within the CAS and the prover is called upon at specific times during the exploration [1].

Ideally, automated theorem provers would be called from within a CAS whenever the user made a conjecture about the functions they were defining. However, this will take increased sophistication in the automated theorem provers and is unlikely to happen in the short term. If the aim of the integration of mathematical systems is to generate *conjectures*, rather than theorems about the functions

J. Calmet et al. (Eds.): AISC-Calculemus 2002, LNAI 2385, pp. 259–274, 2002.

being explored using a CAS, then it is possible to put a positive spin on the relative differences between CAS and ATP. Rather than stating that a disadvantage of ATPs is their limited abilities with concepts of a higher complexity, we note that an advantage of ATPs is that they can be used to prove theorems from first principles, i.e., directly from the axioms of a domain. Furthermore, these theorems are less likely to be of interest to the user than those which cannot be proved by an ATP system. Therefore, in a conjecture-making context, we can use ATP systems to prune conjectures which are provably true from the definitions of the functions, thus improving the quality of the conjectures produced.

We assume a plausible 4-step model of progress in pure mathematics:

1. Some functions are defined in a particular context
2. The functions are calculated over a set of input values
3. The input/output pairs are examined in order to highlight patterns
4. Any observed patterns are stated as conjectures and proved or disproved

We note that, in general, the second step can be automated by computer algebra systems and the fourth step can be automated by theorem provers. Automating the third step — thus providing a possible bridge between CAS and ATP — is the subject of this paper. The making of conjectures necessitates a certain amount of concept formation, as sophisticated conjecture making involves not only making conjectures about the given functions, but also about closely related functions. Hence, we will also be automating the first step and closing a cycle of theory formation.

We use the HR theory formation system [4] to produce conjectures about a set of computer algebra functions provided by the user. In particular, we will use functions from the Maple CAS [16] and we will use the Otter theorem prover [13] to prove some of HR's conjectures, so that we can discard them. In §2, we describe the core functionality behind HR which enables it to make conjectures. In §3 we describe the additional functionality implemented for this application to the generation of conjectures about Maple functions. In §4, we describe a session using HR to generate conjectures about some Maple functions from number theory, and in appendix A, we prove one of the results which HR discovered during the session.

2   Conjecture Making in HR

Much of HR's functionality was used for this application to making conjectures about Maple functions. Each functionality can be broadly characterised as part of one of five tasks: (i) inventing concepts (ii) making conjectures (iii) finding counterexamples (iv) proving theorems and (v) reporting results.

2.1   Inventing Concepts

HR forms theories about a set of *objects of interest*, which are integers in number theory, graphs in graph theory, groups in group theory, etc. It is given background information which describes the objects of interest, namely some initial concepts. As discussed in §3 and §4 below, the objects of interest in the session described in this paper are integers, and the background information is supplied in the form of Maple functions. From the background information, HR uses ten production rules to produce a new concept from one (or two) old concepts. The production rules are described in more detail in [5], and we concentrate here on four:

- The *compose* production rule composes functions using conjugation
- The *disjunct* production rule joins concepts using disjunction
- The *exists* production rule introduces existential quantification
- The *split* production rule instantiates objects

As an example construction, we suppose that HR is given the background concepts of the `isprime(n)` Maple function, which checks whether n is prime and the `tau(n)` Maple function, which calculates the number of divisors of n. Using the compose production rule, HR invents the concept of pairs of integers, $[a, b]$ for which $b = tau(a)$ and *isprime(b)*. Following this, it uses the exists production rule to define the concept of integers, a, for which there exists such a b, i.e., $[a] : \exists\, b\ (tau(a) = b\ \&\ isprime(b))$. Hence HR has invented the concept of integers which have a prime number of divisors, a concept which we discuss further later. This construction is represented in figure 1. We say that the *complexity* of a concept is the number of concepts (including itself) in the construction path of the concept, as explained further in [5]. Hence, the complexity of the concept in figure 1 is the number of boxes, i.e., four.

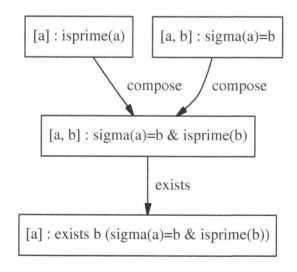

Fig. 1. Construction of a number theory concept

2.2   Making Conjectures

HR has a number of ways to make conjectures, both by noticing empirical patterns and by extracting conjectures from others. Firstly, whenever HR invents a concept, it checks two things empirically:

(i) whether the concept has no examples whatsoever, in which case it makes a non-existence conjecture, i.e., that the definition of the concept is inconsistent with the axioms of the domain. For example, if HR invented the concept of square numbers which are prime, it would find no examples, and make the conjecture that none exist on the number line.

(ii) whether the concept has exactly the same examples as a previous one, in which case, it makes a conjecture that the definitions of the new and old concepts are logically equivalent. For example, if HR invented the concept of integers for which the number of divisors is 2, it would make the conjecture that the new concept is equivalent to the concept of integers which are prime.

If the concept has a non-empty set of examples which differs from all previous concepts, the concept is new and is added to the theory. When added to the theory, HR determines which concepts the new concept *subsumes*, i.e., which concepts have a proper subset of the examples for the new concept. For each old concept that the new concept subsumes, HR makes the implication conjecture that the old definition implies the new definition. Similarly, HR determines which old concepts subsume the new concept, and makes the appropriate implication conjectures. From each subsumption conjecture, HR extracts implicate conjectures. For instance if it made the implication conjecture that $f(a)$ & $g(a) \rightarrow h(a)$ & $x(a)$, it would extract two implicate conjectures: $f(a)$ & $g(a) \rightarrow h(a)$ and $f(a)$ & $g(a) \rightarrow x(a)$.

HR also extracts implicate conjectures from equivalence conjectures and non-existence conjectures. For instance if HR made the equivalence conjecture that $f(a)$ & $g(a) \leftrightarrow h(a)$ & $x(a)$, it would extract four implicates from this:

$$f(a) \ \& \ g(a) \rightarrow h(a)$$
$$f(a) \ \& \ g(a) \rightarrow x(a)$$
$$h(a) \ \& \ x(a) \rightarrow f(a)$$
$$h(a) \ \& \ x(a) \rightarrow g(a)$$

Similarly, if HR made the non-existence conjecture that $\not\exists \, a \, (f(a) \ \& \ g(a))$, it would extract two implicate conjectures: $f(a) \rightarrow \neg g(a)$ and $g(a) \rightarrow \neg f(a)$. We enabled HR to extract implicates, as these are often easier to comprehend than the conjectures from which they are extracted. Often, as in the case in §4, we instruct HR to discard all but the implicates. Note that HR checks whether a new implicate has already been added to the theory, to avoid redundancy.

From implicates, HR can also extract prime implicates, which are such that no proper subset of the premises implies the goal. To do this, it tries to prove that each subset of the premises of an implicate imply the goal, starting with the singleton subsets and trying ever larger subsets. For instance, if starting with the implicate: $f(a)$ & $g(a) \rightarrow h(a)$, HR tries the two prime implicates:

$$f(a) \rightarrow h(a)$$
$$g(a) \rightarrow h(a)$$

If Otter can prove either of these conjectures, then they are added to the set of prime implicates, because clearly no proper subset of the premises imply the goal. The prime implicates represent some of the fundamental truths in a domain. To summarise, HR makes non-existence, equivalence and subsumption conjectures empirically, then extracts implicates from these and prime implicates, where possible, from the implicates.

2.3   Finding Counterexamples

The user can specify that certain objects of interest are given to HR to form a theory with, and others are held back in order to use for counterexamples. Then, whenever HR makes a conjecture, the held-back set is searched in order to find a counterexample. An advantage to this is an increase in efficiency, as often only a fraction of the objects of interest will find their way into the theory as counterexamples, thus whenever HR invents a concept, it will have less work to do to calculate the example set for the concept. Taking this to the extreme, in §4, we give HR only the number 1 to start with, but allow it access to the numbers 2 to 30 in order to find counterexamples to false conjectures. This not only increases efficiency, but it is also instructive to look at the false conjectures HR makes for which each counterexample is introduced. In algebraic domains, HR can also use the MACE model generator [15] to find counterexamples, but discussion of this is beyond the scope of this paper.

2.4   Proving Theorems

HR has some built-in abilities to decide when a conjecture it makes is trivially true, e.g., it can tell that the conjecture $f(a)$ & $g(a) \leftrightarrow g(a)$ & $f(a)$ is true. It also keeps a record of which concepts it generates are functions, so that it can tell that the conjecture $\nexists a(f(a) = k_1$ & $f(a) = k_2)$ is true, where k_1 and k_2 are different ground instances. In fact, it uses its primitive theorem proving to avoid inventing concepts such as this, because it knows in advance that the concept will have no examples, leading to a dull non-existence conjecture. If HR had any more sophisticated theorem proving, then we would, to a certain extent, be re-inventing the wheel, as there are many good theorem provers available for HR to use. In particular, HR invokes the Otter theorem prover to attempt to prove the conjectures it makes. HR has been interfaced to Otter via MathWeb [9,17], but the application here was undertaken using a simple file interface.

2.5   Reporting Results

HR is able to prune the conjectures it produces and order those remaining in terms of measures of interestingness. In particular, for this application, we use

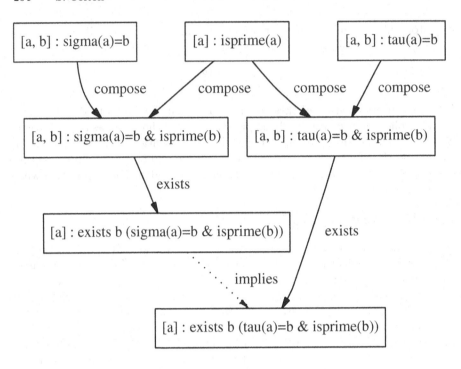

Fig. 2. Construction of an implicate conjecture

HR to keep only implicates extracted from equivalence, non-existence and subsumption conjectures, as these are usually easier to understand. We also instruct HR to discard any conjectures which Otter can prove, as these are likely to follow from the definitions of the Maple functions provided and be fairly uninteresting.

Of those implicates remaining, we use two measures of interestingness to order them. Firstly, each implicate comprises a concept which implies a single clause. The *applicability* of a concept gives an indication of the scope of the conjecture, where the applicability is measured as the proportion of objects of interest in the theory which have non-trivial examples for the concept. The applicability of an implicate conjecture is taken as the applicability of the concept on the LHS of the conjecture. For instance, if HR had the integers 1 to 30 as objects of interest, then the concept of prime numbers would score 10/30 for applicability, as there are 10 prime numbers between 1 and 30, namely 2, 3, 5, 7, 11, 13, 17, 19, 23 and 29. Hence implicate conjectures where the concept which makes up the premises is the concept of prime numbers will score 1/3 for applicability. Conjectures with very low applicability tend to be uninteresting, so sorting the conjectures in terms of decreasing applicability can be useful.

Secondly, equivalence and subsumption conjectures relate two concepts from the theory. HR measures the *surprisingness* of these conjectures as the proportion of concepts in the construction path of either concept which are in the construction path of just one concept. If two concepts conjectured to be related share many concepts in their construction paths, their definitions are likely to be similar, and the relationship between them will be less surprising, so they score less for surprisingness. For example, in figure 2, there are 7 concepts involved in the construction history of the conjecture relating the two concepts joined by a dotted line. Only one of these is shared by the two concepts in the conjecture, hence the conjecture scores 6/7 for surprisingness. The implicates extracted from equivalence and subsumption conjectures inherit the surprisingness value from their parent, so that these too can be measured in terms of surprisingness.

3   Additional Functionality

Each new application of HR necessitates some new functionality. In this case, we have extended HR's functionality in all the five areas discussed in §2 above. In terms of improved concept formation, HR is now able to communicate with Maple. At present, it does this in the same way as it does with Otter, by reading a file, invoking Maple in such a way that it outputs answers to a file, and then reading that file. HR, Maple and Otter are already part of the MathWeb software bus [9] and we have been successful in enabling HR to invoke Otter (and other provers) via MathWeb [17]. We can see no problem in enabling the communication between HR and Maple on a more sophisticated level via MathWeb, and we hope to do this soon.

HR calls Maple at the start of a session to get the initial data for the background concepts. For instance, if the user decides to start HR with the integers 1 to 10 and the Maple number theory functions of tau(n) and sigma(n), (with tau(n) being the number of divisors of n and sigma(n) being the sum of divisors of n), then HR will use Maple to calculate tau(1)=1, ..., tau(10)=4, doing likewise for sigma. HR also calls Maple during concept formation, for instance, if HR used its compose rule to invent the concept of tau(sigma(n)), then it would need to calculate tau(sigma(10)) which is tau(18)=6. In future, we envisage a more sophisticated interface between Maple and HR, in particular, enabling HR to write conjectures in a format Maple can read, then using Maple to check them empirically (over a large set of integers, or graphs, etc.). This interface would improve the efficiency of checking the conjectures, as HR is not optimised like Maple for performing lengthy calculations.

We also improved the way in which HR writes definitions, so that the conjectures about the concepts would be easier to read for the user (intended to be a mathematician). In particular, in order to make the definitions of functions which have been composed more understandable, HR was given the ability to collate and remove existential variables where possible. For example, when HR invents a concept with, say, the definition:

$$[a] : \exists\ b\ (f(a) = b\ \&\ \exists\ c\ (g(b) = c\ \&\ h(a) = c))$$

it first collects together the existential variables thus:

$$[a] : \exists\ b, c\ (f(a) = b\ \&\ g(b) = c\ \&\ h(a) = c),$$

then removes the existential variables b and c thus:

$$[a] : g(f(a)) = h(a)$$

It has done this by both substituting $f(a)$ for b and by removing c by equating $g(b)$ and $h(a)$. As a concrete example, HR rewrites the definition for integers with a prime number of divisors described in §2.1 above in this way:

$$[a] : \exists\ b\ (tau(a) = b\ \&\ isprime(b))\quad \text{becomes}\quad [a] : isprime(tau(a))$$

which is easier to understand. This functionality is also useful for an application to constraint generation [7].

In terms of improved conjecture making and reporting, HR can now make applicability conjectures, which state that a concept is restricted to having only a small number of examples. For instance, when HR invents the concept of integers which are equal to their number of divisors, it notices that this property is only true for integers 1 and 2. It then adds concept formation steps to the agenda which invent (a) the concept of an integer being the number 1 (b) the concept of being the number 2 and (c) the concept of being either 1 or 2. We call such concepts *instantiation* concepts, as they are basically the instantiation of a single object of interest (or a disjunction of similar instantiations). Having invented concept (c) using the disjunct production rule, HR then makes the conjecture that an integer is equal to the number of divisors if and only if it is equal to 1 or 2. HR is then able to identify the conjectures which involve instantiation concepts and discard them, as they are, in general, not particularly interesting.

In terms of improved theorem proving, we gave HR the ability to pass Otter ground instances of the Maple functions. For example, in §4, we describe a session with HR using the Maple `tau(n)` function, which counts the number of divisors of n. Because during that session, HR makes instantiations, it will eventually discover conjectures such as $\forall\ a, ((a = 1 \vee a = 2) \rightarrow tau(a) = a)$. As HR uses Maple to calculate ground instances such as `tau(1)` = 1 and `tau(2)` = 2, etc., and HR gives Otter these ground instances, Otter is able to prove the above theorem and HR discards it because it is unlikely to be interesting.

Furthermore, the user is now able to act as a theorem prover and tell HR that certain conjectures are true and should be given to Otter as additional axioms for future proof attempts. For instance, in the session described in §4 below, HR identifies the conjecture that $isprime(n) \rightarrow tau(n) = 2$. This follows from the definitions, and we told HR to use this as an axiom of the domain. With that information, it was able to prove many more theorems. This also means that, to a certain extent, the user does not have to worry about specifying the axioms of the domain in advance, as HR will come across (some of) them. In fact, for the application in §4, we gave HR no axioms of number theory in advance.

In terms of improved counterexample finding, we enabled the user to step in and check whether certain objects of interest are counterexamples to a particular conjecture HR has made. In number theory, the user can specify a lower and upper bound on a set of integers, and HR checks if any integer in the set breaks the conjecture. To perform the check, HR invokes Maple to calculate the user-given Maple functions for each integer. Using this information, HR calculates examples of the concepts in the conjecture for each integer and tests whether the conjecture still holds. This functionality is useful once HR has identified the interesting conjectures in a session, as the user can choose one and test it empirically before attempting a proof (as we do in §4).

4   Results

In §5 we discuss a planned application of HR to discovery in pure mathematics, for which the interface with Maple will be very important. Our aims for this paper were to show that the pruning measures discussed above are effective and that it is possible to find interesting conjectures about CAS functions using HR.

We give details here of a session with HR in number theory, where HR was given as background knowledge three functions from the Maple numtheory package. The three functions were tau(n), which calculates the number of divisors of n, sigma(n), which calculates the sum of divisors of n, and isprime(n), which tests whether or not n is a prime number. We gave HR only the number 1 to start with, but gave it access to the numbers 2 to 30 from which to find counterexamples to false conjectures. Using a complexity limit of 6, we ran a breadth first search to completion using the compose, exists and split production rules. We also enabled applicability conjecture making, so that HR could make applicability conjectures when concepts applied to 2 or fewer objects of interest. This meant that the disjunct production rule was also used to produce concepts. We specified that HR should produce conjectures through equivalence checking, non-existence checking and subsumption checking. We also specified that it should extract implicates from these conjectures and that it should keep only the implicates. Finally, we specified that it should use Otter to try to prove any implicates it produced. After experimentation, we decided not to extract prime implicates, as this was computationally expensive and mostly fruitless.

The session took around 2 minutes on a Pentium 500Mhz processor, and lasted for 378 theory formation steps. HR produced 48 concepts. Due to the composition of functions, HR called Maple on 120 occasions, to calculate isprime, tau, and sigma for integers ranging from 1 to 195 (which is the sum of the divisors of 72). HR also introduced the numbers 2, 3, 4, 5, 6, 9 and 16 as counterexamples to false conjectures. These false conjectures were made in this order, given with the counterexample HR found to disprove them:

```
all a b (((tau(a)=b) <-> (sigma(a)=b))) [counterexample = 2]
all a b (((tau(a)=b) <-> (tau(a)=b & tau(b)=a))) [3]
all a b (((sigma(a)=b) <-> (sigma(a)=b & tau(b)=a))) [4]
all a ((isprime(a)) <-> ((a=2 | a=3))) [5]
```

```
all a ((a=2 | a=4) <-> (isprime(sigma(a)))) [9]
all a b (tau(a)=b <-> tau(a)=b & tau(sigma(b))=b) [6]
all a b (tau(a)=b & isprime(b) -> tau(a)=b & tau(sigma(b))=b) [16]
```

In the session, HR produced 137 implicate conjectures. Of these, 43 had already been proven by Otter, including ones which followed from a calculation on particular integers, such as:

```
(68) all a (((((a=2 | a=3)) -> (tau((sigma(a)))=a)))
```

Otter could prove this because HR gave it ground instances such as `tau(3)=2` and `sigma(2)=3`. There were also theorems which didn't follow from calculations, but were still obviously true, such as:

```
(56) all a b (tau(a)=b & sigma(b)=a & isprime(b) -> tau(sigma(b))=b)
```

Of the 94 conjectures which remained unsolved, we looked through the first 10 which were produced and added these 9 as axioms:

```
(0) all a (((exists b (tau(a)=b))))
(1) all a (((tau(a)=1) -> (a=1)))
(3) all a (((isprime(a)) -> (tau(a)=2)))
(4) all a (((tau(a)=2) -> (isprime(a))))
(5) all a (((exists b (sigma(a)=b))))
(7) all a (((sigma(a)=1) -> (a=1)))
(8) all a b (((tau(a)=b & sigma(a)=b) -> (tau(b)=a)))
(9) all a b (((tau(a)=b & sigma(a)=b) -> (sigma(b)=a)))
(10) all a b (((sigma(a)=b & sigma(b)=a) -> (tau(a)=b)))
```

Note that conjectures (2) and (6) were proved, hence not in the list of those unsolved conjectures that HR presented to us. The conjecture we did not add from the first 10 unsolved ones was:

```
(11) all a b (((tau(a)=b & isprime(a)) -> (isprime(b))))
```

which we thought might follow from the other axioms, so we left it out. We see that HR has identified the definition of prime numbers in conjectures (3) and (4): `all a (isprime(a) <-> tau(a)=2)`. We also looked through the unsolved conjectures which were instantiations, and added these three as axioms:

```
(15) all a b (sigma(a)=b & sigma(b)=a -> a=1)
(21) all a (tau(tau(a))=a -> (a=1 | a=2))
(135) all a (a=3 -> isprime(sigma(sigma(a))))
```

Having given HR the additional axioms, we then asked it to attempt to re-prove all the unsolved conjectures. This was very effective, and reduced the number of unsolved conjectures from 94 to just 22. We looked at the 17 unsolved conjectures which were not instantiations, and ordered these in terms of a measure of interestingness which was obtained by averaging the normalised applicability and normalised surprisingness. At the top of the ordered list was conjecture number 46, which we found very interesting:

```
(46) all a (isprime(sigma(a)) -> isprime(tau(a)))
```

Paraphrased, this states that, if you take a number and add up the divisors, with the result being a prime number, then the coefficient of divisors you have just added up will also be a prime. We used HR to check this conjecture empirically for the numbers 1 to 100, and it used Maple to perform the appropriate calculations. The empirical test was positive, so we tried to prove this conjecture, which we managed, as reported in appendix A. We then added conjecture 46 as an axiom and asked HR to attempt to prove the remaining unsolved conjectures in the light of this theorem. This reduced the unsolved non-instantiation conjectures to just the following 10, ordered in terms of the interestingness measure mentioned above:

```
(127) all a (tau(tau(a))=a -> tau(sigma(sigma(a)))=sigma(a))
(129) all a (tau(tau(a))=a -> tau(sigma(a))=a)
(130) all a (tau(sigma(a))=a & tau(sigma(sigma(a)))=sigma(a) ->
      tau(tau(a))=a)
(64) all a b (sigma(a)=b & isprime(a) & isprime(b) -> tau(sigma(b))=b)
(111) all a b (sigma(a)=b & isprime(sigma(b)) -> isprime(tau(a)))
(90) all a b (sigma(a)=b & isprime(tau(b)) -> isprime(tau(a)))
(128) all a (tau(sigma(sigma(a)))=sigma(a) -> tau(sigma(tau(a)))=tau(a))
(108) all a b (sigma(a)=b & isprime(sigma(b))  -> tau(b)=a)
(47) all a b (sigma(a)=b & isprime(a) & isprime(b) -> tau(b)=a)
(109) all a b (sigma(a)=b & isprime(sigma(b)) -> isprime(a))
```

We note that conjectures (127) and (129) above should be proved because we gave HR conjecture (21) as an axiom, which states that, given the left hand side of conjecture (127) or (129), then a = 1 or a = 2. However, we found that Otter could not prove either conjecture (with default settings), even when allowed five minutes to prove them. This is an anomaly we are currently investigating. We must also determine the significance – if any – of the other results. However, we feel it is a success that, in such a short session with HR, it managed to find a non-trivial conjecture of enough interest that a generalised theorem (see appendix A) was found and proved with some difficulty. Also, we think that the pruning using Otter and the user to prove easy theorems worked well. In figure 3, we show the decrease in the number of unsolved conjectures at various stages of the session, and we note that the number of unsolved conjectures presented to the user was reduced from 137 to just 11.

5   Conclusions and Further Work

In the final sentence of [4], we state that:

> "... if this technology can be embedded into computer algebra systems, we believe that theory formation programs will one day be important tools for mathematicians." (page 361)

The work presented here represents the first step towards using automated theory formation to enable computer algebra systems to intelligently make research

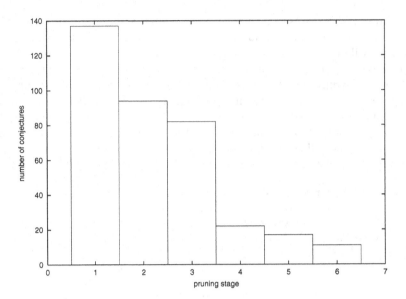

Fig. 3. Pruning of conjectures in stages: stage 1 (all the conjectures), stage 2 (after using Otter to discard trivially true results), stage 3 (after the user chose conjectures to add as axioms), stage 4 (after another round of proving using the additional axioms), stage 5 (after pruning instantiation conjectures), stage 6 (after a final round of proving with a single new axiom added).

conjectures about functions the user is experimenting with. It compliments, but is distinct to, our work datamining the Encyclopedia of Integer Sequences to find conjectures [3,6] which also led to discoveries in number theory.

Other approaches to making research conjectures for mathematicians have either performed an exhaustive search for theorems using the power of an efficient theorem prover, or have required bespoke programs. For instance, in [14], McCune uses an exhaustive search with Otter to find new axiomatisations of group theory and other algebras. Similarly, in [2], Chou used the power of Wu's method to find new constructions in plane geometry. The Graffiti program [8] has produced scores of conjectures which the graph theory community have eagerly proved and disproved, but this is a graph theory specific program which isn't publicly available. To our knowledge, HR is the only program which uses both computer algebra and theorem proving systems to make research conjectures.

We have shown how the HR theory formation system can be used to make conjectures about Maple functions chosen by the user. Given HR's current abilities to form concepts and make conjectures, the main technical difficulty to overcome for this project was to reduce the number uninteresting conjectures produced. To do this, we used much of HR's functionality, including:

[1] Its ability to call Otter to prove theorems from first principles. Such theorems are likely to be uninteresting, and hence can be discarded. In our example

session, this enabled HR to discard 43 such theorems (31% of the 137 implicate conjectures HR produced in total). For this to be effective in number theory, we enabled HR to pass ground instances from Maple to Otter.

[2] A new ability, which allows the user to choose some of HR's conjectures to add as axioms. Subsequent attempts to prove the unsolved conjectures allows more pruning of the theorems because they can be proved from first principles and the (usually simple) axioms added by the user. After giving some of HR's obviously true theorems to Otter as axioms, this reduced the number of unsolved conjectures from 94 to just 22, an acceptable number for the user to look through.

[3] The ability to extract simply stated implicates and order conjectures in terms of measures of interestingness, so that the user can browse the most interesting conjectures first.

The second point above represents a first step towards a more interactive environment for theory development within HR. We hope to pursue such an interactive mode – similar to that employed by Lenat with his AM program [11] – by allowing the user to step in and provide new concepts, conjectures, theorems, proofs and counterexamples at will during the theory formation session. This will be useful for an extended application to mathematical discovery we have planned for HR: the exploration of the domain of Zariski spaces developed by Roy McCasland [12]. Due to the relative complexity of this domain, the interactive mode in HR will be essential. Also, HR's links via MathWeb to various pieces of mathematical software including provers such as Otter, Spass and E, model generators such as MACE, computer algebra systems such as Maple and Gap, and constraint solvers such as Solver, will be essential for this project. Our aim for the HR system is for the theory behind it to encompass more and more abilities, while the tasks reliant on HR's code become fewer, as HR interfaces with more mathematics programs.

The application of HR to finding conjectures about CAS functions is still in its early stages. Our choice of which Maple functions to form conjectures about was inspired by working with these functions in a different project [6], but in general, the user will specify a much larger set of functions. HR must therefore decide which ones to use, possibly discarding some after an initial investigation reveals that there are very few interesting properties about which HR can make conjectures. Furthermore, we need to undertake extended testing of HR to highlight its strengths and limitations when working with CAS functions. Finally, we need to improve the integration of HR and Maple, in terms of (i) the communication between them (i.e., enable HR to write Maple functions so that Maple, rather than HR, can be used to check conjectures empirically) and (ii) the way in which that communication is performed (i.e., by enabling HR to talk to Otter and Maple via the MathWeb software bus).

We hope to have shown here a glimpse of the potential for using HR to discover interesting facts about computer algebra functions and concepts related to them. As the most popular pieces of software within pure mathematics are computer algebra systems, it is essential that HR is able to interact with such programs, and it is a long-term goal of the HR project to embed HR's discovery functionality into computer algebra systems.

Acknowledgments. This work is supported by EPSRC grant GR/M98012. The author is also affiliated to the Department of Computer Science at the University of York. This research was inspired by discussions with Jacques Calmet and Clemens Ballarin during the author's visit to Karlsruhe University, funded by the European Union IHP grant CALCULEMUS HPRN-CT-2000-00102. As with all projects involving HR, the input from Alan Bundy and Toby Walsh has been essential. We would also like to thank the anonymous referees for their interesting comments and suggestions about this work.

References

1. B Buchberger. Theory exploration versus theorem proving. In *Proceedings of Calculemus 99, Systems for Integrated Computation and Deduction*, 1998.
2. S Chou. Proving and discovering geometry theorems using Wu's method. Technical Report 49, Computing Science, University of Austin at Texas, 1985.
3. S Colton. Refactorable numbers - a machine invention. *Journal of Integer Sequences*, 2, 1999.
4. S Colton. *Automated Theory Formation in Pure Mathematics*. PhD thesis, Department of Artificial Intelligence, University of Edinburgh, 2000.
5. S Colton, A Bundy, and T Walsh. Automatic identification of mathematical concepts. In *Machine Learning: Proceedings of the 17th International Conference*, 2000.
6. S Colton, A Bundy, and T Walsh. Automatic invention of integer sequences. In *Proceedings of the Seventeenth National Conference on Artificial Intelligence*, 2000.
7. S Colton and I Miguel. Constraint generation via automated theory formation. In *Proceedings of CP-01*, 2001.
8. S Fajtlowicz. On conjectures of Graffiti. *Discrete Mathematics 72*, 23:113–118, 1988.
9. Andreas Franke and Michael Kohlhase. System description: MATHWEB, an agent-based communication layer for distributed automated theorem proving. In *Proceedings of CADE-16*, pages 217–221, 1999.
10. G Hardy and E Wright. *The Theory of Numbers*. Oxford University Press, 1938.
11. D Lenat. AM: Discovery in mathematics as heuristic search. In D Lenat and R Davis, editors, *Knowledge-Based Systems in Artificial Intelligence*. McGraw-Hill Advanced Computer Science Series, 1982.
12. R McCasland, M Moore, and P Smith. An introduction to Zariski spaces over Zariski topologies. *Rocky Mountain Journal of Mathematics*, 28:1357–1369, 1998.
13. W McCune. The OTTER user's guide. Technical Report ANL/90/9, Argonne National Laboratories, 1990.
14. W McCune. Single axioms for groups and Abelian groups with various operations. *Journal of Automated Reasoning*, 10(1):1–13, 1993.
15. W McCune. A Davis-Putnam program and its application to finite first-order model search. Technical Report ANL/MCS-TM-194, Argonne National Laboratories, 1994.
16. D. Redfern. *The Maple Handbook: Maple V Release 5*. Springer Verlag, 1999.

17. J Zimmer, A Franke, S Colton, and G Sutcliffe. Integrating HR and tptp2x into MathWeb to compare automated theorem provers. Technical report, Division of Informatics, University of Edinburgh, 2001.

A. Proof That $isprime(sigma(n)) \rightarrow isprime(tau(n))$

Lemma
For all n, $\tau(n)$ is prime $\iff n = p^{q-1}$ for primes p and q.

Proof
If $n = p^{q-1}$ then $\tau(n) = q$, hence $\tau(n)$ is prime. Conversely, suppose that the prime factorisation of n is $p_1^{k_1} \ldots p_l^{k_l}$, and that $\tau(n)$ is prime. Now $\tau(n) = (k_1 + 1) \ldots (k_l + 1)$, hence $l = 1$, and n must be of the form p^a for some a. So, $\tau(p^a) = a + 1$, and a must be one less than a prime, q.

Lemma 2
If the prime factorisation of integer n is: $n = \prod_{i=1}^{l} p_i^{k_i}$, then

$$\sigma_m(n) = \prod_{i=1}^{l} \left(\frac{p_i^{m(k_i+1)} - 1}{p_i - 1} \right).$$

(Where $\sigma_m(n)$ is the sum of the mth powers of the divisors of n). For the proof of this result, see thm. 274 of [10]. We also need the following well known identity:

$$\frac{a^b - 1}{a - 1} = 1 + a^2 + \ldots + a^{b-1} = \sum_{i=0}^{b-1} a^i.$$

Theorem
$\forall\, m, n \in \mathbf{N}, \quad \tau(\sigma_m(n)) = 2 \Rightarrow \tau(\tau(n)) = 2.$

Proof
Let the prime factorisation of n be $p_1^{k_1} \ldots p_l^{k_l}$, and let m be an integer. Suppose also that $\tau(\sigma_m(n)) = 2$, i.e. that $\sigma_m(n)$ is prime. We see from lemma 2 that $\sigma_m(n)$ has at least $l+1$ factors (counting 1 as well). Therefore, as $\sigma_m(n)$ is prime, $l = 1$. Hence we can write $n = p^a$ for some prime p and some $a \in \mathbf{N}$. If we assume that $\tau(n)$ is composite, then $\tau(n) = a + 1 = xy$ for some $x, y \in \mathbf{N}, x > 1, y > 1$. Hence $a = xy - 1$. So, using lemma 2 again:

$$
\begin{aligned}
\sigma_m(n) &= \frac{p^{m(a+1)} - 1}{p - 1} = \frac{p^{m(xy-1+1)} - 1}{p - 1} = \frac{p^{mxy} - 1}{p - 1} \\
&= \frac{(p^{mx} - 1)(p^{(y-1)mx} + p^{(y-2)mx} + \ldots + p^{mx} + 1)}{p - 1} \\
&= \frac{p^{mx} - 1}{p - 1} \sum_{i=1}^{y} p^{(y-i)mx} = \left(\sum_{i=0}^{mx-1} p^i \right) \cdot \left(\sum_{j=1}^{y} p^{(y-j)mx} \right)
\end{aligned}
$$

As $x > 1$ and $y > 1$, neither of the factors in this final product equal 1. Hence, this provides a contradiction, because $\sigma_m(n)$ is prime. Hence our assumption that $\tau(n)$ is composite must be false, and we see that $\tau(n)$ is a prime. $\square$

Corollary
Taking $m = 1$ above, we see that: $\forall\, n \in \mathbf{N}, \quad \tau(\sigma(n)) = 2 \Rightarrow \tau(\tau(n)) = 2$, i.e, if the sum of divisors of n is prime, then the number of divisors of n will be prime.

Employing Theory Formation to Guide Proof Planning

Andreas Meier[1], Volker Sorge[2], and Simon Colton[3]*

[1] Fachbereich Informatik, Universität des Saarlandes, Germany,
ameier@ags.uni-sb.de, http://www.ags.uni-sb.de/~ameier
[2] School of Computer Science, University of Birmingham, UK,
V.Sorge@cs.bham.ac.uk, http://www.cs.bham.ac.uk/~vxs,
[3] Division of Informatics, University of Edinburgh, UK,
simonco@dai.ed.ac.uk, http://www.dai.ed.ac.uk/~simonco

Abstract. The invention of suitable concepts to characterise mathematical structures is one of the most challenging tasks for both human mathematicians and automated theorem provers alike. We present an approach where automatic concept formation is used to guide non-isomorphism proofs in the residue class domain. The main idea behind the proof is to automatically identify discriminants for two given structures to show that they are not isomorphic. Suitable discriminants are generated by a theory formation system; the overall proof is constructed by a proof planner with the additional support of traditional automated theorem provers and a computer algebra system.

1 Introduction

In [10] and [11] we present a case study concerned with the automatic classification of residue class sets over the integers into sets of isomorphic structures. The residue class sets have given binary operations and our approach is to prove that two given structures are isomorphic or not in terms of their basic algebraic properties. The necessary proofs are constructed with the MULTI proof planner [13], which forms part of the ΩMEGA theorem proving environment [1]. MULTI uses the guidance of computer algebra and model generation. Particularly hard problems arising in the domain are non-isomorphism proofs, i.e. to show that two given structures are non-isomorphic, since a naïve proof attempt with an exhaustive case analysis is infeasible in most cases. A better approach is to find an *invariant* of the residue class sets (a property which does not change under isomorphism) and show that it differs for a particular pair of residue class sets. We say that the particular invariant found for two structures acts as a *discriminant* for the structures if it has a different value for each. Unfortunately, employing a small set of predefined invariants cannot necessarily deal with all cases which might occur.

---

* The author's work is supported by EPSRC grant GR/M98012 and European Union IHP grant CALCULEMUS HPRN-CT-2000-00102. He is also affiliated with the Department of Computer Science at the University of York.

J. Calmet et al. (Eds.): AISC-Calculemus 2002, LNAI 2385, pp. 275–289, 2002.

In this paper we overcome this dilemma by using the HR automatic theory formation system [3] to automatically detect discriminants for given structures. We then model the discriminants into appropriate concepts and the overall proof is constructed by MULTI with the support of traditional automated theorem provers and a computer algebra system.

The paper is structured as follows: in section 2 we introduce the problem domain and motivate the use of automatic concept formation for non-isomorphism proofs. Section 3 gives a brief overview of the various systems we employ to construct non-isomorphism proofs. Section 4 gives a detailed account of how the proofs are constructed and we present some preliminary results in section 5.

2 Proving Non-isomorphism Problems

In this section we first briefly introduce the problem domain of residue class structures and the proof techniques we have already developed for non-isomorphism proofs as described in [10,11]. Following this, we generalise one of these proof techniques to a new proof scheme which enables the construction of arbitrary discriminants for two algebraic structures that suffice to demonstrate that the structures involved are not isomorphic.

2.1 The Residue Class Domain

We define a residue class set over the integers as the set of all congruence classes modulo an integer n, i.e., $\mathbb{Z}_n$, or as an arbitrary subset of $\mathbb{Z}_n$. More concretely, we are dealing with sets of the form $\mathbb{Z}_3, \mathbb{Z}_5, \mathbb{Z}_3 \backslash \{\bar{1}_3\}, \mathbb{Z}_5 \backslash \{\bar{0}_5\}, \{\bar{1}_6, \bar{3}_6, \bar{5}_6\}$, etc. where $\bar{1}_6$ denotes the congruence class of 1 modulo 6. If c is an integer we also write $cl_n(c)$ for the congruence class of c modulo n. A binary operation $\circ$ on a residue class set is given in λ-function notation, and $\circ$ can be of the form $\lambda xy.x$, $\lambda xy.y$, $\lambda xy.c$ where c is a constant congruence class (e.g., $\bar{1}_3$), $\lambda xy.x\bar{+}y$, $\lambda xy.x\bar{*}y$, $\lambda xy.x\bar{-}y$, where $\bar{+}$, $\bar{*}$, $\bar{-}$ denote addition, multiplication, and subtraction of congruence classes over the integers, respectively. Furthermore, $\circ$ can be any combination of the basic operations with respect to a common modulo factor, e.g., $\lambda xy.(x\bar{+}\bar{1}_3)\bar{-}(y\bar{+}\bar{2}_3)$. We often abbreviate the operations $\lambda xy.x\bar{+}y$, $\lambda xy.x\bar{*}y$ and $\lambda xy.x\bar{-}y$ by $\bar{+}$, $\bar{*}$ and $\bar{-}$, respectively.

For two given structures $(RS_n^1, \circ^1)$ and $(RS_m^2, \circ^2)$ we examine whether or not they are isomorphic; that is, we determine whether or not there is a function $h:(RS_n^1, \circ^1) \to (RS_m^2, \circ^2)$ such that h is injective, surjective, and homomorphic.[1] For proof planning both isomorphism and non-isomorphism proofs, the appropriate guidance for the proof planner is crucial for success. In cases where two structures are isomorphic, it is usually fairly simple to compute an appropriate mapping h with either a computer algebra system or a model generator, and subsequently show that h is indeed an isomorphism. However, when the structures are not isomorphic, it is much more difficult to appropriately guide the necessary non-isomorphism proof.

---

[1] Observe that we avoid confusion between indices and modulo factors by writing indices as superscripts, except in indexed variables such as x_i, y_j as they are clearly distinct from congruence classes of the form $cl_i(x)$.

2.2   Techniques for Non-isomorphism Proofs

In our previous work [10,11], we have implemented several proof techniques for the proof planner MULTI to show that two structures are not isomorphic. These require varying degrees of guidance from computer algebra or model generation:

Testing all possible functions h. Essentially this corresponds to a case split on all possible instantiations for the mapping h and showing in each case that h is not an isomorphism. While this technique does not require any guidance for MULTI, for two structures whose sets have cardinality n, MULTI has to consider n^n possible functions, which becomes infeasible even for relatively small n.

Proof by contradiction. The idea of this technique is to find a pair of distinct elements in one structure that is always mapped to the same image under each homomorphism h. This shows that there exists no injective h and therefore no isomorphism. For this technique, a prospective pair of elements can be computed either with the computer algebra system MAPLE [16] or, more reliably, with the SEM model generator [17]. However, even with this guidance, the subsequent proof process is essentially equational theorem proving, and success is not guaranteed.

Using predefined invariants. An intuitive way to show non-isomorphism is to find an invariant property of one structure that the other structure does not exhibit. We have already implemented a proof planning approach for the following predefined invariants: (1) the structures involved are of different cardinality; (2) the structures form different algebraic entities; e.g., one structure is a group while the other is a semigroup; (3) one of the structures contains an element of some order k and no element in the other structure has order k. For structures without a unit element, we can similarly use the order of traces of elements. MULTI checks these invariants in this order. To compute both orders and traces of elements, MULTI uses the computer algebra system GAP [8]. In the automatic exploration of the residue class domain (see [11]) we usually start with sets of similar algebraic structures of the same cardinality (e.g., quasigroups of order 5). Hence invariant (3) is the only one of relevance, and the predefined criteria are often not sufficient to successfully construct a non-isomorphism proof.

2.3   Systematically Constructing Discriminants

The new proof technique we describe in this paper aims to be a more reliable proof strategy for non-isomorphism proofs. It is essentially a generalisation of the technique presented in section 2.2. Given two structures, we construct an appropriate, bespoke discriminant (i.e., an invariant property that only one of the structures exhibits) to show that the structures are not isomorphic. More formally, for two structures S^1 and S^2 we want to find a property P such that $P(S^1) \land \neg P(S^2)$ holds[2].

For example, consider the pairwise non-isomorphic quasigroups S^1, S^2, S^3 given with their respective multiplication tables in Fig. 1. When comparing the

---

[2] In the remainder of the paper we often use a pair (Set, Op) consisting of a set Set and a binary operation Op to describe a structure S.

$$S^1 = (\mathbb{Z}_5, \bar{-}) \qquad S^2 = (\mathbb{Z}_5, \lambda xy.(\bar{2}_5 \bar{*} x)\bar{\mp}y) \qquad S^3 = (\mathbb{Z}_5, \lambda xy.(\bar{3}_5 \bar{*} x)\bar{\mp}y)$$

S^1	$\bar{0}_5$	$\bar{1}_5$	$\bar{2}_5$	$\bar{3}_5$	$\bar{4}_5$
$\bar{0}_5$	$\bar{0}_5$	$\bar{4}_5$	$\bar{3}_5$	$\bar{2}_5$	$\bar{1}_5$
$\bar{1}_5$	$\bar{1}_5$	$\bar{0}_5$	$\bar{4}_5$	$\bar{3}_5$	$\bar{2}_5$
$\bar{2}_5$	$\bar{2}_5$	$\bar{1}_5$	$\bar{0}_5$	$\bar{4}_5$	$\bar{3}_5$
$\bar{3}_5$	$\bar{3}_5$	$\bar{2}_5$	$\bar{1}_5$	$\bar{0}_5$	$\bar{4}_5$
$\bar{4}_5$	$\bar{4}_5$	$\bar{3}_5$	$\bar{2}_5$	$\bar{1}_5$	$\bar{0}_5$

S^2	$\bar{0}_5$	$\bar{1}_5$	$\bar{2}_5$	$\bar{3}_5$	$\bar{4}_5$
$\bar{0}_5$	$\bar{0}_5$	$\bar{1}_5$	$\bar{2}_5$	$\bar{3}_5$	$\bar{4}_5$
$\bar{1}_5$	$\bar{2}_5$	$\bar{3}_5$	$\bar{4}_5$	$\bar{0}_5$	$\bar{1}_5$
$\bar{2}_5$	$\bar{4}_5$	$\bar{0}_5$	$\bar{1}_5$	$\bar{2}_5$	$\bar{3}_5$
$\bar{3}_5$	$\bar{1}_5$	$\bar{2}_5$	$\bar{3}_5$	$\bar{4}_5$	$\bar{0}_5$
$\bar{4}_5$	$\bar{3}_5$	$\bar{4}_5$	$\bar{0}_5$	$\bar{1}_5$	$\bar{2}_5$

S^3	$\bar{0}_5$	$\bar{1}_5$	$\bar{2}_5$	$\bar{3}_5$	$\bar{4}_5$
$\bar{0}_5$	$\bar{0}_5$	$\bar{1}_5$	$\bar{2}_5$	$\bar{3}_5$	$\bar{4}_5$
$\bar{1}_5$	$\bar{3}_5$	$\bar{4}_5$	$\bar{0}_5$	$\bar{1}_5$	$\bar{2}_5$
$\bar{2}_5$	$\bar{1}_5$	$\bar{2}_5$	$\bar{3}_5$	$\bar{4}_5$	$\bar{0}_5$
$\bar{3}_5$	$\bar{4}_5$	$\bar{0}_5$	$\bar{1}_5$	$\bar{2}_5$	$\bar{3}_5$
$\bar{4}_5$	$\bar{2}_5$	$\bar{3}_5$	$\bar{4}_5$	$\bar{0}_5$	$\bar{1}_5$

Fig. 1. Some quasigroup multiplication tables

tables of S^1 and S^2, one discriminant is fairly obvious: while S^1 has only $\bar{0}_5$ on the main diagonal, all elements on the main diagonal of S^2 are distinct. Thus, the invariant property we can use is $\exists x.\forall y.x = y \circ y$. Things become less obvious when we compare the multiplication tables of S^2 and S^3. Here, one invariant of S^3, which does not hold for S^2, is $\forall x.\forall y.(x \circ x = y) \Rightarrow (y \circ y = x),$.

The generalised proof procedure is as follows: given two structures S^1 and S^2 we have to:

1. find an appropriate discriminant P,
2. show that $P(S^1)$ holds,
3. show that $\neg P(S^2)$ holds, and finally
4. show that $\forall X.\forall Y.P(X) \wedge \neg P(Y) \Rightarrow X \not\simeq Y$ holds [3].

The single proof parts combine to give the following, sketched formal proof:

$$
\cfrac{
\cfrac{\vdots^{?}\, (2) \qquad \vdots^{?}\, (3)}{\cfrac{P(S_1) \quad \neg P(S_2)}{P(S_1) \wedge \neg P(S_2)}\,\wedge Intro}
\qquad
\cfrac{\vdots^{?}\, (4)}{\cfrac{\forall X.\forall Y.P(X) \wedge \neg P(Y) \Rightarrow X \not\simeq Y}{P(S_1) \wedge \neg P(S_2) \Rightarrow S_1 \not\simeq S_2}\,\forall Elim(S_1, S_2)}
}{S_1 \not\simeq S_2}\, ModusPonens
$$

In the following section, we describe how we realize this proof technique with a combination of MULTI and several other heterogenous systems. As we observed in the example, discriminants are not necessarily obvious and can have fairly complicated definitions. Thus, our realization of this technique aims to automatically generate discriminants for an arbitrary pair of structures and provide them to MULTI.

3 The Systems Involved

The proof technique described at the end of the previous section is realized in ΩMEGA's proof planner MULTI. Given two residue class structures $(RS_n^1, \circ^1)$ and

---

[3] While step 4 is fairly obvious for a human mathematician, it is crucial for a formal proof.

$(RS_m^2, \circ^2)$ its task is to show $(RS_n^1, \circ^1) \not\sim (RS_m^2, \circ^2)$ by coordinating the steps 1 to 4 above. To compute a suitable discriminant P, we employ HR [3], a system for theory formation. To obtain a formal proof that P is a discriminant for two arbitrary structures X and Y (i.e., step 4) we use first-order automated theorem provers (ATPs), which we call via TRAMP [9], an interface and transformation system. The proofs that P is a discriminant for the two residue class structures (i.e., that $P(RS_n^1, \circ^1)$ and $\neg P(RS_m^2, \circ^2)$ holds) are done by MULTI itself possibly with the help of the general purpose computer algebra system MAPLE. In this section, we give a short overview of the different systems involved and explain in more detail how the proof planner integrates and coordinates the computations of these systems to assemble an overall proof.

3.1   The HR System

The HR system performs automated theory formation by inventing concepts, making conjectures, proving theorems, and finding counterexamples [3]. The main functionality used for the application to finding discriminants discussed here is concept formation, which is achieved by using production rules which take one (or two) old concepts as input and output a new concept. In particular, we used the following four production rules:

- Compose: this composes functions using conjugation.
- Match: this equates variables in predicate definitions.
- Forall: this introduces existential quantification.
- Exists: this introduces universal quantification.

As an example construction which is further discussed later, consider the concept of there being a single element on the diagonal of the multiplication table of an algebra, as is the case for S^1 in Fig. 1. This concept is constructed by HR using the match, forall and exists production rules, as depicted in Fig. 2. In this scenario, two concepts are supplied by the user, namely the concept of an element of the algebra and the multiplication of two elements to give a third. Using the match production rule with the multiplication concept, it invents the notion of multiplying an element by itself. By using this in the forall production rule, it invents the concept of elements which all other elements multiply by themselves to give. Then, using the exists production rule, HR invents the

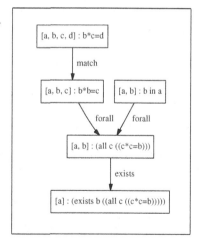

Fig. 2. Example construction

notion of algebras where there is such an element. As we shall see in section 4, this concept can be used to discriminate between the non-Abelian quasigroups in Fig. 1. For more details of how HR forms concepts, see [6].

There are many parameters that users can experiment with in order to get the best performance out of HR for a particular application. For instance, the user can tell HR to ignore concept formation steps if the arity of the function produced will be greater than a certain limit. HR can tell in advance what arity the function will be for a particular concept formation step involving a particular production rule, and the user can specify a limit for individual production rules.

3.2   Calling Automated Theorem Provers via TRAMP

Calls to ATPs are not directly executed but rerouted via the TRAMP system. TRAMP comprises two functionalities: firstly, it is an interface to several ATPs for first-order logic. For a given problem, it produces input in the formats of the connected systems and runs the systems concurrently. Secondly, TRAMP can transform the output of the ATPs into natural deduction (ND) proofs.

TRAMP currently interfaces a variety of first order ATPs and term rewriting systems. For our problems, TRAMP employs OTTER, BLIKSEM, and SPASS. All three systems are based on the resolution principle and can deal with first-order logic problems with equality. In our scenario, it is necessary that the provers can deal with equality, since the discriminant properties found by HR usually contain equations (see the examples discussed in section 4.1).

3.3   The Multiple-Strategies Proof Planner MULTI

Proof planning [2] considers mathematical theorems as planning problems where an *initial partial plan* is composed of the proof *assumptions* and the theorem as an *open goal*. A proof plan is then constructed with the help of abstract planning steps, called *methods*, that are essentially partial specifications of tactics known from tactical theorem proving. In order to ensure correctness, proof plans have to be executed to generate a sound calculus level proof.

In the ΩMEGA system, the traditional proof planning approach is enriched by incorporating mathematical knowledge into the planning process (see [14] for details). That is, methods can encode general proving steps as well as knowledge particular to a mathematical domain. Moreover, *control rules* provide the possibility of introducing mathematical knowledge on how to proceed in the proof planning process by specifying how to traverse the search space. Depending on the mathematical domain or proof situation, they can influence the planner's behaviour at choice points (e.g., which goal to tackle next or which method to apply next).

ΩMEGA's proof planner MULTI additionally provides a strategic level that extends proof planning. For instance, different planning strategies can implement different proof techniques by specifying particular sets of methods and control rules. Thus they enable the same problem to be tackled in different ways. Moreover, different strategies for backtracking and other facilities for refining or modifying proof plans can be specified. When more than one strategy is applicable to a problem, MULTI can reason about which strategy to employ and switch strategies during a proof attempt. In particular, the planner can backtrack from applied strategies and thus perform search on the level of strategies.

3.4   Incorporating HR and TRAMP

One way to incorporate specific knowledge into the planning process is by exploiting knowledge implicitly given in specialised external systems. MULTI supports the access of external systems in methods, control rules, and strategies. In general, computations from external systems can be treated in two ways: as hints or as proof steps. The difference is that the soundness of hints is checked by the subsequent proof planning process, which either fails or succeeds for the given hint. On the other hand, to guarantee the soundness of proof steps, special procedures have to be provided which transform the output of external systems into subproofs that ΩMEGA can check.

In our scenario, we have both: the concepts provided by HR are hints and the output of ATPs are proof steps where we employ TRAMP as a corresponding transformation module. MULTI employs HR in a control rule[4]. When HR succeeds in providing a discriminant P for the problem at hand, the control rule triggers the introduction of this P into the proof plan. Note that MULTI can backtrack on different instantiations of P.

The interface to TRAMP is realized in the method *CallTrampM*. When MULTI applies this method to an open goal, a problem consisting of the formula of the goal and the formulas of the proof assumptions is passed to TRAMP. The method is applicable only if one of the provers interfaced by TRAMP succeeds to prove the problem. Then the goal is simply closed by *CallTrampM* without producing further subgoals. Thus at the method level, the proof of this goal is just one step. However, when the correctness of this method application is checked, its expansion results in the ND proof that TRAMP provides as output.

4   Constructing the Proofs

In this section, we describe the whole proof process and the contributions of the individual systems involved in more detail. We illustrate the procedure with the proof that $(\mathbb{Z}_5, \bar{-}) \not\sim (\mathbb{Z}_5, \lambda xy.\bar{2}_5\bar{*}x\bar{+}y)$ as an example.

The proof procedure is realized with a MULTI proof planning strategy. Among other planning methods, the strategy also contains methods to invoke TRAMP on an appropriately prepared subproblem and a control rule to inject results from HR. When the strategy is applied to a goal of the form $(Set^1, Op^1) \not\sim (Set^2, Op^2)$ MULTI splits this goal into three subgoals of the form:

1. $\mathcal{P}(Set^1, Op^1)$
2. $\neg\mathcal{P}(Set^2, Op^2)$
3. $\forall Set^1, Op^1, Set^2, Op^2.\mathcal{P}(Set^1, Op^1) \wedge \neg\mathcal{P}(Set^2, Op^2)$
$$\Rightarrow [(Set^1, Op^1) \not\sim (Set^2, Op^2)]$$

Here $\mathcal{P}$ is a newly introduced meta-variable (i.e., a placeholder that has to be replaced with some concrete term later in the proof) for the discriminant. Next,

---

[4] Note that the interface between MULTI and HR is currently not automated. That is, the control rule currently allows the user to supply HR's results. However, we intend to make the interface completely automatic.

MULTI receives a suitable discriminant P from HR via the afore-mentioned control rule in order to replace the meta-variable $\mathcal{P}$.

For our example, the discriminant[5] property that HR provides is:
$\lambda Set. \lambda Op. \exists x{:}Set. \forall y{:}Set.\ x = Op(y, y)$.
The concrete instantiated subgoals are therefore of the form:

1. $\exists x{:}\mathbb{Z}_5. \forall y{:}\mathbb{Z}_5. x = y \bar{-} y$
2. $\neg\exists x{:}\mathbb{Z}_5. \forall y{:}\mathbb{Z}_5. x = \bar{2}_5 \bar{*} y \bar{+} y$
3. $\forall Set^1, Op^1, Set^2, Op^2.$
   $[\exists x{:}Set^1. \forall y{:}Set^1. x = Op^1(y, y)] \wedge \neg[\exists x{:}Set^2. \forall y{:}Set^2. x = Op^2(y, y)]$
   $\Rightarrow [(Set^1, Op^1) \not\sim (Set^2, Op^2)]$

Subgoals (1) and (2) are subsequently proved by other MULTI strategies possibly using the computations of the computer algebra system MAPLE. Subgoal (3) is solved with an ATP via TRAMP. In the remainder of this section, we describe the interesting parts of these subproofs in more detail. However, we first explain how we obtain discriminants from HR that are actually useful in our context.

4.1 Obtaining Useful Discriminants from HR

In our scenario, HR needs only to be supplied with the multiplication tables of the example algebras for which it is to find a discriminant, and to be invoked with a flag stating that they are indeed algebraic multiplication tables. Not all possible discriminants are appropriate for our problem. MULTI needs discriminants that are concerned with relations between elements of the sets involved, in particular, existentially or universally quantified statements. For instance, discriminants that involve natural numbers, (e.g., if two structures contain a different number of elements satisfying a certain property) are generally too difficult to handle both for the proof planner and the ATPs. Thus, some of HR's production rules are not appropriate for our application, so they are turned off. In particular, the size production rule that calculates cardinalities of sets – which has been used to find group theory invariants in a different context (see [7] and chapter 12 of [3]) – is not used. We choose only the compose, exists, forall, and match production rules discussed in section 3.1, since they guarantee that only discriminants which are usable in the rest of the proof procedure are found.

HR is run for a given number of concept formation steps and told to output those concepts which both specialise the algebras (e.g., Abelian) and discriminate between the two multiplication tables supplied. The output is then transformed into suitable concepts for MULTI. Currently, the interface between MULTI and HR is not automated. Thus, the discriminants are manually ordered and provided to the control rule that establishes the interface between MULTI and TRAMP. We are currently implementing an automatic interface. The discriminants will then be ordered with respect to their complexity, so that MULTI chooses the least complex first, thereby increasing the chances of success during the subsequent proof planning process and the application of the ATPs.

---

[5] A sorted quantification with a variable $x{:}Set$ states that the variable x ranges over the set Set, only.

For our example from Fig. 1, some of the discriminants HR discovered are:

- $\exists x. \forall y. x = y \circ y$ [There is a single element on the diagonal of the multiplication table (true for S_1, not true for S_2 and S_3).]
- $\forall x, y. (x \circ x = y \Rightarrow y \circ y = x)$ [There is a symmetry on the diagonal, (true for S_1 and S_3, not true for S_2).]
- $\forall x, y. (x \circ y = x \Rightarrow y \circ x = y)$ [If y is a right identity for x, then x is a right identity for y (true for S_1, not true for S_2 and S_3).]

Here $\circ$ stands for the binary operation of the respective structures S_i. When HR passes the properties to MULTI, they are transformed into a lambda abstraction that can be directly inserted for the meta-variable. For the first of the above discriminants, this results in the term $\lambda Set. \lambda Op. \exists x{:}Set. \forall y{:}Set. x = Op(y, y)$, where the quantified variables are restricted to range over the given set Set, only.

4.2   Subproofs with ATPs

The goal we want to prove with an ATP is generally of the form
$$\forall Set^1, Op^1, Set^2, Op^2. P(Set^1, Op^1) \wedge \neg P(Set^2, Op^2)$$
$$\Rightarrow [(Set^1, Op^1) \not\simeq (Set^2, Op^2)]$$
and contains the defined concept of isomorphism. Providing the usual definition of isomorphism to an ATP was often unsuccessful; that is, the ATPs did not find a proof. Instead the following alternative formalisation is better suited: two structures X and Y are isomorphic iff there are two homomorphisms $h{:}X \to Y$ and $j{:}Y \to X$, such that for all x, $h(j(x)) = x$ and $j(h(x)) = x$.

In order to send the problem with the above definition of isomorphism to the ATPs, the original subproblem has to be adequately rewritten. Therefore, MULTI applies an appropriate theorem from ΩMEGA's knowledge base which results in the following new subgoal to be proved:
$$\forall Set^1, Op^1, Set^2, Op^2. P(Set^1, Op^1) \wedge \neg P(Set^2, Op^2) \Rightarrow$$
$$\neg(\exists h. \exists j. hom(h{:}(Set^1, Op^1) \to (Set^2, Op^2)) \wedge hom(j{:}(Set^2, Op^2) \to (Set^1, Op^1))$$
$$\wedge (\forall x : Set^1. j(h(x)) = x \wedge \forall y : Set^2. h(j(y)) = y))$$
After the expansion of the occurrences of the concept hom with the usual definition of homomorphism, the goal can be sent to an ATP.

When trying to close the above subgoal MULTI applies the method *Call-TrampM*. It is applicable when one of the ATPs interfaced by TRAMP succeeds in finding a proof. TRAMP then returns the corresponding ND proof. The formula passed to TRAMP seems to be a higher-order theorem since it contains quantifications on sets, operations, and the functions h and j. However, when TRAMP calls the connected ATPs it first creates a clause normal form of the given theorem and thereby all the higher-order variables become constants (the theorem is negated for clause normalisation). TRAMP does not use any particular settings for the ATPs but calls them all in their default mode.

On the planning level, the subgoal is simply closed by the application of *CallTrampM*. When the plan is executed, the ND proof provided by TRAMP is the expansion of the *CallTrampM* step. Although TRAMP has the ability to translate refutation proofs into direct ND proofs for our domain the proofs are nevertheless indirect since the indirect argument is an inherent part of the problem. Moreover, in our domain, TRAMP's proofs suffer from the fact that they are low level and can become quite lengthy. For our example, TRAMP produces ND proofs containing between 71 (ND proof transformed from SPASS proof) and 104 steps (from BLIKSEM proof). Thus, the expanded subproofs are usually relatively difficult to understand for the user. Nevertheless, TRAMP provides a proof format that ΩMEGA can directly check.

4.3   Subproofs with MULTI

In the remainder of the proof, we have to prove the two goals $P(Set^1, Op^1)$ and $\neg P(Set^2, Op^2)$, where P is the instantiated discriminant. This part is proved by MULTI using one of two proof planning strategies that have already been implemented to prove simple properties of residue class structures such as associativity, existence of unit elements etc. (see [12] for details).

The first strategy implements an exhaustive case analysis. This is possible since in our problems all quantifiers range over finite sets. The strategy proceeds with the case analysis in two different ways, depending on whether (1) a universal or (2) an existential statement has to be proved. In case (1) a split over all the elements in the set involved is performed and the statement in question is proved for every single element separately. In case (2) the single elements of the set involved are examined stepwise until one is found for which the statement holds.

The idea of the second strategy is to use equational reasoning as much as possible to prove properties of residue classes. Instead of checking the validity of the statements for all possible cases, it tries to reduce goals to equations. Then these equations are passed to the computer algebra system MAPLE to check whether the equality actually holds. If an equation contains meta-variables, these are considered as the variables the equation is to be solved for, and they are supplied to MAPLE as additional arguments.

MULTI always tries to apply the equational reasoning strategy first since it is generally faster and produces shorter proofs. If it fails (e.g., if a goal cannot be reduced to equations), an exhaustive case analysis is used, which is applicable to all occurring problems.

In our example, the goals to prove are:

$$\exists x{:}\mathbb{Z}_5 \boldsymbol{.} \forall y{:}\mathbb{Z}_5 \boldsymbol{.}\, x = y \bar{-} y \text{ and } \neg \exists x{:}\mathbb{Z}_5 \boldsymbol{.} \forall y{:}\mathbb{Z}_5 \boldsymbol{.}\, x = \bar{2}_5 \bar{*} y \bar{+} y$$

The latter goal can only be proved with exhaustive case analysis. MULTI first rewrites the goal by pushing the negation inside: $\forall x{:}\mathbb{Z}_5 \boldsymbol{.} \exists y{:}\mathbb{Z}_5 \boldsymbol{.}\, x \neq \bar{2}_5 \bar{*} y \bar{+} y$. Then a case split on the universally quantified variable x over its range $\{\bar{0}_5, \bar{1}_5, \bar{2}_5, \bar{3}_5, \bar{4}_5\}$ is performed. This results in one goal for each of the five elements of the domain. For instance, the resulting goal for $\bar{0}_5$ is $\exists y{:}\mathbb{Z}_5 \boldsymbol{.} \bar{0}_5 \neq \bar{2}_5 \bar{*} y \bar{+} y$. To prove this goal, MULTI tries to find a suitable instantiation for y from the set $\{\bar{0}_5, \bar{1}_5, \bar{2}_5, \bar{3}_5, \bar{4}_5\}$ by inserting each possible instantiation for y in succession and attempting to prove the resulting inequality. In our example, it succeeds for $\bar{1}_5$.

Table 1. Proportion of pairs of algebras for which HR found discriminants

Algebra	Size	Pairs	Successes	Proportion
Abelian Quasigroups	5	3	3	100.0%
Non-Abelian Quasigroups	5	91	90	98.9%
Non-Abelian Semigroups	5	3	3	100.0%
Abelian Magmas	5	14	13	92.8%
Non-Abelian Magmas	5	630	609	96.7%
Non-Abelian Quasigroups	6	1	1	100.0%
Abelian Semigroups	6	6	6	100.0%
Non-Abelian Semigroups	6	28	25	89.3%
Non-Abelian Quasigroups	10	21	21	100.0%
Abelian Semigroups	10	1	1	100.0%
Non-Abelian Semigroups	10	1	1	100.0%
Abelian Magmas	10	15	15	100.0%
Non-Abelian Magmas	10	3	3	100.0%
Total		817	791	96.8%

The second subgoal, $\exists x{:}\mathbb{Z}_5.\forall y{:}\mathbb{Z}_5.x = y\,\bar{-}\,y$, can be proved by the equational reasoning strategy. Thereby, MULTI first decomposes the quantifiers which results in the goal $cl_5(z) = cl_5(c)\,\bar{-}\,cl_5(c)$ where z is a new meta-variable and c is a new constant. Since this goal is a statement on congruence classes, it is transformed into the corresponding goal on integers: $z \bmod 5 = (c - c) \bmod 5$, and passed to MAPLE. MAPLE applies its function `msolve` to compute that the equation holds in general for $z = 0$. This result is used as instantiation for z in the proof plan and the goal is closed.

5   Results

To test HR's effectiveness at finding discriminants for arbitrary pairs of algebraic objects, we used 817 pairs over 6 different algebras (of sizes 5, 6, and 10). For each pair, HR was allowed 500 concept formation steps in order to find discriminants. We allowed HR to use the four production rules previously mentioned, and after some experimentation, we chose the following function arity limits: compose(3), exists(1), forall(2) and match(3). The average session took around 22 seconds on a 500Mhz pentium machine. The results are given in table 1. We note that HR found discriminants for nearly 97% of the pairs given to it, which was a higher success rate than we anticipated given the short session lengths we allowed for these tests. On average, HR found 20 discriminants for each pair, and in total, 517 distinct concepts were used as discriminants.

In the sets of discriminants HR constructed, there were usually some of a complexity similar to those discriminants discussed in section 4.1. In fact, one of the most complex discriminants HR found was for the two non-Abelian magmas M^1 and M^2 given in Fig. 3. HR constructed only two discriminants for M^1 and M^2, including: $\exists x.(x \circ x = x \wedge \forall y.(y \circ y = x \Rightarrow y \circ y = y))$. This states that there exists an idempotent element, x, such that any other element which squares to

$$M^1 = (\mathbb{Z}_5, \lambda xy.(x \bar{-} \bar{1}_5)\bar{*}(y \bar{*} \bar{2}_5)) \qquad M^2 = (\mathbb{Z}_5, \lambda xy.(x \bar{+} \bar{1}_5)\bar{*}(y \bar{*} \bar{2}_5))$$

M^1	$\bar{0}_5$	$\bar{1}_5$	$\bar{2}_5$	$\bar{3}_5$	$\bar{4}_5$
$\bar{0}_5$	$\bar{0}_5$	$\bar{3}_5$	$\bar{1}_5$	$\bar{4}_5$	$\bar{2}_5$
$\bar{1}_5$	$\bar{0}_5$	$\bar{0}_5$	$\bar{0}_5$	$\bar{0}_5$	$\bar{0}_5$
$\bar{2}_5$	$\bar{0}_5$	$\bar{2}_5$	$\bar{4}_5$	$\bar{1}_5$	$\bar{3}_5$
$\bar{3}_5$	$\bar{0}_5$	$\bar{4}_5$	$\bar{3}_5$	$\bar{2}_5$	$\bar{1}_5$
$\bar{4}_5$	$\bar{0}_5$	$\bar{1}_5$	$\bar{2}_5$	$\bar{3}_5$	$\bar{4}_5$

M^2	$\bar{0}_5$	$\bar{1}_5$	$\bar{2}_5$	$\bar{3}_5$	$\bar{4}_5$
$\bar{0}_5$	$\bar{0}_5$	$\bar{2}_5$	$\bar{4}_5$	$\bar{1}_5$	$\bar{3}_5$
$\bar{1}_5$	$\bar{0}_5$	$\bar{4}_5$	$\bar{3}_5$	$\bar{2}_5$	$\bar{1}_5$
$\bar{2}_5$	$\bar{0}_5$	$\bar{1}_5$	$\bar{2}_5$	$\bar{3}_5$	$\bar{4}_5$
$\bar{3}_5$	$\bar{0}_5$	$\bar{3}_5$	$\bar{1}_5$	$\bar{4}_5$	$\bar{2}_5$
$\bar{4}_5$	$\bar{0}_5$	$\bar{0}_5$	$\bar{0}_5$	$\bar{0}_5$	$\bar{0}_5$

Fig. 3. The multiplication tables of two non-Abelian magmas.

give x is itself idempotent. After some thought, it is obvious that this means that there must be an idempotent element which appears only once on the diagonal. Such an element exists in M^2 (the element is $\bar{2}_5$), but not in M^1.

The relative simplicity of the discriminants facilitates the subsequent proofs of MULTI and the ATPs. In particular, for the latter, it is usually hard to predict whether or not they will succeed for a given discriminant. Here, a large set of different discriminants can increase the chance of success, since MULTI can backtrack if the ATPs cannot come up with a proof for one discriminant and instantiate another one. So far, we have tested the planning and ATP side with roughly 100 examples for which HR found discriminants. All tested examples were successfully proved. Among the ATPs, SPASS and BLIKSEM performed best on our problems. In particular, we also proved the example of the two non-Abelian magmas with the discriminant given above. The resulting proof plan consists of 154 steps. The ATP part was proved by BLIKSEM, which was transformed by TRAMP into an ND proof containing 112 steps.

A comparison of attempts to solve problems with structures using $\mathbb{Z}_5$ or $\mathbb{Z}_6$ as opposed to structures using $\mathbb{Z}_{10}$ shows that the incorporated systems depend on the cardinality of the involved residue class structures to different degrees. In our experiments, HR's performance did not vary significantly when applied to problems with $\mathbb{Z}_5$ or $\mathbb{Z}_{10}$. That is, HR solved 100% of the problems of size 10 (see Table 1), but more importantly, it took roughly the same time to provide the solutions. This is because, in this application, HR works with only two structures at a time, which is a much smaller amount of data than it usually works with, so the difference between size 5 and 10 algebras was minimal.

The ATPs were also not affected by the cardinality of the involved structures simply because the subproblem they are responsible for does not contain the residue class structures. The complexity of the problem part for the ATPs depends only on the complexity of the discriminants provided by HR. However, in our experiments, we found that the complexity of HR's discriminants does not depend on the cardinality of the involved residue class structures.

Unlike HR and the ATPs, MULTI's performance depends on both the cardinality of the residue class sets, and the strategy applied. When performing equational reasoning, MULTI solves a subproblem with set $\mathbb{Z}_5$ in the same number of steps as the same subproblem with set $\mathbb{Z}_{10}$. On the other hand, MULTI's performance considerably differs when it applies the exhaustive case analysis strategy. A subproblem with l nested quantifications for $\mathbb{Z}_5$ results in 5^l cases, whereas for $\mathbb{Z}_{10}$, it results in 10^l cases. Although MULTI prefers the application of the equational reasoning, it is never the case that both of MULTI's subproblems, $P(Set^1, Op^1)$ and $\neg P(Set^2, Op^2)$, are solvable by this strategy.

6   A Comparison with a Different Approach

We have also experimented with the model generator SEM [17] to construct discriminants. Here, properties are created in ΩMEGA and SEM tests whether each property is a discriminant. That is, SEM finds models for each of the two involved structures with respect to the created property. A property is a candidate for a discriminant if SEM succeeds for one structure and fails for the other one. The possible properties are constructed as follows. The basic construction element is the equation $x = Op(y, z)$. Associated with this equation are all formulas/properties that result (1) from quantifying the x, y, z either existentially or universally (e.g., $\forall x. \exists y. \exists z. x = Op(y, z)$ or $\exists x. \forall y. \forall z. x = Op(y, z)$ etc.). Further properties are generated by (2) permuting the sequence of the quantifications for x, y, z (e.g., $\forall y. \exists z. \exists x. x = Op(y, z)$ etc.) and by (3) making some variables equal (e.g., $\exists y. \forall x. x = Op(y, x)$). More complicated properties result from the combination of several copies of the basic construction element by $\wedge, \vee, \Rightarrow$, for instance, $x_1 = Op(y_1, z_1) \Rightarrow x_2 = Op(y_2, z_2)$ or $x_1 = Op(y_1, z_1) \wedge (x_2 = Op(y_2, z_2) \Rightarrow x_3 = Op(y_3, z_3))$ etc. The properties associated with such combined equations are again constructed using the mechanisms (1), (2), and (3) etc.

For a residue class structure $(RS_n, \circ)$ and a property p, SEM is passed the multiplication table of the operation of the residue class structure together with the clauses resulting from the normalization of the property p. The multiplication table for RS_n is encoded as a set of n^2 equations of the form $c = Op(a, b)$ where c is the result of $a \circ b$. Then SEM is asked to find a model for this input. We tried to find discriminants for two given residue class structures by either systematically or randomly checking the properties in the space of our construction mechanism. However, it turned out that we could only find very simple discriminants, such as $\lambda Set. \lambda Op. \exists x{:}Set. \forall y{:}Set. x = Op(y, y)$. More complicated discriminants like the discriminant $\exists x. (x * x = x \wedge \forall y. (y \circ y = x \Rightarrow y \circ y = y))$ discussed in section 5 are in our search space but are out of the reach of this approach[6].

The problems of the approach using SEM are twofold. First, we search in an enormous space of possible candidates and second we search blindly since we were not able to guide the search by powerful heuristics. Restricted to the 4 production rules mentioned, HR searches a similar space, but its search mechanism is a little more sophisticated, as it recognises two things. Firstly, if it constructs a concept with no examples, then no further concepts are built from that, as they will also

---

[6] Note that even for the combination of two copies of the basic construction element (e.g., $x_1 = y_1 \ op \ z_1 \Rightarrow x_2 = y_2 \ op \ z_2$) the number of possible properties resulting from the construction mechanisms (1), (2), and (3) is roughly 100000.

have no examples (and hence will be useless as a discriminant). Secondly, if it constructs a concept with the same examples as a previous concept, the new concept is discarded, to reduce redundancy. Such simple empirical checks are very powerful in reducing the amount of search HR undertakes.

HR's success was achieved using only one of many mechanisms available to it: a limited search, where certain search steps are taken off the agenda if they would result in functions of too high an arity. If the problems had been more difficult, there are two additional mechanisms we could have experimented with, namely: (i) a heuristic search, which measures the concepts in various ways and builds new concepts from the most interesting old ones first [5] and (ii) a forward look ahead mechanism, which can tell in advance whether the application of up to three concept formation steps will lead to a concept which achieves a particular categorisation task, in our case finding an invariant, as discussed at length in [6].

7   Conclusion

To produce more sophisticated systems to tackle increasingly difficult problems, it is necessary to combine programs so that the whole is greater than the sum of the parts. We have presented an example here of a fruitful cooperation of heterogeneous mathematical systems to prove non-isomorphism theorems in the residue class domain. The cooperation is essentially orchestrated by a proof planner, MULTI, which uses a theory formation system, HR, to construct appropriate discriminants for given structures. The proof is then planned with the help of automated theorem provers and sometimes a computer algebra system.

The problem of identifying a discriminant for two objects is a machine learning problem, which could, in theory, be solved by a program such as Progol [15], which uses Inductive Logic Programming to identify a concept which correctly categorises a set of positive and negative examples. However, as mentioned in [4], this may be difficult in practice because we only supply a single positive and a single negative example, which would suggest that the amount of compression in a concept would not be high enough to be suggested as a viable solution.

Currently, we call HR manually and supply the proof planner with the appropriate information, but we are working on automating this communication. Nevertheless, we have already tested the cooperation with a number of examples and HR proved successful for the vast majority of these examples. Moreover, HR's discriminants could be handled both by the proof planner and the automated theorem provers successfully even for relatively complicated discriminants.

References

1. C. Benzmüller, L. Cheikhrouhou, D. Fehrer, A. Fiedler, X. Huang, M. Kerber, M. Kohlhase, K. Konrad, E. Melis, A. Meier, W. Schaarschmidt, J. Siekmann, and V. Sorge. ΩMega: Towards a Mathematical Assistant. In *Proceedings of the 14th International Conference on Automated Deduction (CADE-14)*, volume 1249 of *LNAI*, pages 252–255. Springer Verlag, Germany, 1997.

2. A. Bundy. The Use of Explicit Plans to Guide Inductive Proofs. In *Proceedings of the 9th International Conference on Automated Deduction (CADE-9)*, volume 310 of *LNCS*, pages 111–120. Springer Verlag, Germany, 1988.

3. S. Colton. *Automated Theory Formation in Pure Mathematics*. PhD thesis, Department of Artificial Intelligence, University of Edinburgh, 2000.

4. S. Colton. An application-based comparison of automated theory formation and inductive logic programming. *Linkoping Electronic Articles in Computer and Information Science (special issue: Proceedings of Machine Intelligence 17)*, forthcoming, 2002.

5. S. Colton, A Bundy, and T Walsh. On the notion of interestingness in automated mathematical discovery. *International Journal of Human Computer Studies*, 53(3):351–375, 2000.

6. S. Colton, A. Bundy, and T. Walsh. Automatic identification of mathematical concepts. In *Proceedings of the 17th International Conference on Machine Learning (ICML2000)*, pages 183–190. Morgan Kaufmann, USA, 2001.

7. S. Colton, S Cresswell, and A Bundy. The use of classification in automated mathematical concept formation. In *Proceedings of SimCat 1997: An Interdisciplinary Workshop on Similarity and Categorisation*. University of Edinburgh, 1997.

8. The GAP Group, Aachen, St Andrews. *GAP – Groups, Algorithms, and Programming, Version 4*, 1998. http://www-gap.dcs.st-and.ac.uk/~gap.

9. A. Meier. Tramp: Transformation of Machine-Found Proofs into ND-Proofs at the Assertion Level. In *Proceedings of the 17th International Conference on Automated Deduction (CADE-17)*, volume 1831 of *LNAI*, pages 460–464. Springer Verlag, Germany, 2000.

10. A. Meier, M. Pollet, and V. Sorge. Classifying Isomorphic Residue Classes. In *Proceedings of the 8th International Workshop on Computer Aided Systems Theory (EuroCAST 2001)*, volume 2178 of *LNCS*, pages 494–508. Springer Verlag, Germany, 2001.

11. A. Meier, M. Pollet, and V. Sorge. Comparing Approaches to Explore the Domain of Residue Classes. *Journal of Symbolic Computations*, 2002. forthcoming.

12. A. Meier and V. Sorge. Exploring Properties of Residue Classes. In *Proceedings of the CALCULEMUS-2000 Symposium*, pages 175–190. AK Peters, USA, 2001.

13. E. Melis and A. Meier. Proof planning with multiple strategies. In *Proceedings of the First International Conference on Computational Logic*, volume 1861 of *LNAI*. Springer Verlag, Germany, 2000.

14. E. Melis and J. Siekmann. Knowledge-Based Proof Planning. *Artificial Intelligence*, 115(1):65–105, 1999.

15. S. Muggleton. Inverse entailment and Progol. *New Generation Computing*, 13:245–286, 1995.

16. D. Redfern. *The Maple Handbook: Maple V Release 5*. Springer Verlag, Germany, 1999.

17. J. Zhang and H. Zhang. SEM: a System for Enumerating Models. In *Proceedings of the 14th International Joint Conference on Artificial Intelligence (IJCAI)*, pages 298–303. Morgan Kaufmann, USA, 1995.

Unification with Sequence Variables and Flexible Arity Symbols and Its Extension with Pattern-Terms*

Temur Kutsia[1,2]

[1] Research Institute for Symbolic Computation
Johannes Kepler University Linz
A-4040, Linz, Austria
kutsia@risc.uni-linz.ac.at
[2] Software Competence Center Hagenberg
Hauptstrasse 99
A-4232, Hagenberg, Austria
teimuraz.kutsia@scch.at

Abstract. A minimal and complete unification procedure for a theory with individual and sequence variables, free constants and free fixed and flexible arity function symbols is described and a brief overview of an extension with pattern-terms is given.

1 Introduction

We design a unification procedure for a theory with individual and sequence variables, free constants and free fixed and flexible arity function symbols. The subject of this research was proposed by B. Buchberger in [4] and in a couple of personal discussions [5]. The research described in this paper is a part of the author's PhD thesis.

We refer to unification in a theory with individual and sequence variables, free constants and free fixed and flexible arity function symbols shortly as unification with sequence variables and flexible arity symbols, underlining the importance of these two constructs. Sequence variables are variables which can be instantiated by an arbitrary finite (possibly empty) sequence of terms. Flexible arity function symbols can take arbitrary finite (possibly empty) number of arguments. In the literature the symbols with similar property are also referred to as "variable arity", "variadic" or "multiple arity" symbols. Languages with sequence variables and variable arity symbols have been used in various areas. Here we enumerate some of them:

- Knowledge management - Knowledge Interchange Format KIF ([10]) and its version SKIF ([23]) are extensions of first order language with (among other constructs) individual and sequence variables and variable arity function

* Supported by the Austrian Science Foundation (FWF) under Project SFB F1302 and by Software Competence Center Hagenberg (Austria) under MathSoft project.

J. Calmet et al. (Eds.): AISC-Calculemus 2002, LNAI 2385, pp. 290–304, 2002.

symbols. KIF is used to interchange knowledge among disparate computer systems. Another example of using sequence variables and variable arity symbols in knowledge systems is Ontolingua ([8]) - a tool which provides a distributed collaborative environment to browse, create, edit, modify, and use ontologies.

- Databases - sequences and sequence variables provide flexibility in data representation and manipulation for genome or text databases, where much of the data has an inherently sequential structure. Numerous formalisms involving sequences and sequence variables, like Sequence Logic ([11]), Alignment Logic ([12]), Sequence Datalog ([20]), String Calculus ([13],[3]), have been developed for this field.
- Rewriting - variable arity symbols used in rewriting usually come from flattening terms with associative top function symbol. Sequences and sequence variables (sometimes called also patterns), which are used together with variable arity symbols, make the syntax more flexible and expressive, and increase the performance of a rewriting system (see [29], [14]).
- Programming languages - variable arity symbols are supported by many of them. The programming language of Mathematica ([30]) is one of such examples, which uses the full expressive power of sequence variables as well. A relation of Mathematica programming language and rewrite rule languages, and the role of sequence variables in this relation is discussed in [4].
- Theorem proving - the Epilog package ([9]) can be used in programs that manipulate information encoded in Standard Information Format (SIF) - a subset of KIF ([10]) language, containing sequence variables and variable arity symbols. Among the other routines, Epilog includes pattern matchers of various sorts, and an inference procedure based on model elimination.

These applications involve (and in some cases, essentially depend on) solving equations with sequence variables and variable arity symbols. The most used solving technique is matching. However, for some applications, like theorem proving or completion, more powerful solving techniques (unification, for instance) are needed.

The problem whether Knuth-Bendix completion procedure ([16]) can be extended to handle term rewriting systems with function symbols of variable arity, sequences and sequence variables (patterns) is stated as an open problem in [29]. The primary reason why it is an open problem is the absence of appropriate unification algorithm.

In this paper, we make the first step towards solving this problem, providing a unification procedure with individual and sequence variables, fixed and flexible arity function symbols and its extension with pattern-terms. Sequence variables and pattern-terms can be seen as particular examples of the pattern construct of [29]. The term "flexible arity" was suggested by Buchberger ([5]) instead of "variable arity", mainly because of the following reason: variable arity symbols, as they are understood in theorem proving or rewriting, are flattened associative symbols, i.e. flat symbols which take at least two arguments, while flexible arity symbols can have zero or one argument as well and are not necessarily flat. Nonflatness is one of main differences between unification with sequence variables and flexible arity symbols and associative unification: the unification problem $f(x, f(y, z)) \overset{?}{=} f(f(a, b), c)$, with the variables x, y, z and constants a, b, c, has no unifier, if f has a flexible arity, but admits a unifier $\{x \leftarrow a, y \leftarrow b, z \leftarrow c\}$ for associative f. Even when f is a flat flexible arity symbol the problem would not be equivalent to A-unification: the substitution $\{x \leftarrow f(a), y \leftarrow f(b, c), z \leftarrow f()\}$ is a unifier for a flat f, but not for an associative f.

The type of unification with sequence variables and flexible arity symbols in the Siekmann unification hierarchy ([28]) is infinitary: for any unification problem there exists the minimal complete set of unifiers which is infinite for some problems.

It should be mentioned that in the theorem proving context quantification over sequence variables naturally introduces flexible arity symbols and constructs that we call patterns. For instance, Skolemizing the expression $\forall \overline{x} \exists y \Phi[\overline{x}, y]$, where $\overline{x}$ is a sequence variable, y is an individual variable and $\Phi[\overline{x}, y]$ is a formula which depends on $\overline{x}$ and y, introduces a flexible arity Skolem function f: $\forall \overline{x} \Phi[\overline{x}, f(\overline{x})]$. On the other hand, Skolemizing the expression $\forall x \exists \overline{y} \Phi[x, \overline{y}]$ introduces a pattern $h_{1,n(x)}(x)$, which can be seen as an abbreviation of a sequence of terms $h_1(x), \ldots, h_{n(x)}(x)$ of unknown length, where $h_1, \ldots, h_{n(x)}$ are Skolem functions.

The procedure we describe can be used in the theorem proving context in the way similar to [24]: building in equational theories. Although unification with sequence variables and flexible arity symbols is infinitary, special cases can be identified when the procedure terminates.

It is shown in [17] that unification with sequence variables and flexible arity symbols is decidable. Based on the decision procedure, a constraint-based approach to theorem proving with sequence variables and flexible arity symbols can be developed (compare [22], [25]).

Particular instances of unification with sequence variables and flexible arity symbols are word equations ([1],[15], [26]), equations over free semigroups ([19]), equations over lists of atoms with concatenation ([7]), pattern matching.

We have implemented the unification procedure (without decision algorithm) as a Mathematica package and incorporated it into the Theorema system [6], which aims at extending computer algebra systems by facilities for supporting mathematical proving. Currently the package is used in the Theorema Equational Prover. It makes Theorema probably the only system being able to handle equations which involve sequence variables and flexible arity symbols. The package also enhances Mathematica solving capabilities, considering unification as a solving method. We used the package, for instance, to find matches for S-polynomials in non-commutative Gröbner Bases algorithm [21].

The results in this paper are given without proofs. They can be found in [17].

2   Preliminaries

We consider an alphabet consisting of the following pairwise disjoint sets of symbols: the set of individual variables $\mathcal{IV}$, the set of sequence variables $\mathcal{SV}$, the set of object constants $\mathcal{CONST}$, the set of fixed arity function symbols $FFIX$, the set of flexible arity function symbols $\mathcal{FFLEX}$ and a singleton consisting of a binary predicate symbol $\doteq$ (equality).

Let now $\mathcal{V}$ stand for $(\mathcal{IV}, \mathcal{SV})$ (variables), $\mathcal{C}$ - for $(\mathcal{CONST}, \mathcal{FFIX}, \mathcal{FFLEX}, \doteq)$ (a domain of constants) and $\mathcal{P}$ - for $\{(,),,\}$ ("parentheses and comma"). We define terms and equations over $(\mathcal{V}, \mathcal{C}, \mathcal{P})$.

Definition 1 (Term). *The set of terms (over $(\mathcal{V}, \mathcal{C}, \mathcal{P})$) is the smallest set of strings over $(\mathcal{V}, \mathcal{C}, \mathcal{P})$ that satisfies the following conditions:*

- *If $t \in \mathcal{IV} \cup \mathcal{SV} \cup \mathcal{CONST}$ then t is a term.*
- *If $f \in \mathcal{FFIX}$, f is n-ary, $n \geq 0$ and $t_1, \dots, t_n$ are terms such that for all $1 \leq i \leq n$, $t_i \notin \mathcal{SV}$, then $f(t_1, \dots, t_n)$ is a term.*
- *If $f \in \mathcal{FFLEX}$ and $t_1, \dots, t_n$ $(n \geq 0)$ are terms, then so is $f(t_1, \dots, t_n)$.*

f is called the head of $f(t_1, \dots, t_n)$.

Definition 2 (Equation). *The set of equations (over the alphabet $(\mathcal{V}, \mathcal{C}, \mathcal{P})$) is the smallest set of strings over $(\mathcal{V}, \mathcal{C}, \mathcal{P})$ that satisfies the following condition:*

- *If t_1 and t_2 are terms over $(\mathcal{V}, \mathcal{C}, \mathcal{P})$ such that $t_1 \notin \mathcal{SV}$ and $t_2 \notin \mathcal{SV}$, then $\doteq (t_1, t_2)$ is an equation over $(\mathcal{V}, \mathcal{C}, \mathcal{P})$. $\doteq$ is called the head of $\doteq (t_1, t_2)$.*

If not otherwise stated, the following symbols, with or without indices, are used as metavariables: x, y and z - over individual variables, $\overline{x}$, $\overline{y}$ and $\overline{z}$ - over sequence variables, v and u - over (individual or sequence) variables, c - over object constants, f, g and h - over (fixed or flexible arity) function symbols, s and t - over terms. We generalize standard notions of unification theory ([2]) for a theory with sequence variables and flexible arity symbols.

Definition 3 (Substitution). *A substitution is a finite set $\{x_1 \leftarrow s_1, \dots, x_n \leftarrow s_n, \overline{x}_1 \leftarrow t_1^1, \dots, t_{k_1}^1, \dots, \overline{x}_m \leftarrow t_1^m, \dots, t_{k_m}^m\}$ where*

- *$n \geq 0$, $m \geq 0$ and for all $1 \leq i \leq m$, $k_i \geq 0$,*
- *$x_1, \dots, x_n$ are distinct individual variables,*
- *$\overline{x}_1, \dots, \overline{x}_m$ are distinct sequence variables,*
- *for all $1 \leq i \leq n$, s_i is a term, $s_i \notin \mathcal{SV}$ and $s_i \neq x_i$,*
- *for all $1 \leq i \leq m$, $t_1^i, \dots, t_{k_i}^i$ is a sequence of terms and if $k_i=1$ then $t_{k_i}^i \neq \overline{x}_i$.*

Greek letters are used to denote substitutions. The empty substitution is denoted by ε.

Definition 4 (Instance). *Given a substitution θ, we define an instance of a term or equation with respect to θ recursively as follows:*

- $x\theta = \begin{cases} s & \text{if } x \leftarrow s \in \theta, \\ x & \text{otherwise} \end{cases}$
- $\overline{x}\theta = \begin{cases} s_1, \dots, s_m & \text{if } \overline{x} \leftarrow s_1, \dots, s_m \in \theta, \ m \geq 0, \\ \overline{x} & \text{otherwise} \end{cases}$
- $f(s_1, \dots, s_n)\theta = f(s_1\theta, \dots, s_n\theta)$
- $(s_1 \doteq s_2)\theta = s_1\theta \doteq s_2\theta$.

We extend the notion of instance to sequences in a straightforward way - instance of a sequence is a sequence of instances.

Definition 5 (Composition of Substitutions). *Let* $\theta = \{x_1 \leftarrow s_1, \dots, x_n \leftarrow s_n, \overline{x_1} \leftarrow t_1^1, \dots, t_{k_1}^1, \dots, \overline{x_m} \leftarrow t_1^m, \dots, t_{k_m}^m\}$ *and* $\lambda = \{y_1 \leftarrow d_1, \dots, y_n \leftarrow d_l, \overline{y_1} \leftarrow e_1^1, \dots, e_{q_1}^1, \dots, \overline{y_r} \leftarrow e_1^r, \dots, e_{q_r}^r\}$ *be two substitutions. Then the composition of* θ *and* λ *is the substitution, denoted by* $\theta \circ \lambda$, *obtained from the set*

$$\{ x_1 \leftarrow s_1\lambda, \dots, x_n \leftarrow s_n\lambda, \overline{x_1} \leftarrow t_1^1\lambda, \dots, t_{k_1}^1\lambda, \dots, \overline{x_m} \leftarrow t_1^m\lambda, \dots, t_{k_m}^m\lambda, \\ y_1 \leftarrow d_1, \dots, y_l \leftarrow d_l, \overline{y_1} \leftarrow e_1^1, \dots, e_{q_1}^1, \dots, \overline{y_r} \leftarrow e_1^r, \dots, e_{q_r}^r\}$$

by deleting

- *all the elements* $x_i \leftarrow s_i\lambda$ *($1 \le i \le n$) for which* $x_i = s_i\lambda$,
- *all the elements* $\overline{x_i} \leftarrow t_1^i\lambda, \dots, t_{k_i}^i\lambda$ *($1 \le i \le m$) for which* $k_i = 1$ *and* $\overline{x_i} = t_1^i\lambda$,
- *all the elements* $y_i \leftarrow d_i$ *($1 \le i \le l$) such that* $y_i \in \{x_1, \dots, x_n\}$,
- *all the elements* $\overline{y_i} \leftarrow e_1^i, \dots, e_{q_i}^i$ *($1 \le i \le r$) such that* $\overline{y_i} \in \{\overline{x_1}, \dots, \overline{x_m}\}$.

Example 1. Let $\theta = \{x \leftarrow f(y), \overline{x} \leftarrow \overline{y}, x, \overline{y} \leftarrow \overline{y}, \overline{z}\}$ and $\lambda = \{y \leftarrow g(c,c), \overline{x} \leftarrow c, \overline{z} \leftarrow\}$. Then $\theta \circ \lambda = \{x \leftarrow f(g(c,c)), y \leftarrow g(c,c), \overline{x} \leftarrow \overline{y}, c, \overline{z} \leftarrow\}$.

These versions of the notions of substitution, composition, and instance have the same important properties as the standard versions of the same notions:

Theorem 1. *For a term t and substitutions θ and λ $t\theta \circ \lambda = t\theta\lambda$.*

Theorem 2. *For any substitutions θ, λ and σ, $(\theta \circ \lambda) \circ \sigma = \theta \circ (\lambda \circ \sigma)$.*

3   Equational Theory with Sequence Variables and Flexible Arity Symbols

A set of equations E (called representation) defines an equational theory, i.e. the equality of terms induced by E. We use the term E-theory for the equational theory defined by E. We will write $s \doteq_E t$ for $s \doteq t$ modulo E. Solving equations in an E-theory is called E-unification. The fact that the equation $s \doteq_E t$ has to be solved is written as $s \overset{?}{\doteq}_E t$. A finite system of equations $\langle s_1 \overset{?}{\doteq}_E t_1, \dots, s_n \overset{?}{\doteq}_E t_n \rangle$ is called an E-unification problem. Some examples of E-theories are:

1. Free theory ($\emptyset$): $E = \emptyset$;
2. Flat theory (F): $E = \{f(\overline{x}, f(\overline{y}), \overline{z}) \doteq f(\overline{x}, \overline{y}, \overline{z})\}$.
3. Restricted flat theory (RF): $E = \{f(\overline{x}, f(\overline{y_1}, x, \overline{y_2}), \overline{z}) \doteq f(\overline{x}, \overline{y_1}, x, \overline{y_2}, \overline{z})\}$.
4. Orderless theory (O): $E = \{f(\overline{x}, x, \overline{y}, y, \overline{z}) \doteq f(\overline{x}, y, \overline{y}, x, \overline{z})\}$.
5. FO: $E = \{f(\overline{x}, f(\overline{y}), \overline{z}) \doteq f(\overline{x}, \overline{y}, \overline{z}), f(\overline{x}, x, \overline{y}, y, \overline{z}) \doteq f(\overline{x}, y, \overline{y}, x, \overline{z})\}$.
6. RFO:

$$E = \{f(\overline{x}, f(\overline{y_1}, x, \overline{y_2}), \overline{z}) \doteq f(\overline{x}, \overline{y_1}, x, \overline{y_2}, \overline{z}), f(\overline{x}, x, \overline{y}, y, \overline{z}) \doteq f(\overline{x}, y, \overline{y}, x, \overline{z})\}.$$

Definition 6 (Unifier). *A substitution θ is called an E-unifier of an E-unification problem $\langle s_1 \overset{?}{=}_E t_1, \dots, s_n \overset{?}{=}_E t_n \rangle$ iff $s_i\theta \doteq_E t_i\theta$ for all $1 \leq i \leq n$.*

Definition 7 (More General Substitution). *A substitution θ is more general than a substitution σ on a finite set of variables Var modulo a theory E (denoted $\theta \ll_E^{Var} \sigma$) iff there exists a substitution λ such that*

- *for all $\overline{x} \in Var$,*
  - *$\overline{x} \leftarrow \notin \lambda$;*
  - *there exist terms $t_1, \dots, t_n, s_1, \dots, s_n$, $n \geq 0$ such that $\overline{x}\sigma = t_1, \dots, t_n$, $\overline{x}\theta \circ \lambda = s_1, \dots, s_n$ and for each $1 \leq i \leq n$, either t_i and s_i are the same sequence variables or $t_i \doteq_E s_i$;*
- *for all $x \in Var$, $x\sigma \doteq_E x\theta \circ \lambda$.*

Example 2. $\{\overline{x} \leftarrow \overline{y}\} \ll_\emptyset^{\{\overline{x}, \overline{y}\}} \{\overline{x} \leftarrow a, \overline{z}, \ \overline{y} \leftarrow a, \overline{z}\}$, but not $\{\overline{x} \leftarrow \overline{y}\} \ll_\emptyset^{\{\overline{x}, \overline{y}\}} \{\overline{x} \leftarrow, \overline{y} \leftarrow\}$.

Definition 8 (The Minimal Complete Set of Unifiers). *The minimal complete set of E-unifiers of Γ, denoted $MCU_E(\Gamma)$, is an E-minimal set of substitutions with respect to the set of variables Var of Γ, satisfying the following conditions:*

E-Correctness - for all $\theta \in MCU_E(\Gamma)$, θ is an E-unifier of Γ.
E-Completeness - for any E-unifier σ of Γ there exists $\theta \in MCU_E(\Gamma)$ such that $\theta \ll_E^{Var} \sigma$.
E-minimality - for all $\theta, \sigma \in MCU_E(\Gamma)$, $\theta \ll_E^{Var} \sigma$ implies $\theta = \sigma$.

Example 3. Compute the minimal complete set of unifiers in the $\emptyset$, F and RF theories (f and g are free flexible arity symbols, h is flat, rh - restricted flat):

1. $MCU_\emptyset(\langle f(\overline{x}, a) \overset{?}{=}_\emptyset f(a, \overline{x}) \rangle) = \{\{\overline{x} \leftarrow \}, \{\overline{x} \leftarrow a\}, \{\overline{x} \leftarrow a, a\}, \dots \}$.
2. $MCU_\emptyset(\langle f(g(a, g(\overline{y}, c)), \overline{x}) \overset{?}{=}_\emptyset f(\overline{u}, g(b, \overline{v})) \rangle) = \{\{\overline{u} \leftarrow g(a, \overline{x}), \overline{y} \leftarrow b, \overline{v} \leftarrow c\}, \{\overline{x} \leftarrow, \overline{u} \leftarrow g(a), \overline{y} \leftarrow b, \overline{v} \leftarrow c\}, \{\overline{u} \leftarrow g(a, \overline{x}), \overline{y} \leftarrow b, \overline{y}, \overline{v} \leftarrow \overline{y}, c\}, \{\overline{x} \leftarrow, \overline{u} \leftarrow g(a), \overline{y} \leftarrow b, \overline{y}, \overline{v} \leftarrow \overline{y}, c\}\}$.
3. $MCU_F(\langle x \overset{?}{=}_F h(x) \rangle) = \{\{x \leftarrow h(x)\}\}$.
4. $MCU_F(\langle h(\overline{x}) \overset{?}{=}_F h(a) \rangle) = \{\{\overline{x} \leftarrow a\}, \{\overline{x} \leftarrow h(a)\}, \{\overline{x} \leftarrow a, h()\}, \{\overline{x} \leftarrow h(a), h()\}, \{\overline{x} \leftarrow h(), a\}, \{\overline{x} \leftarrow h(), h(a)\}, \{\overline{x} \leftarrow h(), a, h()\}, \dots \}$.
5. $MCU_{RF}(\langle rh(\overline{x}) \overset{?}{=}_{RF} rh(a) \rangle) = \{\{\overline{x} \leftarrow a\}, \overline{x} \leftarrow rh(a)\}$.

Below in this paper we consider only the $\emptyset$-theory, although the results that are valid for arbitrary E-theories are formulated in a general setting.

4   General Unification Procedure in the Free Theory with Sequence Variables and Flexible Arity Symbols

In this section we design a unification procedure to solve general unification problem of the form $t_1 \overset{?}{=}_\emptyset t_2$[1], built over the alphabet which consists of sequence and individual variables, free flexible arity function symbols, free constants and free fixed arity function symbols. We denote it as $GUP_\emptyset$. The unification procedure is a tree generation process based on two basic steps: projection and transformation.

4.1   Projection

The idea of projection ([1]) is to eliminate some sequence variables from the given unification problem UP. Let $\Pi(UP)$ be the following set of substitutions: $\{\{\overline{x} \leftarrow | \overline{x} \in S\} \mid S \subseteq vars(UP) \cap \mathcal{SV}\}$, where $vars(UP)$ is a set of variables of UP. $\Pi(UP)$ is called the set of projecting substitutions for UP. Each $\pi \in \Pi$ replaces some sequence variables from UP with the empty sequence. The projection rule is shown in Figure 1.

Projection: $s \overset{?}{=}_\emptyset t \rightsquigarrow \langle\langle s\pi_1 \overset{?}{=}_\emptyset t\pi_1, \pi_1\rangle, \ldots,$ where $\{\pi_1, \ldots, \pi_k\} = \Pi(s \overset{?}{=}_\emptyset t)$.
    $\langle s\pi_k \overset{?}{=}_\emptyset t\pi_k, \pi_k\rangle\rangle$

Fig. 1. Projection rule.

4.2   Transformation

Each of the transformation rules for unification have one of the following forms: $UP \rightsquigarrow \bot$ or $UP \rightsquigarrow \langle\langle SUC_1, \sigma_1\rangle, \ldots, \langle SUC_n, \sigma_n\rangle\rangle$ where each of the successors SUC_i is either $\top$ or a new unification problem.

The full set of transformation rules are given on Figure 2. It consists of four family of rules: Success, Failure, Elimination and Splitting. Note the usage of widening techniques (similar to [18],[24],[27],[28]) in the elimination rules for sequence variables.

4.3   Unification Procedure – Tree Generation

In [15] and [26] tree generation construction is used for solving word equations. We use the similar idea for unification procedure with sequence variables and flexible arity symbols. Projection and transformation can be seen as single steps

---

[1] In the case of $\emptyset$-theory it is enough to consider single equations instead of systems of equations in unification problems, because $\langle s_1 \overset{?}{=}_\emptyset t_1, \ldots, s_n \overset{?}{=}_\emptyset t_n \rangle$ has the same set of unifiers as $f(s_1, \ldots, s_n) \overset{?}{=}_\emptyset f(t_1, \ldots, t_n)$, where f is a free flexible arity symbol.

Success: $t \stackrel{?}{\doteq}_\emptyset t \rightsquigarrow \langle\langle \top,\ \varepsilon\rangle\rangle.$

$x \stackrel{?}{\doteq}_\emptyset t \rightsquigarrow \langle\langle \top,\ \{x \leftarrow t\}\rangle\rangle,$ if $x \notin vars(t)$.

$t \stackrel{?}{\doteq}_\emptyset x \rightsquigarrow \langle\langle \top,\ \{x \leftarrow t\}\rangle\rangle,$ if $x \notin vars(t)$.

Failure: $c_1 \stackrel{?}{\doteq}_\emptyset c_2 \rightsquigarrow \bot,$ if $c_1 \neq c_2$.

$x \stackrel{?}{\doteq}_\emptyset t \rightsquigarrow \bot,$ if $t \neq x$ and $x \in vars(t)$.

$t \stackrel{?}{\doteq}_\emptyset x \rightsquigarrow \bot,$ if $t \neq x$ and $x \in vars(t)$.

$f_1(\tilde{t}) \stackrel{?}{\doteq}_\emptyset f_2(\tilde{s}) \rightsquigarrow \bot,$ if $f_1 \neq f_2$.

$f() \stackrel{?}{\doteq}_\emptyset f(t_1, \tilde{t}) \rightsquigarrow \bot.$

$f(t_1, \tilde{t}) \stackrel{?}{\doteq}_\emptyset f() \rightsquigarrow \bot.$

$f(\overline{x}, \tilde{t}) \stackrel{?}{\doteq}_\emptyset f(s_1, \tilde{s}) \rightsquigarrow \bot,$ if $s_1 \neq \overline{x}$ and $\overline{x} \in vars(s_1)$.

$f(s_1, \tilde{s}) \stackrel{?}{\doteq}_\emptyset f(\overline{x}, \tilde{t}) \rightsquigarrow \bot,$ if $s_1 \neq \overline{x}$ and $\overline{x} \in vars(s_1)$.

$f(t_1, \tilde{t}) \stackrel{?}{\doteq}_\emptyset f(s_1, \tilde{s}) \rightsquigarrow \bot,$ if $t_1 \stackrel{?}{\doteq}_\emptyset s_1 \rightsquigarrow \bot$.

Eliminate: $f(t_1, \tilde{t}) \stackrel{?}{\doteq}_\emptyset f(s_1, \tilde{s}) \rightsquigarrow \langle\langle g(\tilde{t}\sigma) \stackrel{?}{\doteq}_\emptyset g(\tilde{s}\sigma),\ \sigma\rangle\rangle,$ if $t_1 \stackrel{?}{\doteq}_\emptyset s_1 \rightsquigarrow \langle\langle \top,\ \sigma\rangle\rangle.$

$f(\overline{x}, \tilde{t}) \stackrel{?}{\doteq}_\emptyset f(\overline{x}, \tilde{s}) \rightsquigarrow \langle\langle g(\tilde{t}) \stackrel{?}{\doteq}_\emptyset g(\tilde{s}),\ \varepsilon\rangle\rangle.$

$f(\overline{x}, \tilde{t}) \stackrel{?}{\doteq}_\emptyset f(s_1, \tilde{s}) \rightsquigarrow$ if $s_1 \notin \mathcal{SV}$ and $\overline{x} \notin vars(s_1)$,

$\qquad \langle\langle g(\tilde{t}\sigma_1) \stackrel{?}{\doteq}_\emptyset g(\tilde{s}\sigma_1),\ \sigma_1\rangle,$ where $\sigma_1 = \{\overline{x} \leftarrow s_1\}$,

$\qquad \langle g(\overline{x}, \tilde{t}\sigma_2) \stackrel{?}{\doteq}_\emptyset g(\tilde{s}\sigma_2),\ \sigma_2\rangle\rangle,$ $\sigma_2 = \{\overline{x} \leftarrow s_1, \overline{x}\}$.

$f(s_1, \tilde{s}) \stackrel{?}{\doteq}_\emptyset f(\overline{x}, \tilde{t}) \rightsquigarrow$ if $s_1 \notin \mathcal{SV}$ and $\overline{x} \notin vars(s_1)$,

$\qquad \langle\langle g(\tilde{s}\sigma_1) \stackrel{?}{\doteq}_\emptyset g(\tilde{t}\sigma_1),\ \sigma_1\rangle,$ where $\sigma_1 = \{\overline{x} \leftarrow s_1\}$,

$\qquad \langle g(\tilde{s}\sigma_2) \stackrel{?}{\doteq}_\emptyset g(\overline{x}, \tilde{t}\sigma_2),\ \sigma_2\rangle\rangle,$ $\sigma_2 = \{\overline{x} \leftarrow s_1, \overline{x}\}$.

$f(\overline{x}, \tilde{t}) \stackrel{?}{\doteq}_\emptyset f(\overline{y}, \tilde{s}) \rightsquigarrow$ where

$\qquad \langle\langle g(\tilde{t}\sigma_1) \stackrel{?}{\doteq}_\emptyset g(\tilde{s}\sigma_1),\ \sigma_1\rangle,$ $\sigma_1 = \{\overline{x} \leftarrow \overline{y}\},$

$\qquad \langle g(\overline{x}, \tilde{t}\sigma_2) \stackrel{?}{\doteq}_\emptyset g(\tilde{s}\sigma_2),\ \sigma_2\rangle,$ $\sigma_2 = \{\overline{x} \leftarrow \overline{y}, \overline{x}\},$

$\qquad \langle g(\tilde{t}\sigma_3) \stackrel{?}{\doteq}_\emptyset g(\overline{y}, \tilde{s}\sigma_3),\ \sigma_3\rangle\rangle,$ $\sigma_3 = \{\overline{y} \leftarrow \overline{x}, \overline{y}\}.$

Split: $f(t_1, \tilde{t}) \stackrel{?}{\doteq}_\emptyset f(s_1, \tilde{s}) \rightsquigarrow$ if $t_1, s_1 \notin \mathcal{IV} \cup \mathcal{SV}$ and

$\qquad \langle\langle f(r_1, \tilde{t}\sigma_1) \stackrel{?}{\doteq}_\emptyset f(q_1, \tilde{s}\sigma_1),\ \sigma_1\rangle, \ldots,$ $t_1 \stackrel{?}{\doteq}_\emptyset s_1 \rightsquigarrow \langle\langle r_1 \stackrel{?}{\doteq}_\emptyset q_1,\ \sigma_1\rangle, \ldots,$

$\qquad \langle f(r_k, \tilde{t}\sigma_k) \stackrel{?}{\doteq}_\emptyset f(q_k, \tilde{s}\sigma_k),\ \sigma_k\rangle\rangle$ $\langle r_k \stackrel{?}{\doteq}_\emptyset q_k,\ \sigma_k\rangle\rangle.$

Fig. 2. Transformation rules. $\tilde{t}$ and $\tilde{s}$ are possibly empty sequences of terms. $f, f_1, f_2 \in \mathcal{FFIX} \cup \mathcal{FFLEX}$. $g \in \mathcal{FFLEX}$ is a new symbol, if in the same rule $f \in \mathcal{FFIX}$. Otherwise $g = f$.

in a tree generation process. Each node of the tree is labeled either with a unification problem, $\top$ or $\bot$. The edges of the tree are labeled by substitutions. The nodes labeled with $\top$ or $\bot$ are terminal nodes. The nodes labeled with unification problems are non-terminal nodes. The children of a non-terminal node are constructed in the following way:

Given a nonterminal node, let UP be a unification problem attached to it. First, we decide whether UP is unifiable. If the answer is negative, we replace UP with the new label $\bot$. If UP is unifiable, we apply projection or transformation on UP and get $\langle\langle SUC_1, \sigma_1\rangle, \dots, \langle SUC_n, \sigma_n\rangle\rangle$. Then the node UP has n children, labeled respectively with $SUC_1, \dots, SUC_n$ and the edge to the SUC_i node is labeled with σ_i $(1 \le i \le n)$. The set $\{\sigma_1, \dots, \sigma_n\}$ is denoted by $SUB(UP)$.

We design the general unification procedure as a breadth first (level by level) tree generation process. Let $GUP_\emptyset$ be a unification problem. We label the root of the tree with $GUP_\emptyset$ (zero level). First level nodes (the children of the root) of the tree are obtained from the original problem by projection. Starting from the second level, we apply only a transformation step to a unification problem of each node, thus getting new successor nodes. The branch which ends with a node labeled by $\top$ is called a successful branch. The branch which ends with a node labeled by $\bot$ is a failed branch. For each node in the tree, we compose substitutions (top-down) displayed on the edges of the branch which leads to this node and attach the obtained substitution to the node together with the unification problem the node was labeled with. The empty substitution is attached to the root. For a node N, the substitution attached to N in such a way is called the associated substitution of N. Let $\Sigma(GUP_\emptyset)$ be the set of all substitutions associated with the $\top$ nodes. We call the tree a unification tree for $GUP_\emptyset$ and denote it $UT(GUP_\emptyset)$.

Example 4. Figure 3 shows development of successful branches of the unification tree for $GUP_\emptyset = f(x, b, \overline{y}, f(\overline{x})) \doteq_\emptyset^? f(a, \overline{x}, f(b, \overline{y}))$. $\Sigma(GUP_\emptyset) = \{\{x \leftarrow a, \overline{x} \leftarrow b, \overline{x}, \overline{y} \leftarrow \overline{x}\}, \{x \leftarrow a, \overline{x} \leftarrow b, \overline{y} \leftarrow\}\}$.

A stronger notion than minimality – disjointness – is used to prove Theorem 3 below. Formally, disjointness is defined as follows:

Definition 9. *A set of substitutions Σ is called disjoint modulo E with respect to a set of variables Var iff for all $\theta, \sigma \in \Sigma$, if there exist substitutions λ_1, λ_2 such that*

- *for all sequence variables $\overline{x} \in Var$,*
  - *$\overline{x} \leftarrow \notin \lambda_1$,*
  - *$\overline{x} \leftarrow \notin \lambda_2$,*
  - *there exist terms $t_1, \dots, t_n, s_1, \dots, s_n$, $n \ge 0$ such that $\overline{x}\theta \circ \lambda_1 = t_1, \dots, t_n$, $\overline{x}\sigma \circ \lambda_2 = s_1, \dots, s_n$ and for all $1 \le i \le n$, either t_i and s_i are the same sequence variables or $t_i \doteq_E s_i$ and*
- *for all individual variables $x \in Var$,*
  - *$x\theta \circ \lambda_1 \doteq_E x\sigma \circ \lambda_2$,*

then $\theta = \sigma$.

The main result of this paper says that $\Sigma(GUP_\emptyset)$ is a minimal complete set of free unifiers of $GUP_\emptyset$:

Theorem 3. $\Sigma(GUP_\emptyset) = MCU_\emptyset(GUP_\emptyset)$.

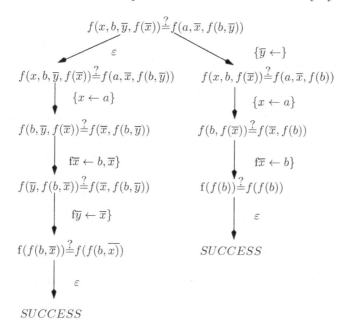

Fig. 3. Successful branches of $UT(f(x, b, \overline{y}, f(\overline{x})) \overset{?}{=}_\emptyset f(a, \overline{x}, f(b, \overline{y})))$.

Proof. Here we briefly sketch the idea. Details can be found in [17].

Completeness follows from the fact that for every unifier ϕ of $GUP_\emptyset$ there exists a branch β in $UT(GUP_\emptyset)$ such that for every substitution θ associated with a unification problem in β we have $\theta \ll_\emptyset^{vars(GUP_\emptyset)} \phi$.

Minimality is implied by disjointness of $\Sigma(GUP_\emptyset)$, which itself follows from the facts that for each non-terminal node UP in $UT(GUP_\emptyset)$ the set $SUB(UP)$ is a disjoint set of substitutions and every projecting or transforming substitution preserves disjointness. □

It is clear that the unification procedure terminates if $GUP_\emptyset$ contains no sequence variables (in this case the problem can be considered as a Robinson unification). Another terminating case is when one of the terms to be unified is ground. It yields to the following result:

Theorem 4. *Matching in a theory with individual and sequence variables, free constants, free fixed and flexible arity function symbols is finitary.*

We can add a cycle-checking method to the procedure: stop with failure if a unification problem attached to a node of unification tree coincides with a unification problem in the same branch of the tree. Then the following theorem holds:

Theorem 5. *The unification procedure with cycle-checking for $GUP_\emptyset$ terminates if no individual and sequence variables occur more than twice in $GUP_\emptyset$.*

If $GUP_\emptyset = t_1 \overset{?}{=} t_2$ has the property that sequence variables occur only as arguments of t_1 or t_2, then we can weaken the condition of the previous theorem:

Theorem 6. *The unification procedure with cycle-checking for $GUP_\emptyset = t_1 \overset{?}{=} t_2$, where sequence variables occur only as arguments of t_1 or t_2, terminates if no sequence variable occurs more than twice in $GUP_\emptyset$.*

The following termination condition does not require cycle-checking and does not depend on the number of occurrences of sequence variables. Instead, it requires for a unification problem of the form $f(\overline{x}) \overset{?}{=}_\emptyset f(t_1, \dots, t_n)$, $n > 1$, to check whether $\overline{x}$ occurs in $f(t_1, \dots, t_n)$. We call it *the sequence variable occurrence checking*. We can tailor this checking into the unification tree generation process as follows: if in the tree a successor of the unification problem of the form $f(\overline{x}) \overset{?}{=}_\emptyset f(t_1, \dots, t_n)$, $n > 1$, has to be generated, perform the sequence variable occurrence checking. If $\overline{x}$ occurs in $f(t_1, \dots, t_n)$, label the node with $\bot$, otherwise proceed in the usual way (projection or transformation).

Theorem 7. *If $GUP_\emptyset$ is a unification problem such that all sequence variables occurring in $GUP_\emptyset$ are only the last arguments of the term they occur, then the unification procedure with the sequence variable occurrence checking terminates.*

The fact that in most of the applications sequence variables occur precisely only at the last position in terms, underlines the importance of Theorem 7. The theorem provides an efficient method to terminate unification procedure in many practical applications.

5   Extension with Pattern-Terms

In this section we give a brief informal overview of an extension of the theory with patterns. Detailed exposition can be found in [17]. Pattern is an extended construct of the form $h_{m,k}(t_1, \dots, t_n)$, where h is a fixed or flexible arity function symbol, m and k are linear polynomials with integer coefficients with the special types of variables - called index variables, which are disjoint from sequence and individual variables. Instances of patterns are: $h_{1,vn+3}(\overline{x}, y)$, $f_{vm,vk}(a)$, etc., where vn, vm and vk are index variables. The intuition behind patterns is that they abbreviate term sequences of unknown length: $h_{vm,vk}(t)$ abbreviates $h_{vm}(t), \dots, h_{vk}(t)$. Patterns can occur as arguments in terms with flexible arity heads only. Such terms are called pattern-terms (P-terms). A minimal and complete unification procedure with patterns, individual and sequence variables, free constants, fixed and flexible arity function symbols is described in [17]. The procedure enumerates substitution/constraint pairs which constitute the minimal complete set of solutions of the problem.

Example 5. Let $\Gamma = f(\overline{x}, \overline{y}) \overset{?}{=}_\emptyset f(h_{vm,vk}(z))$. Then the unification procedure returns the set of substitution/constraint pairs

$$\{ \ \{\{\overline{x} \leftarrow , \overline{y} \leftarrow h_{vm,vk}(z)\}, 1 \leq vm \wedge vm \leq vk\},$$
$$\{\{\overline{x} \leftarrow h_{vm,vk}(z), \overline{y} \leftarrow \}, 1 \leq vm \wedge vm \leq vk\},$$
$$\{\{\overline{x} \leftarrow h_{vm,vn}(z), \overline{y} \leftarrow h_{vn+1,vk}(z)\}, 1 \leq vm \wedge vm \leq vn \wedge vn + 1 \leq k\} \ \},$$

with the property that each integer solution of a constraint, applied to the corresponding substitution, generates an element of the minimal complete set of solutions. For instance, the solution $vm = 1$, $vn = 3$, $vk = 4$ of the constraint $1 \leq vm \wedge vm \leq vn \wedge vn + 1 \leq vk$ applied on the substitution $\{\overline{x} \leftarrow h_{vm,vn}(z), \overline{y} \leftarrow h_{vn+1,vk}(z)\}$ gives a substitution $\{\overline{x} \leftarrow h_{1,3}(z), \overline{y} \leftarrow h_{4,4}(z), vm \leftarrow 1, vn \leftarrow 3, vk \leftarrow 4\}$ which belongs to the minimal complete set of solutions of Γ. In the expanded form the substitution looks like $\{\overline{x} \leftarrow h_1(z), h_2(z), h_3(z), \overline{y} \leftarrow h_4(z), vm \leftarrow 1, vn \leftarrow 3, vk \leftarrow 4\}$.

As we have already mentioned in the Introduction, patterns naturally appear in the proving context, when one wants to Skolemize, for instance, the expression $\forall x \exists \overline{y} (g(x) \doteq g(\overline{y}))$. Here $\overline{y}$ should be replaced with a sequence of terms $f_1(x), \ldots, f_{n(x)}(x)$, where $f_1, \ldots, f_{n(x)}$ are Skolem functions. The problem is that we can not know in advance the length of such a sequence. Note that in the unification we use an index variable vn instead of $n(x)$. This is because, given a unification problem UP in which $n(x)$ occurs, we can do a variable abstraction on $n(x)$ with a fresh index variable vn and instead of UP consider UP' together with the constraint $vn = n(x)$, where UP' is obtained from UP by replacing each occurrence of $n(x)$ with vn. One of the tasks for unification with patterns is to find a proper value for vn, if possible.

6   Applications

We have implemented the unification procedure (without the decision algorithm) as a Mathematica package and incorporated it into the Theorema system, where it is used by the equational prover. Besides using it in the proving context, the package can be used to enhance Mathematica solving capabilities, in particular, the Solve function of Mathematica. Solve has a rich arsenal of methods to solve polynomial and radical equations, equations involving trigonometric or hyperbolic functions, exponentials and logarithms. The following example shows, for instance, how a radical equation is solved:

$\text{In}[1]:= \mathbf{Solve}\left[\mathbf{x}^{1/3} + \sqrt{\mathbf{x}} == 1, \mathbf{x}\right]$

$\text{Out}[1]= \left\{\left\{x \rightarrow -\dfrac{2}{3} - \dfrac{11}{3}\left(\dfrac{2}{101 + 15\sqrt{69}}\right)^{1/3} + \dfrac{1}{3}\left(\dfrac{1}{2}\left(101 + 15\sqrt{69}\right)\right)^{1/3}\right\}\right\}$

However, it is unable to solve a symbolic equation like $f(x, y) = f(a, b)$:

$\text{In}[2]:= \mathbf{Solve}[\mathbf{f}[\mathbf{x}, \mathbf{y}] == \mathbf{f}[\mathbf{a}, \mathbf{b}], \{\mathbf{x}, \mathbf{y}\}]$
Solve::"dinv": The expression $\mathbf{f}[\mathbf{x}, \mathbf{y}]$ involves unknowns in more than one argument, so inverse functions cannot be used
$\text{Out}[2]= \text{Solve}[\mathbf{f}[\mathbf{x}, \mathbf{y}] == \mathbf{f}[\mathbf{a}, \mathbf{b}], \{\mathbf{x}, \mathbf{y}\}]$

Also, Solve can not deal with equations involving sequence variables.

On the basis of the unification package, we implemented a function called SolveExtended, which has all the power of Solve and, in addition, deals with symbolic equations which can be solved using unification methods. An equation like $f(x, y) = f(a, b)$ becomes a trivial problem for SolveExtended:

In[3]:= **SolveExtended**[**f**[**x**, **y**] == **f**[**a**, **b**], {{**x**, **y**}, {}}]

> Answer 1
>
> $\{x \rightarrow a, y \rightarrow b\}$
>
> The procedure terminated

Out[3]= $\{\{x \rightarrow a, y \rightarrow b\}\}$

The function is able to solve (a system of) equations involving sequence variables as well. In the following example x is an individual variable and X and Y are sequence variables:

In[4]:= **SolveExtended**[**f**[**x**, **b**, **Y**, **f**[**X**]] == **f**[**a**, **X**, **f**[**b**, **Y**]], {{**x**}, {**X**, **Y**}}]

> Answer 1
>
> $\{x \rightarrow a, X \rightarrow b, Y \rightarrow \text{Sequence}[]\}$
>
> Answer 2
>
> $\{x \rightarrow a, X \rightarrow \text{Sequence}[b, X], Y \rightarrow X\}$
>
> The procedure terminated

Out[4]= $\{\{x \rightarrow a, X \rightarrow b, Y \rightarrow \text{Sequence}[]\}, \{x \rightarrow a, X \rightarrow \text{Sequence}[b, X], Y \rightarrow X\}\}$

All the problems Solve deals with can also be solved by SolveExtended, e.g.:

In[5]:= **SolveExtended**$\left[x^{1/3} + \sqrt{x} == 1, x \right]$

Out[5]= $\left\{ \left\{ x \rightarrow -\frac{2}{3} - \frac{11}{3} \left(\frac{2}{101 + 15\sqrt{69}} \right)^{1/3} + \frac{1}{3} \left(\frac{1}{2} \left(101 + 15\sqrt{69} \right) \right)^{1/3} \right\} \right\}$

Options allow SolveExtended to switch on/off cycle-detecting and last sequence variable checking modes. Printing answers as they are generated is useful when the problem has infinitely many solutions and the procedure does not terminate, but this facility can also be switched off, if the user wishes so. It is possible to terminate execution after generating a certain number of solutions.

We found another application of the unification package in computing minimal matches for non-commutative Gröbner basis procedure. We show it on the following example: let p and q be non-commutative polynomials $a*a*b*a+a*b$ and $a*b*a*a+b*a$. In order to find S-polynomials of p and q one needs to compute all minimal matches between the leading monomials $a*a*b*a$ and $a*b*a*a$, which, in fact, is a problem of solving equations (for instance, of the form $a*a*b*a*R1 = L2*a*b*a*a$ for the variables $R1$ and $L2$) in a free semigroup. Solving equations in a free semigroup is a particular case of unification with sequence variables and flexible arity symbols. Even more, the equations we need to solve to find the matches, belong to a case when the unification procedure terminates. The example below demonstrates how the function MinimalMatches, based on the unification package, computes matches for $a*a*b*a$ and $a*b*a*a$:

In[6]:= **MinimalMatches**[**a** * **a** * **b** * **a**, **a** * **b** * **a** * **a**]

Out[6]={{a ∗ a ∗ b ∗ a ∗ R1, L2 ∗ a ∗ b ∗ a ∗ a, {L2 → a, R1 → a}},

{a ∗ a ∗ b ∗ a ∗ R1, L2 ∗ a ∗ b ∗ a ∗ a, {L2 → a ∗ a ∗ b, R1 → b ∗ a ∗ a}},

{L1 ∗ a ∗ a ∗ b ∗ a, a ∗ b ∗ a ∗ a ∗ R2, {L1 → a ∗ b, R2 → b ∗ a}},

{L1 ∗ a ∗ a ∗ b ∗ a, a ∗ b ∗ a ∗ a ∗ R2, {L1 → a ∗ b ∗ a, R2 → a ∗ b ∗ a}}}

7   Conclusion

We considered a unification problem for the equational theory with sequence and individual variables, free fixed and free flexible arity function symbols and gave a unification procedure which enumerates the minimal complete set of unifiers. Several sufficient termination conditions have been established. We gave a brief overview of a theory extended with constructs called patterns, which are used to abbreviate sequences of unknown lengths of terms matching a certain "pattern". Unification procedure for the extended theory enumerates substitution/constraint pairs which constitute the minimal complete set of solutions of the problem. Computer algebra related applications have been discussed.

Acknowledgments. I wish to thank Prof. Bruno Buchberger for supervising the work and for many helpful discussions.

References

1. H. Abdulrab and J.-P. Pécuchet. Solving word equations. *J. Symbolic Computation*, 8(5):499–522, 1990.
2. F. Baader and W. Snyder. Unification theory. In A. Robinson and A. Voronkov, editors, *Handbook of Automated Reasoning*, volume I, pages 445–532. Elsevier Science, 2001.
3. M. Benedikt, L. Libkin, T. Schwentick, and L. Segoufin. String operations in query languages. In *Proceedings of the 20th ACM SIGACT-SIGMOD-SIGART Symposium on Principles of Database Systems*, 2001.
4. B. Buchberger. Mathematica as a rewrite language. In T. Ida, A. Ohori, and M. Takeichi, editors, *Proceedings of the 2nd Fuji International Workshop on Functional and Logic Programming)*, pages 1–13, Shonan Village Center, Japan, 1–4 November 1996. World Scientific.
5. B. Buchberger. Personal communication, 2001.
6. B. Buchberger, C. Dupre, T. Jebelean, F. Kriftner, K. Nakagawa, D. Vasaru, and W. Windsteiger. The Theorema project: A progress report. In M. Kerber and M. Kohlhase, editors, *Symbolic Computation and Automated Reasoning (Proceedings of CALCULEMUS 2000)*, pages 98–113, St.Andrews, 6–7 August 2000.
7. A. Colmerauer. An introduction to Prolog III. *CACM*, 33(7):69–91, 1990.
8. A. Farquhar, R. Fikes, and J. Rice. The Ontolingua Server: A tool for collaborative ontology construction. *Int. J. of Human-Computer Studies*, 46(6):707–727, 1997.
9. M. R. Genesereth. Epilog for Lisp 2.0 Manual. Epistemics Inc., Palo Alto, 1995.
10. M. R. Genesereth and R. E. Fikes. Knowledge Interchange Format, Version 3.0 Reference Manual. Technical Report Logic-92-1, Computer Science Department, Stanford University, Stanford, June 1992.

11. S. Ginsburg and X. S. Wang. Pattern matching by Rs-operations: Toward a unified approach to querying sequenced data. In *Proceedings of the 11th ACM SIGACT-SIGMOD-SIGART Symposium on Principles of Database Systems*, pages 293–300, San Diego, 2–4 June 1992.
12. G. Grahne, M. Nykänen, and E. Ukkonen. Reasoning about strings in databases. In *Proceedings of the Thirteenth ACM SIGACT-SIGMOD-SIGART Symposium on Principles of Database Systems*, pages 303–312, Minneapolis, 24–26 May 1994.
13. G. Grahne and E. Waller. How to make SQL stand for string query language. *Lecture Notes in Computer Science*, 1949:61–79, 2000.
14. M. Hamana. Term rewriting with sequences. In *Proceedings of the First International Theorema Workshop*, Hagenberg, Austria, 9–10 June 1997.
15. J. Jaffar. Minimal and complete word unification. *J. of ACM*, 37(1):47–85, 1990.
16. D. E. Knuth and P. B. Bendix. Simple word problems in universal algebras. In J. Leech, editor, *Computational Problems in Abstract Algebra*, pages 263–298, Oxford, 1967. Pergamon Press. Appeared 1970.
17. T. Kutsia. Unification in a free theory with sequence variables and flexible arity symbols and its extensions. Technical report, RISC-Linz, 2002. Available from ftp://ftp.risc.uni-linz.ac.at/pub/people/tkutsia/reports/unification.ps.gz.
18. A. Lentin. Equations in free monoids. In M. Nivat, editor, *Automata, Languages and Programming*, pages 67–85, Paris, 3–7 July 1972. North-Holland.
19. G. S. Makanin. The problem of solvability of equations on a free semigroup. *Math. USSR Sbornik*, 32(2), 1977.
20. G. Mecca and A. J. Bonner. Sequences, Datalog and transducers. In *Proceedings of the Fourteenth ACM SIGACT-SIGMOD-SIGART Symposium on Principles of Database Systems*, pages 23–35, San Jose, 22–25 May 1995.
21. F. Mora. Gröbner bases for non-commutative polynomial rings. In J. Calmet, editor, *Proceedings of the 3rd International Conference on Algebraic Algorithms and Error-Correcting Codes (AAECC-3)*, volume 229 of *LNCS*, pages 353–362, Grenoble, July 1985. Springer Verlag.
22. R. Nieuwenhuis and A. Rubio. Theorem proving with ordering and equality constrained clauses. *Journal of Symbolic Computation*, 19:321–351, 1995.
23. P.Hayes and C. Menzel. A semantics for the knowledge interchange format. Available from http://reliant.teknowledge.com/IJCAI01/HayesMenzel-SKIFIJCAI2001.pdf, 2001.
24. G. Plotkin. Building in equational theories. In B. Meltzer and D. Michie, editors, *Machine Intelligence*, volume 7, pages 73–90. Edinburgh University Press, 1972.
25. A. Rubio. Theorem proving modulo associativity. In *Proceedings of the Conference of European Association for Computer Science Logic*, LNCS, Paderborn, Germany, 1995. Springer Verlag.
26. K. U. Schulz. Word unification and transformation of generalized equations. *J. Automated Reasoning*, 11(2):149–184, 1993.
27. J. Siekmann. String unification. Research paper, Essex University, 1975.
28. J. Siekmann. Unification and matching problems. Memo, Essex University, 1978.
29. M. Widera and C. Beierle. A term rewriting scheme for function symbols with variable arity. TR 280, Prakt. Informatik VIII, FernUniversitaet Hagen, 2001.
30. S. Wolfram. *The Mathematica Book*. Cambridge University Press and Wolfram Research, Inc., fourth edition, 1999.

Combining Generic and Domain Specific Reasoning by Using Contexts

Silvio Ranise*

Université H. Poincaré-Nancy 2 & LORIA-INRIA-Lorraine
615, rue du Jardin Botanique, BP 101,
54602 Villers Les Nancy Cedex, France
Silvio.Ranise@loria.fr

Abstract. The most effective theorem proving systems (such as PVS, Acl2, and HOL) provide a kind of two-level reasoning, where the knowledge of a given domain is treated by a special purpose reasoner and a generic reasoning module is used for the actual problem specification. To obtain an effective integration between these two levels of reasoning is far from being a trivial task. In this paper, we propose a combination of Window Reasoning and Constraint Contextual Rewriting to achieve an effective integration of such levels. The former supports hierarchical reasoning for arbitrarily complex expressions. The latter provides the necessary theorem proving support for domain specific reasoning. The two levels of reasoning cooperate by building and exploiting a context, i.e. a set of facts which can be assumed true while transforming a given subexpression. We also argue that the proposed combination schema can be useful for building sound simplifiers to be used in computer algebra systems.

1 Introduction

In the verification arena, the most effective theorem proving systems (such as PVS, Acl2, and HOL) provide some kind of two-level reasoning, where the knowledge of a given domain (or theory) is separated and treated by a special purpose reasoner (called the *background reasoner*) and a generic reasoning module (called the *foreground reasoner*) is used for the actual problem specification. As already experienced by Boyer and Moore in their seminal work [8], to obtain an effective integration between these two levels of reasoning is far from being a trivial task. This is so mainly for two reasons.

- First, it is difficult to design an effective interface between the two levels of reasoning which is independent on the theory handled by the background reasoner. In fact, most incorporation schemas available in state-of-the-art

* This work is partly supported by the European Union "CALCULEMUS Project" (HPRN-CT-2000-00102). The author would like to thank A. Armando and M. Rusinowitch for many helpful discussions on issues related to this paper. The anonymous referees helped improving the paper.

J. Calmet et al. (Eds.): AISC-Calculemus 2002, LNAI 2385, pp. 305–318, 2002.

systems (see e.g. [8,7]) apply to decision procedures for specific theories only and therefore they are not immediately applicable to decision procedures for different theories. On the other hand, formal verification requires a high degree of automation for a wide variety of domain specific reasoning tasks. So, a generic interface between the two levels of reasoning which allows for the "plug-and-play" of a wide class of background reasoners is required.

– Second, it is often the case that only a tiny portion of the proof obligations arising in practical applications fall exactly into the domain the background reasoners are designed to solve. This is so because specifications include both symbols with their usual mathematical meaning as well as additional function symbols that correspond to entities in the system being specified. For example, in a program verification scenario, facts like $L \neq [] \implies min(L) \leq max(L)$ involve the symbol $\leq$ with its usual meaning of (weak) "less-than" relation over integers in addition to the functions min and max appearing in the program of which some behaviour is expressed and which are supposed to return the minimum and the maximum of the non-empty list (cf. $L \neq []$) of integers L. As a consequence, either the available decision procedures need to be supplemented with mechanisms capable of widening their scope or dedicated background reasoners should be built from scratch.

Similar problems are experienced in the computer algebra community. To illustrate, consider the problem of computing the definite integral $\int_0^1 \sqrt{x^2} dx$. The following simplification is one of the most important steps towards the desired result:

$$\int_0^1 \sqrt{x^2} dx = \int_0^1 x dx.$$

The justification is that $\sqrt{x^2} = x$ for each x s.t. $0 \leq x \leq 1$. We can regard $0 \leq x \leq 1$ as an additional assumption which justifies the simplification $\sqrt{x^2} = x$. Here, it is correct to assume $0 \leq x \leq 1$ since it is a consequence of the context in which the expression $\sqrt{x^2}$ is placed (namely, the bounds of the definite integral). In this informal discussion, we have left implicit the fact that a suitable theory of analysis is assumed as the background theory w.r.t. which all the simplification steps are performed. Furthermore, it is crucial to notice that the additional assumption $0 \leq x \leq 1$ only holds in the context (under the theory of analysis) since the variable x is captured by the integral and it cannot be referenced outside its scope. This rather complicated way of building and using the context is almost completely neglected by many computer algebra systems which often perform incorrect simplifications. For example, consider the problem of computing the (slightly different) definite integral $\int_{-1}^1 \sqrt{x^2} dx$. Some versions of Maple will incorrectly return 0 as the result. The simplification $\sqrt{x^2} = x$ is applied regardless the fact that the context of $\sqrt{x^2}$ is changed and we are no more entitled to apply the previous simplification rule (i.e. $\sqrt{x^2} = x$). Also in this situation, two cooperating activities for background and foreground reasoning would allow to cope with the problem of simplifying complex (possibly containing quantifiers) symbolic expressions. In fact, the background reasoner would perform the simplifications in the theory of analysis and the foreground reasoner would soundly build up the context used for simplification.

Given the importance of the problem of obtaining an effective integration between background and foreground reasoning both for automated theorem proving and for computer algebra, this paper will investigate how to obtain such an effective integration between a sophisticated form of contextual rewriting (as the background reasoner) and a form of hierarchical reasoning which builds up a context once a given subexpression is selected (as the foreground reasoner).

1.1 A More Precise Statement of the Problem

In order to provide a flexible incorporation schema of (ground) satisfiability procedures in rewriting, we have proposed (see [3,4]) a generalised form of contextual rewriting [25], called *Constraint Contextual Rewriting* (CCR for short), which allows the available satisfiability procedure to access and manipulate the rewriting context. CCR is independent of the theory handled by the procedure and therefore its applicability is not restricted to procedures for certain theories. Furthermore, CCR features a powerful mechanism capable of extending the information available to the satisfiability procedure with facts encoding properties of symbols the procedure does not know anything about. This mechanism can greatly widen the scope of the satisfiability procedure thereby dramatically improving its effectiveness. The flexibility of CCR is demonstrated in [3] where it is shown that the integration schemas employed in the simplifiers of Acl2 [8] and Tecton [19] are all instances of CCR as well as in [1] where a rational reconstruction of Maple's evaluation process is obtained by suitably instantiating CCR.

Another important feature of CCR (as explained in [3]) is the possibility to make CCR independent of the way the context manipulated by both the rewriter and the satisfiability procedure is built. To do this, it is sufficient to assume the existence of a suitable function which returns a finite conjunction of literals which can be assumed true during the constraint contextual rewriting activity. As an example, let us consider clauses only. Then, the function building the context returns the conjunction of the negation of all literals in the clause but one which is currently being rewritten. Although it is possible to translate any first-order formula into clausal normal form (CNF) by well-known means, this situation is not completely satisfactory since many verification efforts are expressed in larger fragments of full first-order logic and it is well-known that the "quality" of the CNF translation is crucial for any afterwards applied theorem proving method (see [24] for an exhaustive discussion about this issue). Furthermore, the original structure of the formula is partially (if not totally) lost in the translation and a problem-reduction style of reasoning cannot be supported this way. In order to overcome this problem, more sophisticated instances of the process of building the context are required. Window reasoning (WR, for short) [22,15] provides a sophisticated way of obtaining a context in such a way that the original structure of the formula is unaltered and also hierarchical reasoning is supported. More precisely, WR is the activity of transforming a logical expression by restricting to one of its subexpressions (the focus expression) and transforming it. The remainder of the surrounding expression (the context) is left unchanged and it can be assumed true while transforming the focus expression. For example, when transforming ψ into an equivalent formula ψ' in the expression $\phi \wedge \psi$, we can assume that ϕ is true. WR has been found particularly useful for program verification and program refinement (see [15] for a complete discussion of this issue).

This paper describes our experience in building and using a system which features a combination of CCR and WR in order to support the activity of simplifying complex (possibly containing quantifiers) expressions. On the one hand, CCR plays the rôle of the background reasoner. It provides an adequate degree of automation for reasoning in rich background theories as well as the flexibility of fine tuning the domain specific reasoning to the actual verification effort. On the other hand, WR plays the rôle of the foreground reasoner, thereby providing the support for hierarchical top-down reasoning. The two levels of reasoning cooperate via a context built by WR and used during the simplification activity by CCR. The interface between the two levels of reasoning is the process of building the context. In order to validate our approach, we apply a (prototype) implementation of the proposed combination to the task of proving the correctness of Hoare's *Find* algorithm [17] which also requires dealing with quantifiers.

The paper is organised as follows. Sections 2 and 3 briefly describe CCR and WR. In section 4, the combination of CCR and WR is discussed and its implementation is briefly sketched. Then, Section 5 overviews the case study. Finally, section 6 draws some conclusions and sketches some lines of research for future work.

2   Constraint Contextual Rewriting

CCR features a tight integration of three reasoning activities: *contextual rewriting, satisfiability checking,* and *lemma speculation.* In the following, let T_c be a first-order theory whose satisfiability problem is decidable and T be the background theory obtained by closing $T_c \cup R$ under logical implication, where R is a set of facts of the form $\forall x_1, ..., x_m . h_1 \wedge \cdots \wedge h_n \implies c$ and $x_1, ..., x_m$ are all the variables in the formula, $h_1, ..., h_n$ are literals, and c is either an equality or a literal whose top-most function symbol is in the language of T_c. We now describe how CCR can realize the background reasoning activity in the theory T.[1]

In CCR, we assume the availability of a satisfiability procedure for T_c. The procedure works on a data structure (called *constraint store*) representing a conjunction of ground literals which are assumed true during rewriting. Three interface functionalities must be provided by a satisfiability procedure to be plugged in CCR. First, `cs_init(C)` initialises the constraint store C. Second,

---

[1] For lack of space, the description of CCR is not exhaustive but only sufficient to make the paper self-contained. A complete overview of CCR can be found in [3].

$\mathtt{cs\_unsat}(C)$ checks whether C is unsatisfiable in T_c. Third, $P :: C \xrightarrow[\text{cs--extend}]{} C'$ extends the constraint store C with the set of literals P yielding a new constraint store C'.

The activity of constraint contextual rewriting the literal p to p' in context C (in symbols $C :: p \xrightarrow[\text{ccr}]{} p'$) is done in two ways. First, the literal p can be rewritten to $true$ in a given context C if p is entailed by C. In order to check whether a literal p is entailed by C, we can check the unsatisfiability of C and the negation of p. In CCR, this can be done by invoking $\mathtt{cs\_unsat}(C')$, where C' is the extension of C with the negation of the literal p. Similarly, we can rewrite to $false$ the literal p if CCR is capable of checking that the negation of p is entailed by the context C. Second, given a literal $p[l\sigma]$, the rewriting activity of CCR returns $p[r\sigma]$ if the following condition is satisfied; there exists a rewrite rule of the form $\forall x_1, ..., x_m.h_1 \wedge \cdots \wedge h_n \implies l = r$ in R and a ground substitution σ such that each $h_i\sigma$ (for $i = 1, ..., n$) can be simplified to $true$ by recursively invoking the activity of constraint contextual rewriting.

By lemma speculation we mean the activity of finding out and feeding the satisfiability procedure with new facts about function symbols which are otherwise uninterpreted in T_c. Augmentation is a form of lemma speculation which extends the information available to the satisfiability procedure with selected instances of facts in R (which are supposed to encode properties of symbols the satisfiability procedure does not know anything about). We still need to describe how such facts are exploited in CCR. First of all, recall that instances of facts in R are of the form $(h_1 \wedge \cdots \wedge h_n \implies c)\sigma$, where σ is a ground substitution and c is a literal whose topmost predicate symbol is in the language of T_c. In order to add $c\sigma$ to the actual constraint store, the hypotheses $h_1\sigma, ..., h_n\sigma$ must preliminary be relieved. This problem is solved by trying to constraint contextually rewrite each $h_i\sigma$ (for $i = 1, ..., n$) to $true$.

3   Window Reasoning

In [15], window reasoning transforms an expression of interest (the focus) F under some preorder relation R (such as, e.g. implication). A *window* is an object of the form

$$
\begin{array}{l}
! \; \Gamma_1 \\
\quad \cdots \\
! \; \Gamma_n \\
\rho \star F
\end{array}
\tag{1}
$$

where $\Gamma_1, ..., \Gamma_n$ are (sets of) assumptions, F is the focus expression, and ρ is the relation to be preserved. The window (1) states the intention to prove $\bigcup_{i=1}^{n} \Gamma_i \vdash F \rho F'$ for some resulting expression F', where $\vdash$ is the derivability relation in (say) the Natural Deduction calculus [21].

There are four basic operations in WR. The first is to make an initial window for a specific focus F and relation ρ. The second is to transform a window by replacing the focus expression $F[f]$ with $F[f']$ whenever $\Gamma' \vdash f \rho' f'$ has been proved for a given relation ρ' and $\Gamma' \supseteq \Gamma$. The third is to open a new window at a selected subexpression. To do this, the child window is pushed on a window stack, the new focus is the selected subexpression of the parent window and the relation in the child window is determined by the opening rule used. The hypotheses of the parent window are inherited by the child window and possibly the opening rule can add extra hypotheses. The last operation is to close back to a parent window by replacing the original subexpression in the parent window which was selected for opening. The child window is then popped from the window stack.

A usable mechanisation of WR requires to hide the explicit choice of opening rules. It must be possible to specify an arbitrary position in the formula and the window opening mechanism should be capable of implicitly composing many rules so as to open a window at the desired position. Indeed, a set of basic rules must be identified. These rules can be found by considering the definition of (first-order) formula and finding out at which position a window can be opened. Hence, there will be opening rules for the connectives and for the universal and the existential quantifiers. In addition to the basic window rules, others are required to make WR usable in practice. Such additional rules preserve domain specific relations such as $\leq$.

4   Combining Constraint Contextual Rewriting and Window Reasoning

In our combination, the background reasoning activity is provided by CCR and the foreground reasoning by WR. These two activities cooperate via a context which is built by WR and used by CCR. In fact, WR builds the context while breaking down the formula to literal level. Once at the literal level, CCR is capable of simplifying the literal in a suitable context w.r.t. a given background theory.

As already observed in Section 3, the task of breaking down a formula is straightforward as soon as we consider the inductive definition of the set of first-order formulae. The concurrent activity of building a context is also simple since the notion of truth (in first order logic) is recursively defined over the structure of the formulae (see, e.g., [12]). The only difficulty in doing this is posed by the requirements which the resulting context and focus expression are supposed to satisfy in order to be suitably exploited by CCR. Let us analyse these requirements. First, the focus expression and the context must be ground. Second, the context C must be such that $(\bigwedge C \implies (l \iff l')) \implies (\phi[l]_u \iff \phi[l']_u)$ holds in the background theory for a formula ϕ.

To see how these two requirements can be easily satisfied, consider the set $\mathcal{L}$ of first order literals built in the usual way over the signature in which the actual verification problem is expressed. Then, consider the set $\mathcal{H}$ containing $\mathcal{L}$ as well as the (finite) conjunctions of literals in $\mathcal{L}$. We restrict our attention to the sentences in the (inductively defined) set $\mathcal{VC}$ of formulae such that (i) $\mathcal{L} \subseteq \mathcal{VC}$, (ii) if G_1 and G_2 are in $\mathcal{VC}$ then $G_1 \wedge G_2$ is in $\mathcal{VC}$, (iii) if H is in $\mathcal{H}$ and G is in $\mathcal{VC}$ then

$H \Longrightarrow G$ is in $\mathcal{VC}$, and (iv) if G is in $\mathcal{VC}$ then $\forall x.G$ is in $\mathcal{VC}$. If a sentence ϕ in $\mathcal{VC}$ contains the universally quantified variables $x_1, ..., x_n$, then replace each x_i ($i = 1, ..., n$) by a "fresh" (i.e. not occurring in ϕ) constant. In the following, we assume that all sentences in $\mathcal{VC}$ are transformed in this way to eliminate all universal quantifiers. As a consequence, we consider only ground formulae. We are now in the position to define the function $\mathsf{Cxt}(G, u)$ building the context C for any sentence G in $\mathcal{VC}$ and literal position u in G as follows: $\mathsf{Cxt}(L, \epsilon) := \{L\}$ when L is in $\mathcal{L}$, $\mathsf{Cxt}(G_1 \wedge G_2, i.u) := \mathsf{Cxt}(G_i, u)$ where $i = 1, 2$, and $\mathsf{Cxt}(H \Longrightarrow G, 2.u) := \{h | h$ is a literal in the conjunction $H\} \cup \mathsf{Cxt}(G, u)$; otherwise, the function Cxt returns the empty set.[2] (We notice that WR can easily mechanise the function Cxt.) With this definition of the function building the context, it is easy to check that both requirements above are satisfied. Furthermore, if the activity of CCR is complete w.r.t. the background theory T (i.e. for any context C and literal p we have that $C :: p \xrightarrow[\text{ccr}]{} true$ iff $T \cup C \models p$), then the proposed combination of WR and CCR is complete for the set of sentences in $\mathcal{VC}$ (i.e. a sentence α in $\mathcal{VC}$ is simplified to $true$ iff $T \models \alpha$). We can exploit a complete implementation of CCR whenever we have a background theory $T = T_c$ for which a satisfiability procedure is available (and hence $R = \emptyset$ since T is the deductive closure of $T_c \cup R$). In this case, as already described in Section 2, any focus literal p is rewritten to $true$ ($false$) in $\mathsf{Cxt}(G, u)$ where u is the position of G at which p is located by checking the unsatisfiability of the ground set of literals $\{\neg p\} \cup \mathsf{Cxt}(G, u)$ ($\{p\} \cup \mathsf{Cxt}(G, u)$, respectively).

Unfortunately, in some cases, the set of sentences in $\mathcal{VC}$ is not sufficient to express the verification conditions arising in practice. Often, the reason for this is that invariants involving existential quantification are required. So, we extend the set $\mathcal{VC}$ to include the formulae with existential quantifiers as follows. First, we consider the set $\mathcal{H}^e \supseteq \mathcal{H}$ to be such that if H is in $\mathcal{H}^e$ then $\forall x.H$ is in $\mathcal{H}^e$. Then, we define (by induction) the set $\mathcal{VC}^e$ to be such that (i) $\mathcal{VC}^e \supseteq \mathcal{H}^e$, (ii) $\mathcal{VC}^e \supseteq \mathcal{VC}$, (iii) if G is in $\mathcal{VC}^e$ then $\exists x.G$ is in $\mathcal{VC}^e$, and (iv) if H is in $\mathcal{H}$ then $\forall x.H$ is in $\mathcal{VC}^e$. (Notice that the universal quantifier in $\forall x.H \Longrightarrow G$ encodes an existential choice.) The situation is now more complicated. In fact, for sentences in $\mathcal{VC}^e$, WR can generate contexts containing non-ground literals. As a consequence, we need to extend the background reasoner to handle non-ground literals. From an operational point of view, this requires to lift rewriting to narrowing and the activity of checking the satisfiability of ground literals to handling (and possibly instantiating) variables in search of an inconsistency in the background theory. To understand how difficult this problem is, let the background theory be (finitely) axiomatised by a set of equations E, the context be the singleton set containing the literal $P(u_1, ..., u_n)$ and $P(v_1, ..., v_n)$ be the focus literal. Then, consider the situation in which we want to rewrite the focus literal to $true$ by checking the unsatisfiability of the conjunction of its negation (namely $\neg P(v_1, ..., v_n)$) with the context. In this scenario, the background reasoner must be capable of answering the following question: given a set E of equations and

---

[2] The uniqueness of the function Cxt is an immediate consequence of the recursion theorem (see, e.g. [12]).

some pairs of terms (u_i, v_i), is there a substitution σ such that for all equations $(u_i = v_i)\sigma$ holds that $E\sigma \cup \{(u_i = v_i)\sigma\}$ is unsatisfiable? This problem can be shown equivalent (see, e.g. [6]) to the rigid E-unification problem [14], which is known to be decidable and NP-complete.[3] So, ensuring completeness for the set of sentences in $\mathcal{VC}^e$ is quite expensive from a computational point of view. Furthermore, the situation is much more complicated as soon as we consider theories which are not axiomatised by a finite set equations (see [7] for a discussion of this problem in the context of a refutation procedure).

Instead of handling existential quantifiers in a general way by using narrowing in place of rewriting and constraint solving in place of satisfiability checking, we take a more pragmatic (although incomplete) approach. There are two cases to be considered depending on the fact that the existential quantification is implicit or explicit. Firstly, when the existential quantification is encoded by a universal quantification whose scope is the antecedent of an implication, the implication is added to the set R of background facts and augmentation is asked to (possibly) find a suitable instance for the universally quantified variables in the fact (see Section 5.1 for an example of this). Secondly, when the existential quantification is explicit in the formula, we extend the function Cxt building the context with a ground substitution for the existential variables in the formula as an extra argument. Indeed, this allows us to easily obtain a suitable ground version of the context and the focus literal. In both cases, CCR can be applied without modifications. (Notice that also this version of the function Cxt can easily be mechanised by WR.)

4.1 Generality of the Approach

So far, we have been able to give completeness results of the combination between CCR and WR for some restricted classes of formulae, namely $\mathcal{VC}$ and $\mathcal{VC}^e$. This characterization, although important from a practical point of view since many verification conditions fall in these two classes, is not completely satisfactory. In [10], Corella investigates the problem of determining what sentences can be assumed true when proving $s = t$ so to establish—by the principle of substitution of equals for equals—that $C[s] = C[t]$ for a suitable context $C[\cdot]$ and some set of assumptions Γ. He notices that there are some sentences related to the context $C[\cdot]$ which can be assumed to hold although they are not in Γ and that there exists a closed form characterization of such a set of sentences in Church's system [9] (which combines the simple theory of types with Church's lambda-calculus). To obtain a similar closed form solution for a natural deduction calculus for first-order logic [21], the introduction of an "if-then-else" construct is required due to a lack of expressive power of first-order logic. Let $\mathcal{A}(\Gamma, C[\cdot])$ be the set of sentences which can be assumed true in order to prove that $\Gamma \vdash C[s] = C[t]$ for a suitable derivability relation $\vdash$. In general, it is undecidable whether a formula α is in $\mathcal{A}(\Gamma, C[\cdot])$ or not; but it is possible to identify useful sufficient conditions

---

[3] Notice that the related problem of simultaneous rigid E-unification is undecidable [11].

for α to be an element of $\mathcal{A}(\Gamma, C[\cdot])$ which are decidable, as we have done for $\mathcal{VC}$ and $\mathcal{VC}^e$. So, it is natural to investigate more sophisticated WR rules for building the context which can be afterwards used by CCR. For example, we think that a characterization of some classes of formulae (in the same spirit of $\mathcal{VC}$ and $\mathcal{VC}^e$) can be given by considering proof obligations arising in typical computer algebra sessions. To illustrate, consider the task of computing definite integrals. We can think of deriving the elements of $\mathcal{A}(\Gamma, C[\cdot])$ from the function which is going to be integrated (more precisely from its definition) and the bounds of the integral. For example, consider the following integral: $\int_0^1 |x| dx$. From the definition of absolute value (namely, $|x| = x$ if $x \geq 0$ and $|x| = -x$ if $x < 0$) which is assumed to be in Γ and the interval considered for integration (namely, $[0, 1]$), it is not difficult to derive the following sentence in $\mathcal{A}(\Gamma, C[\cdot])$: $\forall x.(0 \leq x \wedge x \leq 1 \implies |x| = x)$. This fact together with some form of extensionality (which again we assume to be in Γ) allow us to perform the desired simplification $\int_0^1 |x| dx = \int_0^1 x dx$.

Finally, it is important to notice (as shown in [10]) the incompleteness of the method of showing that $\Gamma \vdash C[s] = C[t]$ by proving that $s = t$ follows from Γ and $\mathcal{A}(\Gamma, C[\cdot])$. This means that for some context $C[\cdot]$ and some set of assumptions Γ, there exists an equality $s = t$ which is not in $\mathcal{A}(\Gamma, C[\cdot])$ but is such that $\Gamma \vdash C[s] = C[t]$.

4.2   Implementation

We have implemented the proposed combination schema on top of **RDL**(acronym for **R**ewrite and **D**ecision procedure **L**aboratory) [2], an automatic theorem prover based on CCR. We have extended the input set of formulae to the system to full first-order logic and we have implemented the various operations to support WR (see Section 3). This task has been routine since the implementation language of **RDL** is Prolog and this allows the programmer to easily manipulate symbolic expressions. Furthermore, since CCR provide the uniform interface to background reasoning, there has been no need to implement many domain specific window inference rules preserving a variety of relations such as $\leq$. This is in contrast with previous systems based on WR. For example, the window library distributed with the HOL system consists of about 40 rules, whereas our implementation consists of 11 rules to handle the connectives and the quantifiers as well as to combine them. An additional rule allows to invoke CCR on the focus literal of the active window in the context obtained by collecting the hypotheses of the window. For the sentences in $\mathcal{VC}$, the WR mechanism implemented in our system gives exactly the implementation of function $\mathsf{Cxt}(\phi, u)$ defined in Section 4 when ϕ is a sentence in $\mathcal{VC}$. As a consequence, our system becomes complete for sentences in $\mathcal{VC}$ as soon as a complete satisfiability procedure for the background theory is available. This is the case for the theory of arrays with extensionality and permutation for which a satisfiability procedure has been obtained by coupling the equational theorem prover E [23] to our system, following the methodology described in [5] (see Section 5 for more details on this).

5   The Proof of Correctness of Hoare's Find

In [17], Hoare shows the correctness and termination of his algorithm *Find* [16]. Given an array a of comparable elements and an element $sel(a, f)$ at position f of a, *Find* consists in reorganising the elements of a in such a way that all elements on the left side of f are smaller than or equal to $sel(a, f)$ and all elements on the right side of f are greater than or equal to $sel(a, f)$. Hoare's proof is based on the use of invariants. Proof obligations are made explicit at each step and proved as lemmas in a hierarchical (top-down) way. In order to prove most lemmas, reasoning in the background theory of comparable elements (and arrays) is essential.

Since WR supports hierarchical reasoning and CCR provides the background reasoning in a theory of comparable elements and arrays, Hoare's correctness proof appears as an interesting case study for our system. Furthermore, *Find* is a non-trivial program.

Both the indexes and the elements of the array are modeled as integers so that the available (in **RDL**) satisfiability procedure for Universal Presburger Arithmetic can be used. Arrays are axiomatised by using the following axioms:

$$\forall A, I, E. sel(store(A, I, E), I) = E \qquad (2)$$
$$\forall A, I, J, E. I \neq J \implies sel(store(A, I, E), J) = sel(A, J), \qquad (3)$$

where $store(A, I, E)$ is the array whose I-th element is E and whose J-th component for $I \neq J$ is $sel(A, J)$. These axioms are stored in the set R thereby specifying an extension of Universal Presburger Arithmetic as the background theory.

We consider the proof of the lemmas listed in Hoare's paper [17]. Each lemma states that a given invariant of the algorithm is satisfied at a certain point in the algorithm. Lemmas 1–9 can be easily proved by our system by invoking the augmentation mechanism only to find suitable instances of the existentially quantified variable encoded in the universal quantification whose scope is the antecedent of an implication. There is no need of finding instantiations of the axioms for arrays, since lemmas 1–9 encode the correctness of operations which perform comparisons on the elements of the array but do not alter it, thus only reasoning in Universal Presburger Arithmetic is necessary. Lemmas 10–13 require the augmentation mechanism also to find suitable instances of the axioms for arrays. In fact, these four lemmas express the correctness of exchanging two elements in the array, thus the fundamental properties of arrays are necessary.

Lemmas 14–18 concern the proof of termination of the algorithm. Notice that these lemmas require an explicit existential quantification in the conclusion of the formula. As already said in Section 4, our system takes a ground term as input in order to instantiate the explicitly existentially quantified variable of the formulae. All the required terms are extracted from Hoare's paper and the proofs are then automatically completed by our system.

There are still two important aspects of the correctness proof to be considered. The former is proving that all indices used in the program are within the bounds of the array. Although mentioned, the problem is not formalised in Hoare's paper. The lemmas encoding these properties can be easily stated in Universal Presburger Arithmetic and automatically proved in our system thanks to the available satisfiability procedure for such a theory. The second aspect amounts to proving that the initial elements stored in the array are preserved by the program. In [17], Hoare informally proves this property by observing that the program only performs exchanges of pairs of elements and, as a consequence, the elements in the array after the execution of the program are disposed in a permutation of the original array. Hoare also points out that the formalisation of this argument is quite involved and it implies tedious proofs. In [13], Filliâtre formalises the above argument in Coq [18] by using inductive definitions for the concept of permutation and he claims that the resulting proofs obligations are easy to prove. Since our system does not provide facilities to handle inductive definition, we take another approach to prove the property of preservation of the algorithm.

In [5], a methodology to build satisfiability procedures for certain equational theories is described and it is applied to the theory of arrays considered here. The satisfiability procedure for such a theory roughly consists of the exhaustive application of the rules of a superposition calculus [20] to the axioms of the theory and the ground literals whose satisfiability we want to check. Now, consider the following definition of permutation (expressed in second-order logic, see [12]):

$$\forall M, N.(Perm(M, N) \Longleftrightarrow \\ \exists p.(Bijective(p) \wedge \forall I.(sel(M, I) = sel(N, p(I))))) \tag{4}$$

where $Bijective(p)$ encodes the property that the function p is a bijection.[4] Notice that given two arrays denoted by two (first order) ground terms m and n formula (4) can be reduced to the following first-order formula:

$$Perm(m, n) \Longleftrightarrow \forall I.(p(q(I)) = q(p(I)) \wedge sel(M, I) = sel(N, p(I)) \tag{5}$$

where q is the inverse of the function p. This definition of $Perm$ is sufficient for our case study since we need to handle the (un-)satisfiability of ground literals only. The satisfiability procedure for arrays described in [5] can be easily extended to handle lemmas of the form (5).[5] Thus a superposition based theorem prover can be used to rapid prototype a satisfiability procedure for the theory of arrays with permutations. To this end, we have used the theorem prover E [23]. A simple interface between our system and E has been implemented so that the satisfiability procedure can be readily plugged into our system. At this point, the proof obligations expressing the preservation of the elements of the array are automatically proved.

---

[4] A permutation is formally defined as a bijection.

[5] This is so because (roughly) no superpositions are possible between the right hand sides of formulae of the form (5) and the axioms of arrays.

5.1   A Worked Out Example

Let us consider the following verification condition corresponding to Lemma 10 of [17]:

$$(m \leq i \leq j \wedge \forall p.(1 \leq p < i \Longrightarrow sel(a,p) \leq r)) \Longrightarrow \\ (m \leq i \wedge \forall p.(1 \leq p < i \Longrightarrow sel(store(store(a,i,sel(a,j)),j,sel(a,i)),p) \leq r)) \tag{6}$$

where $A \leq B < C$ abbreviates $A \leq B \wedge B < C$, and m, i, j, a, and r are constants. (Notice that (6) is a sentence in $\mathcal{VC}^e$ and $\forall p.(1 \leq p < i \Longrightarrow sel(a,p) \leq r)$ in (6) encodes an existential quantification.)

Our system begins the proof by opening a window on the literal $m \leq i$ (occurring in the conclusion of the implication) and CCR is invoked on it in a suitable context as follows: $\{m \leq i, ...\} :: m \leq i \xrightarrow[ccr]{} true$, since $m \leq i$ is readily found entailed by the context thanks to the available satisfiability procedure for Universal Presburger Arithmetic. Then, we close the window and we go back to its parent where the literal we originally opened the sub-window on (namely $m \leq i$) is replaced by $true$, thereby obtaining the following formula (equivalent to (6)):

$$(m \leq i \leq j \wedge \forall p.(1 \leq p < i \Longrightarrow sel(a,p) \leq r)) \Longrightarrow \\ (true \wedge \forall p.(1 \leq p < i \Longrightarrow sel(store(store(a,i,sel(a,j)),j,sel(a,i)),p) \leq r)) \tag{7}$$

Then the system opens a sub-window on the literal $sel(store(store(a,i,sel(a,j)),j,sel(a,i)),\overline{p}) \leq r$ (where $\overline{p}$ is a fresh constant) of (7) and we invoke CCR on it:

$$\{\overline{p} < i, i \leq j, ...\} :: \\ sel(store(store(a,i,sel(a,j)),j,sel(a,i)),\overline{p}) \leq r \xrightarrow[ccr]{} \\ sel(store(store(a,i,sel(a,j))),\overline{p}) \leq r \xrightarrow[ccr]{} sel(a,\overline{p}) \leq r \xrightarrow[ccr]{} true \tag{8}$$

(notice also that $\forall p.(1 \leq p < i \Longrightarrow sel(a,p) \leq r)$ is added to R) since axiom (3) is instantiated to

$$\overline{p} \neq j \Longrightarrow sel(store(store(a,i,sel(a,j)),j,sel(a,i)),\overline{p}) = sel(store(a,i,sel(a,j)),\overline{p})$$

and the focus literal can be rewritten to $sel(store(a,i,sel(a,j))),\overline{p}) \leq r$ (the condition $\overline{p} \neq j$ is rewritten to $true$ since it is entailed by the context), then axiom (3) is instantiated to

$$\overline{p} \neq i \Longrightarrow sel(store(a,i,sel(a,j))),\overline{p}) = sel(a,\overline{p})$$

and the focus literal can be rewritten to $sel(a,\overline{p}) \leq r$ (the condition $\overline{p} \neq i$ is rewritten to $true$ since it is entailed by the context). Finally, the literal $sel(a,\overline{p}) \leq r$ is rewritten to $true$ since augmentation instantiates the fact $\forall p.(1 \leq p < i \Longrightarrow sel(a,p) \leq r)$—which, we recall, has been added to the set R—to $1 \leq \overline{p} < i \Longrightarrow sel(a,\overline{p}) \leq r$. The conditions of this fact are easily found to be entailed by the context and the conclusion is added to the context. At this point, the literal $sel(a,\overline{p}) \leq r$ can be rewritten to $true$ by detecting that it is obviously entailed by the (extended) context.

Now, we close the sub-window and we replace the rewritten focus expression in the parent, thereby obtaining the following formula:

$$(m \leq i \leq j \land \forall p.(1 \leq p < i \implies sel(a, p) \leq r)) \implies (true \land true)$$

which is easily transformed to *true* by applying some obvious window inferences.

6 Conclusions

We have presented a combination of generic and domain specific reasoning which allowed us to easily prototype a reasoning system which supports the activity of verification in a flexible way and with a high degree of automation. The generic reasoning module is based on WR which is particularly suited for program verification and refinement. The domain specific reasoning is mechanised in a uniform way by CCR. The two reasoning activities communicate via a context. We have also seen how the combined reasoning activity allows CCR to handle a class of formulae containing existentially quantified variables, not only ground formulae. The prototype system has been successfully used for proving the correctness of Hoare's *Find* algorithm, thereby confirming the validity of our approach. A key to the flexibility and high degree of automation provided by the proposed combination schema is the ability to extend and to build satisfiability procedures for rich background theories. In our case study, this is exemplified by the procedure for arrays with permutations. An interesting line of research is to extend the system by incorporating a module which analyses the axioms in the background theory and automatically proves the termination property required to obtain a satisfiability procedure along the lines given in [5] and finally invokes an equational theorem prover.

Another promising development of this work is to systematically investigate classes of proof obligations arising in computer algebra sessions for which the method proposed in the paper can be proved complete. We think that in order to do this in a first-order setting, we should investigate how to cope with some form of extensionality which seems to be the key ingredient to handle simplifications taking place in the scope of quantifiers.

References

1. A. Armando and C. Ballarin. Maple's Evaluation Process as Constraint Contextual Rewriting. In *Proc. of the 2001 Int. Symp. on Symbolic and Algebraic Computation (ISSAC-01)*, pages 32–37, New York, July 22–25 2001. ACMPress.
2. A. Armando, L. Compagna, and S. Ranise. **RDL**—Rewrite and Decision procedure Laboratory. In *Proceedings of the International Joint Conference on Automated Reasoning (IJCAR 2001)*, Siena, Italy, June 2001.
3. A. Armando and S. Ranise. Constraint Contextual Rewriting. In *Proc. of the 2nd Intl. Workshop on First Order Theorem Proving (FTP'98)*, 1998.
4. A. Armando and S. Ranise. Termination of Constraint Contextual Rewriting. In *Proc. of the 3rd Intl. W. on Frontiers of Comb. Sys.'s (FroCos'2000)*, LNCS 1794, 2000.

5. A. Armando, S. Ranise, and M. Rusinowitch. Uniform Derivation of Decision Procedures by Superposition. In *Computer Science Logic (CSL01), Paris, France, 10-13 September*, 2001.

6. P. Baumgartner. An Ordered Theory Resolution Calculus. In *Logic Programming and Automated Reasoning*, number 624 in LNAI, pages 119–130, 1992.

7. N. Bjørner. *Integrating Decision Procedures for Temporal Verification*. PhD thesis, Stanford University, 1999.

8. R.S. Boyer and J S. Moore. Integrating Decision Procedures into Heuristic Theorem Provers: A Case Study of Linear Arithmetic. *Machine Intelligence*, 11:83–124, 1988.

9. A. Church. A Formulation of the Simple Theory of Types. *J. of Symbolic Logic*, 5(1):56–68, 1940.

10. F. Corella. What Holds in a Context? *J. of Automated Reasoning*, 10:79–93, 1993.

11. A. Degtyarev and A. Voronkov. The Undecidability of Simultaneous Rigid E-Unification. *Theoretical Computer Science*, 166(1–2):291–300, 1996.

12. H. B. Enderton. *A Mathematical Introduction to Logic*. Academic Pr., 1972.

13. J.-C. Filliâtre. Formal Proof of a Program: Find. *Science of Computer Programming*, To appear.

14. J. H. Gallier, S. Ratz, and W. Snyder. Theorem Proving Using Rigid E-Unification: Equational Matings. In *Logic in Computer Science (LICS'87)*, Ithaca, New York, 1987.

15. Jim Grundy. *A Method of Program Refinement*. PhD thesis, University of Cambridge, Computer Laboratory, New Museums Site, Pembroke Street, Cambridge CB2 3QG, England, November 1993. Technical Report 318.

16. C. A. R. Hoare. Algorithm 65, Find. *Comm. of the ACM*, 4(7):321, July 1961.

17. C. A. R. hoare. Proof of a Program: FIND. *Comm. of the ACM*, 14(1):39–45, January 1971.

18. G. Huet, G. Kahn, and C. Paulin-Mohring. The Coq Proof Assistant: a tutorial. Technical Report 204, INRIA-Rocquencourt, 1997.

19. D. Kapur and X. Nie. Reasoning about Numbers in Tecton. Technical report, Department of Computer Science, State University of New York, Albany, NY 12222, March 1994.

20. R. Nieuwenhuis and A. Rubio. Paramodulation-based theorem proving. In A. Robinson and A. Voronkov, editors, *Hand. of Automated Reasoning*. 2001.

21. D. Prawitz. Natural Deduction. Acta Universitatis Stockholmiensis, Stockholm Studies in Philosophy 3. Almqvist & Wiksell, Stockholm, 1965.

22. P. J. Robinson and J. Staples. Formalising the Hierarchical Structure of Practical Mathematical Reasoning. *Journal of Logic and Computation*, 3(1):47–61, February 1993.

23. S. Schulz. System Abstract: E 0.61. In R. Goré, A. Leitsch, and T. Nipkow, editors, *Proc. of the 1st IJCAR, Siena*, number 2083 in LNAI, pages 370–375. Springer, 2001.

24. C. Weidenbach and A. Nonnengart. Small clause normal form. In A. Robinson and A. Voronkov, editors, *Hand. of Automated Reasoning*. 2001.

25. H. Zhang. Contextual Rewriting in Automated Reasoning. *Fundamenta Informaticae*, 24(1/2):107–123, 1995.

Inductive Theorem Proving and Computer Algebra in the MathWeb Software Bus

Jürgen Zimmer[1]* and Louise A. Dennis[2]**

[1] Division of Informatics, University of Edinburgh,
jzimmer@mathweb.org
[2] School of Computer Science and Information Technology, University of Nottingham,
lad@cs.nott.ac.uk

Abstract. Reasoning systems have reached a high degree of maturity in the last decade. However, even the most successful systems are usually not general purpose problem solvers but are typically specialised on problems in a certain domain. The MathWeb Software Bus (MathWeb-SB) is a system for combining *reasoning specialists* via a common software bus. We describe the integration of the λ-Clam system, a reasoning specialist for proofs by induction, into the MathWeb-SB. Due to this integration, λ-Clam now offers its theorem proving expertise to other systems in the MathWeb-SB. On the other hand, λ-Clam can use the services of any reasoning specialist already integrated. We focus on the latter and describe first experiments on proving theorems by induction using the computational power of the MAPLE system within λ-Clam.

1 Introduction

Reasoning systems have reached a high degree of maturity in the last decade.[1] However, these systems are usually not general purpose problem solvers but should be seen as *reasoning specialists* with individual expertise in a certain domain. Automated Theorem Provers (ATPs), for instance, are specialised in equality proofs or proof by induction. Some Computer Algebra Systems (CASs) are, for example, specialised on group theory, others in computations in polynomial rings.

The MathWeb Software Bus (MathWeb-SB) [FK99] provides the means for inter-operability of reasoning specialists. It connects a wide-range of reasoning systems (*MathWeb services*), such as ATPs, CASs, model generators (MGs), (semi-)automated proof assistants, constraint solvers (CSs), human interaction units, and automated concept formation systems, by a common *mathematical*

---

* The author is supported by the European Union IHP grant CALCULEMUS HPRN-CT-2000-00102.
** The author was funded by EPSRC grant Gr/M45030.

[1] Throughout this paper, by *reasoning system* we mean both logic-based Deduction Systems and computation-based systems such as Computer Algebra Systems or Constraint Solvers.

J. Calmet et al. (Eds.): AISC-Calculemus 2002, LNAI 2385, pp. 319–331, 2002.

software bus. Reasoning systems integrated in the MathWeb-SB can therefore offer new functionality to the pool of services, and can in turn use all services offered by other systems. The communication between systems in the Math-Web-SB is based on the standards OPENMATH [CC98] and OMDoc [Koh00] for encoding mathematical content.

A crucial idea behind frameworks like the MathWeb-SB is that opening existing reasoning specialists and combining them via a common software bus can lead to synergetic effects. By using the expertise of other reasoning specialists, a reasoning system can solve problems that are way beyond its previous problem solving horizon. A typical example for this kind of synergy is the use of external reasoning systems, like CASs, MGs, CSs, and ATPs, in the mathematical assistant system ΩMEGA [BCF$^+$97]. The use of external reasoners allows ΩMEGA to solve problems in many different domains, such as group theory, analysis, and economics.

In the last three years we have steadily increased the list of reasoning specialists integrated in the MathWeb-SB. In this paper, we describe our work in progress on the integration of the λ-Clam [RSG98] system, a proof planning system specialised on proof by induction. The advantages of integrating λ-Clam into the MathWeb-SB are twofold: i) λ-Clam can offer its expertise in inductive theorem proving to other reasoning systems, e.g., to the ΩMEGA proof planner which does not have the expertise to perform inductive proofs, and ii) λ-Clam can use all reasoning specialists already available in the MathWeb-SB.

For our work we have investigated both directions, λ-Clam offering services and λ-Clam using external reasoning systems. Due to space limitations we focus on the latter in this paper, especially on the use of CAS computation in the λ-Clam proof planner. For our first experiments with λ-Clam using CAS we studied theorems proved by the systems *Analytica* [CZ92,BCZ98] and CLAM-Lite [Wal00]. *Analytica* is a theorem prover implemented in the programming language of the CAS *Mathematica*. CLAM-Lite is a simple re-implementation of the proof planner CLAM [BvHHS90] in MAPLE's programming language. Thus, both systems are based on the (re-)implementation of a theorem prover in the programming language of a CAS. Preliminary results of our experiments show that, in principle, we can prove the same theorems as *Analytica* and CLAM-Lite using a far more general approach which is even easier to implement, namely the combination of existing reasoning systems via a software bus. In our case, the MAPLE system was already available in the MathWeb-SB which reduced the task to the integration of λ-Clam and the implementation of services (*phrase-books*) for the translation of OPENMATH and OMDoc from and to λ-Clam's logic.

The structure of this paper is as follows. We first give brief descriptions of the MathWeb-SB and the λ-Clam system in sections 2 and 3 respectively. In section 4, we describe the integration of λ-Clam into the MathWeb-SB. We also describe the reasoning services offered by λ-Clam and discuss the use of external reasoning systems in inductive theorem proving in λ-Clam. In section 5 we present first results from experiments with λ-Clam using the CAS MAPLE.

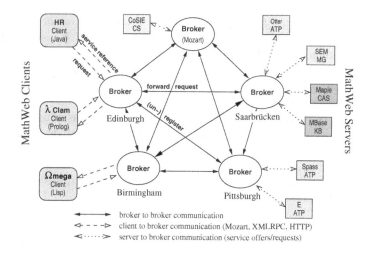

Fig. 1. The MathWeb Software Bus

We draw a conclusion in section 6 and list various directions for future work in section 8.

2 The MathWeb Software Bus

The MathWeb Software Bus [FK99] is a platform for distributed automated reasoning that supports the connection of a wide range of mathematical services by a common software bus.[2] The MathWeb-SB provides the functionality for turning existing reasoning systems into mathematical services that are homogeneously integrated into a networked proof development environment. Each system benefits from being able to use the features of the other components in the MathWeb-SB. The MathWeb-SB is implemented in the multi-paradigm programming language Mozart which enables easy distribution of applications over a LAN or the Internet [Smo95,Con]. This allows the MathWeb-SB to offer a stable network of distributed mathematical services that is in every day use. Client applications can access 23 different reasoning systems, for instance, the CASs MAPLE, MAGMA, and CoCoA, the CS $\mathcal{CoSIE}$, mediators, and MGs such as Mace and Satchmo. Moreover, the MathWeb-SB integrates nine first order ATPs, such as Otter, Spass, Bliksem, and E.

The current architecture of the MathWeb system is depicted in Fig. 1. Local *brokers* provide routing and authentication information to the mathematical services. MathWeb-SB servers (wrappers) encapsulate the actual reasoning systems and offer the mathematical services to their local broker. Brokers register to other brokers running remotely and thus build a dynamic web of brokers. Client

---

[2] Further information about the MathWeb-SB is available at
http://www.mathweb.org/mathweb/.

applications such as the ΩMEGA system, HR, or λ-Clam, connect to one of the MathWeb brokers and request services. If the requested service is not offered by a local reasoning system, the broker forwards the request to all known remote brokers until the service is found. If the requested service is found, the client application receives a reference to a newly created *service object*. The client can then send messages directly to this service object.

The MathWeb-SB originated in the use of external reasoning specialists in the ΩMEGA system. It turned out that the use of external reasoners in proof planning can give valuable support to a human user constructing a proof plan and can significantly extend the problem solving horizon of an automated proof planner. Currently, ΩMEGA's use of external reasoning systems via the Math-Web-SB can be classified into the following categories:

- the use of CAS computations during proof planning
- controlling the proof planning process with the help of CAS computation
- sending subgoals to first order (higher order) ATPs
- finding witness terms for existentially quantified variables using model generators and constraint solvers

The successful use of these reasoning systems in ΩMEGA was one of the main motivations for the work described in this paper, namely the integration λ-Clam into the MathWeb-SB. On the one hand, λ-Clam can now use all reasoning specialists available in the MathWeb-SB. On the other hand, λ-Clam can offer its expertise in inductive theorem proving to other reasoning systems, e.g., the ΩMEGA proof planner which does not provide the proof planning methods and the control knowledge needed for proofs by induction.

In the following, we briefly describe the λ-Clam system and its main reasoning features.

3   The λ-Clam Proof Planner

λ-Clam [RSG98] is a higher-order proof planning system. It is a descendant of the CIAM [BvHHS90] series and is specialised for proof by induction, but it is also intended to allow the rapid prototyping of automated theorem proving strategies. It is implemented in Teyjus λ-PROLOG which is a higher order extension of standard PROLOG.

Proof Methods. λ-Clam works by using depth-first planning with *proof methods.* *Atomic methods* encode more or less complex reasoning steps ranging from quantifier elimination to the step case of an induction proof or difference reduction between a goal and its hypotheses. Atomic methods can be composed to *compound methods* using *methodicals* [RS01]. Compound methods offer the means to define search strategies at various levels of abstraction. A proof plan produced by λ-Clam is a tree of the subgoals generated during proof labelled with the atomic method applications applied to that subgoal. The leaves of the tree are the trivial subgoals: *true*. These proofs are not fully formal but are closer to the formality of proofs found in mathematical textbooks.

Rippling is one of the most widely used and most powerful proof methods. It is used for difference reduction, particularly in inductive proof planning. It was first introduced in [BSvH+93]. The implementation of rippling in λ-Clam is based on the theory presented by Smaill & Green [SG96] who proposed a version that naturally coped with higher-order features. The main advantages of rippling are that it allows an equation to be treated as a rewrite in both directions without loss of termination and provides useful information for automatically patching failed proof attempts.

Proof Critics. The methodical expressions which make up compound methods provide a guide to which proof methods should be chosen at any given stage of the proof. Knowing which method is expected to apply gives additional information in case the system generating the plan fails to apply it. Since heuristics are employed in the generation of proof plans it is possible for a proof planning system to fail to find a plan even when one exists. To this end *proof critics* [IB96] can be employed to analyse the reasons for failure and propose alternative choices. The proof critics currently implemented in λ-Clam allow for generalising the original goal, modifying the chosen induction scheme or speculating a missing lemma (rewrite rule).

4 λ-Clam in the MathWeb-SB

We integrated λ-Clam into the MathWeb-SB using the methodology that has been successfully used for the integration of other systems like the ΩMEGA proof planner or the higher order theorem prover TPS: A MathWeb-SB *wrapper* implements the interface to the services offered by λ-Clam and has full control over λ-Clam by simulating user input using socket communication. λ-Clam itself can use the wrapper to access other services as external reasoning systems.

4.1   Services Offered by λ-Clam

All services offered and used by λ-Clam are based on the OPENMATH [CC98] and the OMDoc [Koh00] standard. We implemented a translation service (a phrase-book) to translate incoming formulae, definitions, and conjectures into λ-Clam's higher order abstract syntax. For the sake of efficiency, this translation is performed by the MathWeb-SB wrapper. λ-Clam currently offers two services to the MathWeb-SB:

planProblem. This service takes an OMDoc document, containing a single conjecture, as an argument. The service starts the λ-Clam proof planning mechanism on the conjecture. In our current implementation, the service expects the conjecture to be about natural number arithmetic. We plan an extension of the service such that clients can also provide the theory the conjecture is defined in.

Client applications using the `planProblem` service can use optional arguments to determine which proof strategy (compound method) λ-Clam should use for the planning attempt, and to give a time limit in seconds. In the current implementation, the service simply returns the OPENMATH symbol *true* if λ-Clam could find a proof plan within the given time limit, and *false* if no proof plan could be found.

`ripple`. We argued in section 3 that rippling is one of the most successful heuristics for guiding (inductive) proof planning. Therefore, λ-Clam offers its rippling mechanism as a separate service to the MathWeb-SB. The service is given a single input formatted using the OMDoc standard. The OMDoc must contain a non-empty set of rewrite rules, formalised as lemmas, and a goal sequent $H \vdash \phi$ as a conjecture. The `ripple` service tries to reduce the difference between ϕ and the best suitable hypothesis in H using the rewrite rules. The `ripple` service also tries to apply fertilisation to reduce the goal ϕ to the trivial goal *true*. As a result, the service `ripple` returns an OMDoc which contains the resulting proof planning goal as a sequent $H \vdash \phi'$.

The services offered by λ-Clam can be used by other reasoning systems via the MathWeb-SB. First experiments with the use of the `planProblem` service and the `ripple` service within the ΩMEGA proof planner were successful. We formalised problems in the natural number theory of ΩMEGA and implemented proof planning methods that call the `planProblem` and the `ripple` service of λ-Clam to close open subgoals or to reduce a subgoal to a simpler one, respectively.

One advantage of passing lemmas from ΩMEGA's theories as rewrites to λ-Clam's `ripple` service is that the rewriting process is completely independent of λ-Clam's theories. Thus, λ-Clam can be used as an abstract rewriting engine whose termination is guaranteed. However, the current version of λ-Clam does not maintain a trace of the positions of the subterms to which a rewrite rules was applied. The latter would allow the `ripple` service to tell a client application, like ΩMEGA, exactly which rewrite has been applied to which subterm of the planning goal not just the rewrite rule that had been applied to that goal. ΩMEGA could then use this information to construct a natural deduction proof for the reasoning steps performed during rippling.

4.2 λ-Clam Using MathWeb Services

For the use of external reasoning systems, we extended λ-Clam by a module which abstracts from socket communication details and offers a convenient interface to MathWeb-SB wrapper. Furthermore, we implemented a module to translate λ-Clam's formulae into the OPENMATH standard using the core OPEN-MATH Content Dictionaries (CDs). Thus, the module forms the second half of an OPENMATH phrase-book for λ-Clam. Using the `mathweb` module, λ-Clam can now access every reasoning system available in the MathWeb-SB. We are currently thinking of the following applications of external reasoners in λ-Clam:

Using ATPs to prove simple subgoals: This approach has already been used in the ΩMEGA proof planner to restrict the search space. In some cases, even higher order problems can be transformed to first order problems and then sent to one or more first order ATPs in the MathWeb-SB.

Using ATPs on the control level: It is a known shortcoming of λ-Clam that it does not check the consistency of hypotheses after performing a case split. This leads to λ-Clam missing easy proofs by inconsistency. Modern ATPs are very efficient and could detect trivial inconsistencies in a few milliseconds. We therefore try to prune inconsistent search paths in proof planning with the help of ATPs like OTTER.

CAS computation in proof planning: Due to our positive experience with CAS computations in many proof planning domains, we think that the use of CASs in inductive proof planning can enable λ-Clam to solve problems it couldn't solve before. The rewriting capabilities of a CAS can complement the rewriting of λ-Clam and can thus enhance the reasoning capabilities of λ-Clam.

In our work, we followed all three applications of external reasoning system but focused on the last application, namely the use of MAPLE's simplification in inductive proof planning. On the MAPLE side, we therefore extended the simplifyTerm service. This service now accepts a term (formula) in OPENMATH representation. It returns the OPENMATH representation of the term (formula) resulting from the call of the simplify function of MAPLE. Before this extension, simplifyTerm accepted only MAPLE syntax. To allow the access of external reasoning systems in the proof planner, we implemented two new proof methods for λ-Clam. The method otter can be used to send an open subgoal to the ATP OTTER. The maple_simplify method calls the simplifyTerm service of MAPLE on the current subgoal and introduces the resulting formula as a new subgoal. λ-Clam treats both these systems as oracles and simply labels the appropriate node in the proof plan with the name of the Oracle.

In the following section we present first experience with the use of MAPLE in λ-Clam.

5   Using Maple in λ-Clam

Our experiments should rather be seen as a *proof of concept* than as an extensive case study. So far, we have focused on theorems taken from two works: The *Analytica* theorem prover and the CLAM-Lite proof planner. All theorems have been formalised in λ-Clam's theories and will be available in version 3.2 of λ-Clam. In the following, we describe the theorems we have been looking at so far and give examples for the rewriting (simplification) performed by MAPLE in the different cases. As we mentioned above, the communication between λ-Clam and MAPLE is fully based on the OPENMATH standard, but for the sake of readability we use a convenient mathematical notation here.

5.1   Analytica Theorems

The *Analytica* system [CZ92,BCZ98] is a theorem prover implemented in the programming language of the CAS *Mathematica*. It is based on rewriting and uses term simplification and other computations available in *Mathematica*. E. Clarke and X. Zhao used their system to prove rather complex theorems in real-valued standard analysis, such as theorems about inequalities, limit theorems, finite sums, and about the convergence of series. *Analytica* can also prove that a function found by Weierstrass is continuous over the real numbers but nowhere differentiable.

Since λ-Clam is specialised to inductive proofs we first tried to prove *Analytica* problems that require induction. We formalised the two induction problems presented in [CZ92] and [BCZ98]. The first theorem gives a closed representation for the Fibonacci numbers:

Theorem 1. *Let $Fib(n)$ be the nth Fibonacci number, i.e. $Fib(0) = 0$, $Fib(1) = 1$, and $\forall n \in \mathbb{N}. n \geq 2 \Rightarrow Fib(n) = Fib(n-1) + Fib(n-2)$. Then the following holds:*

$$\forall n. Fib(n) = \frac{(1+\sqrt{5})^n - (1-\sqrt{5})^n}{2^n \sqrt{5}}.$$

Proving Theorem 1 requires complex reasoning about the square root function and exponentiation which is very hard if not infeasible to encode as rewriting rules in λ-Clam. λ-Clam therefore uses MAPLE's simplification whenever the rewrite rules initially given are not sufficient to perform the required reasoning steps. λ-Clam uses the following induction scheme for the proof planning attempt:

$$\frac{P(0) \land P(s(0)) \land \forall n. (P(n) \land P(s(n)) \Rightarrow P(s(s(n))))}{\forall n. P(n)},$$

where s is the successor function on the natural numbers. In the step case $(P(n) \land P(s(n)) \Rightarrow P(s(s(n))))$ of the inductive proof, λ-Clam uses the definition of the Fibonacci numbers to rewrite the goal until no rewriting is applicable anymore. At this point of the proof λ-Clam tries to apply the `maple_simplify` method which performs the following rewriting:

$$\frac{(1+\sqrt{5})^n - (1-\sqrt{5})^n}{2^n \cdot \sqrt{5}} + \frac{(1+\sqrt{5})^{n+1} - (1-\sqrt{5})^{n+1}}{2^{n+1} \cdot \sqrt{5}} = \frac{(1+\sqrt{5})^{n+2} - (1-\sqrt{5})^{n+2}}{2^{n+2} \cdot \sqrt{5}} \rightsquigarrow$$
$$\frac{1}{10} \cdot 5^{\frac{1}{2}} \cdot 2^{-n} \cdot (3 \cdot (1+5^{\frac{1}{2}})^n - 3 \cdot (1-5^{\frac{1}{2}})^n + (1+5^{\frac{1}{2}})^n \cdot 5^{\frac{1}{2}} + (1-5^{\frac{1}{2}})^n \cdot 5^{\frac{1}{2}} =$$
$$\frac{1}{10} \cdot 5^{\frac{1}{2}} \cdot 2^{-n} \cdot (3 \cdot (1+5^{\frac{1}{2}})^n - 3 \cdot (1-5^{\frac{1}{2}})^n + (1+5^{\frac{1}{2}})^n \cdot 5^{\frac{1}{2}} + (1-5^{\frac{1}{2}})^n \cdot 5^{\frac{1}{2}}.$$

λ-Clam can easily reduce the resulting goal to the trivial goal by applying equality reflexivity.

The second theorem proved by λ-Clam is taken from [BCZ98] and gives a closed form of a modified geometric sum:

Theorem 2. *For all natural numbers $m \neq 1$ the following holds:*

$$\forall n. \sum_{k=0}^{n} \frac{2^k}{1+m^{2^k}} = \frac{1}{m-1} + \frac{2^{n+1}}{1-m^{2^{n+1}}}$$

During the planning for this theorem, one crucial simplification step that cannot be performed by λ-Clam, but by MAPLE, occurs in the step case: $\frac{1}{-1+m} + \frac{2}{1-m^2} \rightsquigarrow \frac{1}{1+m}$.

5.2   CLAM-Lite Theorems

In [WNB92], T. Walsh, A. Nunes and A. Bundy presented automatically synthe-sised closed forms of sums. Recently, T. Walsh implemented the simple "proof planning" shell CLAM-Lite in the CAS MAPLE and used his system to prove the correctness of some of these closed forms using MAPLE computations [Wal00].

We formalised all theorems of [WNB92] in λ-Clam and tried to prove them using MAPLE via the MathWeb-SB. The theorems and results are shown in Fig. 2.

No	Closed Form	CLAM-Lite	λ-Clam using MAPLE
3	$\sum i = \frac{n \cdot (n+1)}{2}$	+	+
4	$\sum i^2 = \frac{2n^3 + 3n^2 + n}{6}$	+	+
5	$\sum i + i^2 = \frac{2n^3 + 6n^2 + 4n}{6}$	+	+
6	$\sum a^i = \frac{a^{n+1} - 1}{a - 1}$	+	+
7	$\sum i a^i = \frac{(n+1)a^{n+1} - a \cdot \frac{a^{n+1} - 1}{a-1}}{a - 1}$	+	-
8	$\sum (i + 1) \cdot a^i = \frac{(n+1)a^{n+1}}{a - 1} - \frac{a^{n+1} - 1}{(a-1)^2}$	+	+
9	$\sum \frac{1}{i(i+1)} = \frac{n}{n+1}$	+	-
10	$\sum Fib(i) = Fib(n + 2) - 1$	-	+

Fig. 2. Closed forms proved by CLAM-Lite and by λ-Clam. All sums are over i from 0 to n, $a \neq 1$

A plus sign in the column for CLAM-Lite or λ-Clam means that the respec-tive system could find a proof plan for the problem at hand. A minus indicates failure.

The proof of Theorem 3 in λ-Clam is very similar to the proof presented in [Wal00] and MAPLE performs the same rewriting in the step case of the proof. The reason for λ-Clam's failure on Theorems 7 and 9 is more a technical than a conceptual problem: The normal form of terms returned by MAPLE is sometimes much bigger than the original term and due to a bug in Teyjus λ-PROLOG (the implementation language of λ-Clam), λ-Clam can currently not read in terms that exceed a certain size.

Interestingly, CLAM-Lite could not find a proof plan for Theorem 10 al-though it had access to the same version of MAPLE that λ-Clam is using. On the other hand, λ-Clam could find a proof plan, using MAPLE to perform the following rewrite at the end of the step case:

$$(Fib(n) + Fib(n + 1) - 1) + Fib(n + 1) = (Fib(n + 1) + (Fib(n) + Fib(n + 1))) - 1$$
$$\leadsto 2 \cdot Fib(n + 1) + Fib(n) - 1 = 2 \cdot Fib(n + 1) + Fib(n) - 1,$$

which again is difficult to perform solely with rewriting in λ-Clam because we would have to define a special term ordering to guaranty the termination of the rewriting system. Walsh names the simplicity of CLAM-Lite's planning methods as a reason for CLAM-Lite's failure on Theorem 10. The fact that λ-Clam could prove the theorem underlines our thesis that it is a better choice to combine existing deduction systems via protocols instead of re-implementing them on top of a Computer Algebra System.

6   Conclusion

We describe the integration of the λ-Clam proof planner that is specialised to inductive theorem proving into the MathWeb-SB. We described the advantages of this integration: λ-Clam can now offer its expertise in proof by induction to other reasoning systems in the MathWeb-SB and can use the reasoning services already offered in the MathWeb-SB. We focused on the latter and described first experiments with the use of the CAS MAPLE in λ-Clam to prove theorems about Fibonacci numbers and closed forms of sums. λ-Clam could *not* find proof plans for these theorems without MAPLE's computational power. This paper describes work in progress and we hope to present more results in the final version.

7   Related Work

Several frameworks for the combination of reasoning systems have been developed. The Logic Broker Architecture (LBA) [AZ00] is most similar to the Math-Web-SB. It is a platform for the distributed automated reasoning based on the distribution facilities of the CORBA middle-ware and the OPENMATH standard.

The PROSPER project [DCN+00] aims at developing the technology needed to deliver the benefits of formal specification and verification to system designers in industry. The central idea of the PROSPER project is that of a *proof engine* (a custom built verification engine) which can be accessed by applications via an application programming interface (API). This API is supposed to support the construction of CASE-tools.

Our experiments were strongly inspired by our positive experience with the use of CAS computations in the ΩMEGA proof planner [KKS96]. Similar work has been done for an integration of CASs in the higher order theorem prover HOL [HT93] and for the MAPLE-PVS interface described in [DGKM01].

8   Future Work

The preliminary results we got from our experiments with the use of CAS in inductive proof planning were promising and in the near future we will extend our experiments to a fully-fledged case study. We are going to formalise more theorems about closed forms of summations and try to find proof plans for these theorems. We plan a more detailed comparison of our work with [Wal00] also on the basis of quantitative results (runtime comparisons).

λ-Clam offers the means to formalise and prove theorems in non-standard analysis (NSA) [MFS02]. Using NSA, λ-Clam could already find proof plans for the limit theorems LIM+ and LIM-TIMES and some other analysis theorems, e.g., the mean-value theorem. In contrast to the classical ϵ-δ proofs, NSA proofs tend to be much shorter and more intuitive. Hence, it would be interesting to use λ-Clam's NSA theory to construct alternative proofs for some of the complex theorems proved by the *Analytica* system using the computational power of CASs available in the MathWeb-SB.

We hope in future to experiment, where appropriate, with the communication of plans between systems. This would allow λ-Clam's proof plans to be replayed in an LCF-style theorem prover to produce fully formal proofs. It would also allow the output of oracles such as MAPLE to be expanded within λ-Clam to alleviate concerns about the combinations of different logics etc.

The proper use of context in the communication between reasoning systems is still an open research question. Context can not only reduce the amount of information that has to be transfered between systems, it is also crucial to establish more complex forms of collaboration and coordination between reasoning systems. The λ-Clam proof planner, for instance, offers the powerful *critics* mechanism which analyses the failure of a proof attempt and gives feedback to the user about possible ways of correcting the proof. This feedback can include generalising the original goal, modifying the chosen induction scheme, or speculating a new rewrite rule. Potentially this feedback could also be given to another reasoning system using λ-Clam and this would involve far more complex interactions in terms of context. For instance, if λ-Clam were to suggest modifying the induction scheme this new proposed scheme might have to be transmitted back to the client system for verification in terms of its own logic. To enable this form of fine-grained interaction between reasoning systems we plan to develop a general notion of context in inter-system communication and to use λ-Clam as a prototypical reasoning system that builds up a context.

References

[AZ00]   A. Armando and D. Zini. Towards Interoperable Mechanized Reasoning Systems: the Logic Broker Architecture. In A. Poggi, editor, *Proceedings of the AI*IA-TABOO Joint Workshop 'From Objects to Agents: Evolutionary Trends of Software Systems'*, Parma, Italy, May 2000.

[BCF+97]   C. Benzmüller, L. Cheikhrouhou, D. Fehrer, A. Fiedler, X. Huang, M. Kerber, K. Kohlhase, A. Meirer, E. Melis, W. Schaarschmidt, J. Siekmann, and V. Sorge. ΩMEGA: Towards a mathematical assistant. In W. McCune, editor, *Proc. of the 14th Conference on Automated Deduction*, volume 1249 of *LNAI*, pages 252–255. Springer Verlag, 1997.

[BCZ98]   A. Bauer, E. Clarke, and X. Zhao. Analytica — an Experiment in Combining Theorem Proving and Symbolic Computation. *Journal of Automated Reasoning (JAR)*, 21(3):295–325, 1998.

[BSvH+93] A. Bundy, A. Stevens, F. van Harmelen, A. Ireland, and A. Smaill. Rippling: A heuristic for guiding inductive proofs. *AI*, 62:185–253, 1993. Also available from Edinburgh as DAI Research Paper No. 567.

330   J. Zimmer and L.A. Dennis

[BvHHS90] A. Bundy, F. van Harmelen, C. Horn, and A. Smaill. The Oyster-Clam system. In M. E. Stickel, editor, *Proc. of the 10th International Conference on Automated Deduction*, pages 647–648. Springer-Verlag, 1990. LNAI No. 449. Also available from Edinburgh as DAI Research Paper 507.

[CC98] O. Caprotti and A. M. Cohen. Draft of the Open Math standard. The Open Math Society, http://www.nag.co.uk/projects/OpenMath/omstd/, 1998.

[Con] The Mozart Consortium. The mozart programming system. http://www.mozart-oz.org/.

[CZ92] E. Clarke and X. Zhao. Analytica – A Theorem Prover for Mathematica. Technical Report CMU//CS-92-117, Carnegie Mellon University, School of Computer Science, October 1992.

[DCN+00] L. A. Dennis, G. Collins, M. Norrish, R. Boulton, K. Slind, G. Robinson, M. Gordon, and T. Melham. The prosper toolkit. In *Proc. of the 6th International Conference on Tools and Algorithms for the Construction and Analysis of Systems, TACAS-2000*, LNCS, Berlin, Germany, 2000. Springer Verlag.

[DGKM01] M. Dunstan, H. Gottliebsen, T. Kelsey, and U. Martin. A maple-pvs interface. *Proc. of the Calculemus Symposium 2001*, 2001.

[FK99] A. Franke and M. Kohlhase. System description: MATHWEB, an agent-based communication layer for distributed automated theorem proving. In Harald Ganzinger, editor, *Proc. of the 16th Conference on Automated Deduction*, volume 1632 of *LNAI*, pages 217–221. Springer Verlag, 1999.

[HT93] J. Harrison and L. Théry. Extending the HOL Theorem Prover with a Computer Algebra System to Reason About the Reals. In C.-J. H. Seger J. J. Joyce, editor, *Higher Order Logic Theorem Proving and its Applications (HUG '93)*, volume 780 of *LNCS*, pages 174–184. Springer Verlag, 1993.

[IB96] A. Ireland and A. Bundy. Productive use of failure in inductive proof. *JAR*, 16(1–2):79–111, 1996. Also available as DAI Research Paper No 716, Dept. of Artificial Intelligence, Edinburgh.

[KKS96] M. Kerber, M. Kohlhase, and V. Sorge. Integrating Computer Algebra with Proof Planning. In Jaques Calmet and Carla Limongelli, editors, *Design and Implementation of Symbolic Computation Systems; International Symposium, DISCO '96, Karlsruhe, Germany, September 18-20, 1996; Proc.*, volume 1128 of *LNCS*. Springer Verlag, 1996.

[Koh00] M. Kohlhase. OMDOC: An open markup format for mathematical documents. Seki Report SR-00-02, Fachbereich Informatik, Universität des Saarlandes, 2000. http://www.mathweb.org/omdoc.

[MFS02] E. Maclean, J. Fleuriot, and A. Smaill. Proof-planning non-standard analysis. In *Proc. of the 7th International Symposium on Artificial Intelligence and Mathematics*, 2002.

[RS01] J. Richardson and A. Smaill. Continuations of proof strategies. In Maria Paola Bonacina and Bernhard Gramlich, editors, *Proc. of the 4th International Workshop on Strategies in Automated Deduction*, Siena, Italy, June 2001.

[RSG98] J.D.C. Richardson, A. Smaill, and I. Green. System description: Proof planning in higher-order logic with lambda-clam. In C. Kirchner and H. Kirchner, editors, *Proc. of the 15th International Conference on Automated Deduction*, volume 1421 of *LNCS*, pages 129–133. Springer-Verlag, 1998.

[SG96]   A. Smaill and I. Green. Higher-order annotated terms for proof search. In Joakim von Wright, Jim Grundy, and John Harrison, editors, *Theorem Proving in Higher Order Logics: 9th International Conference, TPHOLs'96*, volume 1275 of *LNCS*, pages 399–414, Turku, Finland, 1996. Springer-Verlag. Also available as DAI Research Paper 799.

[Smo95]   G. Smolka. The Oz programming model. In Jan van Leeuwen, editor, *Computer Science Today*, volume 1000 of *LNCS*, pages 324–343. Springer-Verlag, Berlin, 1995.

[Wal00]   T. Walsh. Proof Planning in Maple. In *Proc. of the CADE-17 workshop on Automated Deduction in the Context of Mathematics*, 2000.

[WNB92]   T. Walsh, A. Nunes, and A. Bundy. The use of proof plans to sum series. In D. Kapur, editor, *Proc. of the 11th Conference on Automated Deduction*, volume 607 of *LNCS*, pages 325–339, Saratoga Spings, NY, USA, 1992. Springer Verlag.

YACAS: A Do-It-Yourself Symbolic Algebra Environment

Ayal Z. Pinkus[1] and Serge Winitzki[2]

[1] 3e Oosterparkstraat 109-III, Amsterdam, The Netherlands (`apinkus@xs4all.nl`)
[2] Tufts Institute of Cosmology, Department of Physics and Astronomy, Tufts University, Medford, MA 02155, USA (`serge@cosmos.phy.tufts.edu`)

Abstract. We describe the design and implementation of YACAS, a free computer algebra system currently under development. The system consists of a core interpreter and a library of scripts that implement symbolic algebra functionality. The interpreter provides a high-level weakly typed functional language designed for quick prototyping of computer algebra algorithms, but the language is suitable for all kinds of symbolic manipulation. It supports conditional term rewriting of symbolic expression trees, closures (pure functions) and delayed evaluation, dynamic creation of transformation rules, arbitrary-precision numerical calculations, and flexible user-defined syntax using infix notation. The library of scripts currently provides basic numerical and symbolic functionality. The main advantages of YACAS are: free (GPL) software; a flexible and easy-to-use programming language with a comfortable and adjustable syntax; cross-platform portability and small resource requirements; and extensibility.

1 Introduction

YACAS is a computer algebra system (CAS) which has been in development since the beginning of 1999. The goal was to make a small system that allows to easily prototype and research symbolic mathematics algorithms. A secondary future goal is to evolve YACAS into a full-blown general purpose CAS.

YACAS is primarily intended to be a research tool for easy exploration and prototyping of algorithms of symbolic computation. The main advantage of YACAS, besides being free software, is its rich and flexible scripting language. The language is closely related to LISP [WH89] but has a recursive descent infix grammar parser [ASU86], includes expression transformation (term rewriting), and supports defining infix operators at run time similarly to Prolog [B86].

The YACAS language interpreter comes with a library of scripts that implement a set of computer algebra features. The YACAS script library is in active development and at the present stage does not offer the rich functionality of industrial-strength systems such as MATHEMATICA or MAPLE. Extensive implementation of algorithms of symbolic computation is one of the future development goals.

YACAS handles input and output in plain ASCII, either interactively or in batch mode. (A graphical interface is under development.) There is also an op-

J. Calmet et al. (Eds.): AISC-Calculemus 2002, LNAI 2385, pp. 332–336, 2002.

tional plugin mechanism whereby external libraries can be linked into the system to provide extra functionality.

YACAS currently (at version 1.0.49) consists of approximately 22000 lines of C++ code and 13000 lines of script code, with 170 functions defined in the C++ kernel and 600 functions defined in the script library.

2   Basic Design

YACAS consists of a "core engine" (kernel), which is an interpreter for the YACAS scripting language, and a library of script code.

The YACAS engine has been implemented in a subset of C++ which is supported by almost all C++ compilers. The design goals for YACAS core engine are: portability, self-containment (no dependence on extra libraries or packages), ease of implementing algorithms, code transparency, and flexibility. The YACAS system as a whole falls into the "prototype/hacker" rather than into the "axiom/algebraic" category, according to the terminology of Fateman [F90]. There are relatively few specific design decisions related to mathematics, but instead the emphasis is made on extensibility.

The kernel offers sufficiently rich but basic functionality through a limited number of core functions. This core functionality includes substitutions and rewriting of symbolic expression trees, an infix syntax parser, and arbitrary precision numerics. The kernel does not contain any definitions of symbolic mathematical operations and tries to be as general and free as possible of predefined notions or policies in the domain of symbolic computation.

The plugin inter-operability mechanism allows to extend the YACAS kernel or to use external libraries, e.g. GUI toolkits or implementations of special-purpose algorithms. A simple C++ API is provided for writing "stubs" that make external functions appear in YACAS as new core functions. Plugins are on the same footing as the YACAS kernel and can in principle manipulate all YACAS internal structures. Plugins can be compiled either statically or dynamically as shared libraries to be loaded at runtime from YACAS scripts.

The script library contains declarations of transformation rules and of function syntax (prefix, infix etc.). The intention is that almost all symbolic manipulation algorithms and definitions of mathematical functions should be held in the script library and not in the kernel.

For example, the mathematical operator "+" is an infix operator defined in the library scripts. To the kernel, this operator is on the same footing as any other function defined by the user and can be redefined. The YACAS kernel itself does not store any properties for this operator. Instead it relies entirely on the script library to provide transformation rules for manipulating expressions involving the operator "+". In this way, the kernel does not need to anticipate all possible meanings of the operator "+" that users might need in their calculations.

3   Advantages of YACAS

The "policy-free" kernel design means that YACAS is highly configurable through its scripting language. It is possible to create an entirely different symbolic manipulation engine based on the same kernel, with different syntax and different naming conventions, by loading another script library. An example of the flexibility of the YACAS system is a sample script wordproblems.ys. It contains a set of rule definitions that make YACAS recognize simple English sentences, e.g. "Tom has 3 chairs" or "Jane gave an apple to Jill", as valid YACAS expressions. YACAS can then "evaluate" these sentences to True or False according to the semantics of the described situation.

This example illustrates a major advantage of YACAS—the flexibility of its syntax. Although YACAS works internally as a LISP-style interpreter, the script language has a C-like grammar. Infix operators defined in the script library or by the user may contain non-alphabetic characters such as "=" or "#". This means that the user works with a comfortable and adjustable infix syntax, rather than a LISP-style syntax. The user can introduce such syntactic conventions as are most convenient for a given problem. For example, algebraic expressions can be entered in the familiar infix form such as

```
(x+1)^2 - (y-2*z)/(y+3) + Sin(x*Pi/2)
```

The same syntactic flexibility is available for defining transformation rules. Suppose the user needs to reorder expressions containing non-commutative operators of quantum theory. It takes about 20 rules to define an infix operation "**" to express non-commutative multiplication with the appropriate commutation relations and to automatically reorder all expressions involving both non-commutative and commutative factors. Thanks to the YACAS parser, the rules for the new operator can be written in a simple and readable form. Once the operator "**" is defined, the rules that express distributivity of this operation with respect to addition may look like this:

```
15 # (_x + _y) ** _z <-- x ** z + y ** z;
15 # _z ** (_x + _y) <-- z ** x + z ** y;
```

Here, "15 #" is a specification of the rule precedence, "_x" denotes a pattern-matching variable x and the expression to the right of "<--" is to be substituted instead of a matched expression on the left hand side.

Rule-based and functional programming can be freely combined with procedural programming style when the latter is more appropriate for reasons of efficiency or simplicity. Standard patterns of procedural programming, such as subroutines that return values, with code blocks and temporary local variables, the familiar if / else construct and For(), ForEach() loop functions are defined in the script library for the convenience of users. The YACAS interpreter is sufficiently powerful to define these functions in the script library itself rather than in the kernel. This power is fully given to the user, since the library scrips are on the same footing as any user-defined code. Many library functions are intended mainly as tools available to a YACAS user to make algorithm implementation more comfortable.

4   The YACAS Kernel Functionality

YACAS script is a functional language based on various ideas that seemed useful for an implementation of a CAS: list-based data structures, object properties, functional programming (à la LISP); term rewriting [BN98] with pattern matching somewhat along the lines of MATHEMATICA; user-defined infix operators á la PROLOG; delayed evaluation of expressions; and arbitrary precision arithmetic. Garbage collection is implemented through reference counting.

The kernel provides three basic data types: numbers, strings, and atoms and two container types: list and static array (for speed). Additional container or data types ("generic objects") can be made available through C++ plugins. Atoms are implemented as strings that can be assigned values and evaluated. Boolean values are simply atoms **True** and **False**. Hash tables, stacks, and closures (pure functions) are implemented using nested lists. Kernel primitives are available for arbitrary precision arithmetic, string and list manipulation, control flow, defining transformation rules, and declaring function syntax. Expression trees are internally represented by nested lists. Expressions can be "tagged" (assigned a "property object" à la LISP).

The interpreter engine recursively evaluates expression trees according to the transformation rules from the script library. Evaluation proceeds from the leaves of the tree upwards. The engine tries to apply all existing rules to each subexpression, rewriting leaves or branches of the expression tree, until no more rules apply. This type of semantic matching has been implemented in the past (see, e.g., [C86]). However, the YACAS language includes some advanced features to create a more flexible and powerful term rewriting system.

Rules have predicates that determine whether a rule should be applied to an expression. Predicates can be any YACAS expressions that evaluate to the atoms **True** or **False** and are typically functions of pattern variables.

All rules are assigned a precedence value (a positive integer) and rule matching is attempted in the order of precedence. (Thus YACAS provides somewhat better control over the automatic recursion than e.g. the pattern-matching system of MATHEMATICA which does not allow for rule precedence.) Using rule precedence and predicates, a recursive implementation of the integer factorial function may look like this:

```
10 # Factorial(0) <-- 1;
20 # Factorial(n_IsInteger) _ (n>0) <-- n*Factorial(n-1);
```

The rules have precedence 10 and 20, therefore the first rule will be tried first and the recursion will stop when $n = 0$ is reached.

New rules can be defined dynamically as a side-effect of evaluation. This means that there is no predefined "ranking alphabet" of "ground terms" (in the terminology of [TATA99]). It is possible to define a "rule closure" that defines rules depending on its arguments, or to erase rules. Thus, a (read-only) YACAS script library does not actually represent a fixed tree rewriting automaton; an implementation of machine learning is possible.

The Knuth-Bendix termination algorithm [KB70] is not used because rules in YACAS are not an expression of mathematical equivalence but a programming technique. Termination is the responsibility of the user who has complete control over the order of rule application.

5 Current Status

Currently, the script library implements basic algorithms of computer algebra: manipulation of polynomials and elementary functions, limits, derivatives and (basic) symbolic integration, solution of (simple) equations, and some special-purpose functions. (The primary sources of inspiration were the books [K98], [GG99] and [B86].) The system is free (GNU GPL) software and comes with ample documentation to facilitate cooperative development. The main Internet site for YACAS is http://www.xs4all.nl/~apinkus/.

The main development platform is GNU/Linux, but YACAS runs also under various Unix flavors, Windows environments, Psion organizers (EPOC32), Ipaq PDAs running the Linux kernel, BeOS, and Mac OS X. Creating an executable for another platform (including embedded platforms) should not be difficult.

In the future, YACAS is intended to grow into a general-purpose CAS as well as a repository and a testbed of algorithms. In our opinion, YACAS is a promising research tool for exploring symbolic computation.

References

[ASU86]   A. Aho, R. Sethi and J. Ullman, *Compilers (Principles, Techniques and Tools)*, Addison-Wesley, 1986.

[B86]   I. Bratko, *Prolog (Programming for Artificial Intelligence)*, Addison-Wesley, 1986.

[BN98]   F. Baader and T. Nipkow, *Term rewriting and all that*, Cambridge University Press, 1998.

[C86]   G. Cooperman, *A semantic matcher for computer algebra*, in Proceedings of the symposium on symbolic and algebraic computation (1986), Waterloo, Ontario, Canada (ACM Press, NY).

[F90]   R. Fateman, *On the design and construction of algebraic manipulation systems*, also published as: ACM Proceedings of the ISSAC-90, Tokyo, Japan.

[GG99]   J. von zur Gathen and J. Gerhard, *Modern Computer Algebra*, Cambridge University Press, 1999.

[K98]   D. E. Knuth, *The Art of Computer Programming (Volume 2, Seminumerical Algorithms)*, Addison-Wesley, 1998.

[KB70]   D. E. Knuth and P. B. Bendix, *Simple word problems in universal algebras*, in *Computational problems in abstract algebra*, ed. J. Leech, p. 263, Pergamon Press, 1970.

[TATA99]   H. Comon, M. Dauchet, R. Gilleron, F. Jacquemard, D. Lugiez, S. Tison, and M. Tommasi, *Tree Automata Techniques and Applications*, 1999, online book: http://www.grappa.univ-lille3.fr/tata

[WH89]   P. Winston and B. Horn, *LISP*, Addison-Wesley, 1989.

Focus Windows: A New Technique for Proof Presentation*

Florina Piroi and Bruno Buchberger

Research Institute For Symbolic Computation,
4232 Hagenberg, Austria
{fpiroi,buchberg}@risc.uni-linz.ac.at

Abstract. Whether they are hand written or generated by an auto-mated prover, long proofs may be difficult to understand and follow. The main reason for this is that at some point in the proof formulae that occur lines, paragraphs or even pages before, are used. The proof presentation method proposed here tries to overcome this by showing, in each proof step, exactly the formulae that are relevant for the particular proof step. We describe the implementation of this method in the frame of the *Theorema* system.

1 Introduction

Proofs in mathematical publications are linear texts. We view them as sequences of proof steps, i.e. at each step a new formula is derived from formulae appearing in earlier steps, by some proof technique. In long proofs, the used formulae may occur a couple of lines, paragraphs, or even pages distant from the place where they are used. These formulae are usually referred by labels and the reader has to jump back and forth between the referenced formulae and the proof step in which they are needed. This is unpleasant and makes understanding proofs quite difficult even if the proofs are nicely structured and well presented.

Most automated theorem provers do not put emphasis on producing proofs that are easy to read and understand. (A telling illustration of this is given in [7].) Even those which provide tools for studying proofs (as, for example, the Omega system [5]) have the problem described above.

From the outset, in *Theorema* [2] we tried to emphasize on attractive proof presentation. *Theorema* proofs are designed to resemble proofs generated by humans, i.e. they contain formulae and explanatory text in english. In addi-tion, *Theorema* provides various tools that help the reader in browsing proofs: references to formulae are realized as hyper–links that display the referenced formula in a small auxiliary window; nested brackets at the right–hand margin make contracting entire sub–proofs to just one line possible; various color codes distinguish the (temporary) proof goals from the (temporary) knowledge base

---

* partially supported by the RISC PhD scholarship program of the government of Upper Austria and by the FWF (Austrian Science Foundation) SFB project P1302

J. Calmet et al. (Eds.): AISC-Calculemus 2002, LNAI 2385, pp. 337–341, 2002.

formulae; etc. Still, reading and understanding long linear proofs is difficult even for proofs generated by the typical *Theorema* provers.

Focus windows provide means to overcome this difficulty. The idea of focus windows as a technique for proof presentation was introduced in [1] and is as follows: Starting from the root of a proof object, in each proof step, one analyzes which formulae are used and which ones are produced. Then, a window containing exactly these formulae for the proof step that is being analyzed is composed. The window also contains buttons for moving to, and analyzing the next proof step in the proof. For proof steps that branch into two or more sub–proofs the subsequent windows are displayed in contracted form the user being allowed to decide which one to open next.

In the following we give some comments on the Focus Windows from the user's point of view. In Section 3, we briefly describe the implementation of the Focus Windows technique in *Theorema* and then present the final conclusions.

2   Using Focus Windows

In this section we try to shortly present the Focus Windows from the user's side.

A typical call for starting a *Theorema* prover to work on a proof problem looks like this:

Prove[Lemma["Lm"], using → KnowledgeBase, by → SomeProver,

ProverOptions → {options for the Prover}, **showBy** → **SomeDisplayer**];

The user of *Theorema* can control both the work of SomeProver by setting the ProverOptions and the way the proof is presented by setting the showBy option. By default, *Theorema* will present the proof in a new *Mathematica* notebook as a linear proof text. By setting *showBy* → *FocusWindows* the Focus Windows display method will be invoked. (For a complete description of the options of the Prove command and other details about *Theorema* see [8], [6]).

As mentioned before, the Focus Windows method presents proofs in a step–wise manner. Each step of the proof will be shown to the user in two phases: the attention phase and the transformation phase with the corresponding Attention Window (the formulae inferred at the inspected proof step are not yet shown to the user) and the Transformation Window. Each of these windows has

- a "goal area" in which the current goals are shown,
- an "assumptions area" in which the "relevant" assumptions are shown,
- a "proof tree area" in which the proof tree is displayed in a schematic form,
- an area that presents all the assumptions that are available (the "all assumptions area"),
- and a "navigation area" that allows the user to step forward or backward in the proof, in the order suggested by the prover that generated it.

As a concrete example, the goal area of the Transformation Window in the picture below contains the formulae (2.1) and (3). The latter is the formula that was inferred in the presented proof step, therefore its 'New Goal' heading. The assumptions area contains the definitions of the functions 'class' and 'factor–set' and the predicate 'is–all–nonempty'. If new assumptions would have been derived in this proof step, the corresponding formulae would have been shown under the heading 'New Assumptions'. The area containing all the assumptions that are currently available is shown in a closed cell, following the basic philosophy of the Focus Windows technique, that the user will normally *not* want to see all the assumptions that are available in the proof at that point. If the user is interested to see the contents of it (s)he has to double–click on the respective cell bracket. (The organization of notebooks using cells is a standard *Mathematica* feature, see [9]).

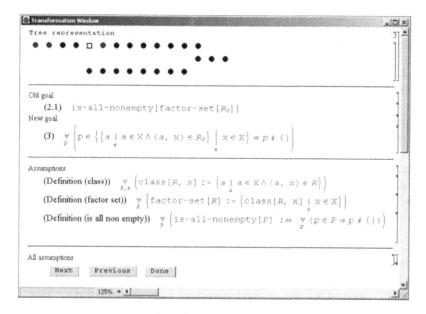

Fig. 1.

The simplified proof representation in the proof tree area, at the top of the above focus window, is not only a graphical representation but it also has some functionality. The nodes of the simplified tree representation are in one–to–one correspondence with the proof steps of the proof object, the current one being high–lighted (□). Clicking any of these nodes will cause the window to shift its focus to the proof step linked to the clicked node. Thus, the user is allowed to read the proof in the order (s)he prefers.

3 Implementation Issues

The proof presentation technique explained in the section above should not be difficult to implement in any existing automated prover, even for systems that do not actually generate proofs automatically but restrict automation to checking proofs generated by humans (like HOL [3], Mizar [4]). The main pre–requisite is that the results of the provers in the system must be formal proof objects that contain sufficient information for extracting the used and inferred formulae, in any particular step.

We implemented the Focus Windows method in *Mathematica* [9], which is also the language we chose for the implementation of *Theorema*. In fact, the implementation was straightway because of two reasons:

• From the outset, the data structure of *Theorema* proof objects was carefully designed in order to give easy access to the relevant formulae in each proof step.

• The front end of *Mathematica* provides convenient programming tools for active objects that, basically, allow to apply the usual *Mathematica* programming style also for programming man–machine interfaces. We use this facility for attaching certain information to the buttons of the navigation area and of the schematic proof tree representation, reducing drastically the time needed for searching information in the proof tree. We give some more details below.

The user actions are taken in via the buttons 'Next', 'Previous' and 'Done' in the navigation area and the schematic proof tree presentation whose nodes are, in fact, buttons. The schematic proof tree representation is a static object in the sense that the data attached to its node buttons does not change during the presentation of the proof by the focus window viewer. In contrast, the buttons 'Next' and 'Previous' are dynamic objects, whose information is used in the following way:

• Suppose that the focus window is presenting the Attention Window of some node n of the proof tree. Then the data attached to the 'Previous' button is a link to the parent node of n. The data attached to the 'Next' button is a link to the node n because when pressing it we want to bring up the Transformation Window of the same node n.

• Suppose that the focus window is presenting the Transformation Window of some node n of the proof tree. Note that such a window may have several branches. Then the data attached to the 'Previous' button in each of the branches is a link to the node n because when pressing it we want to bring up the Attention Window of the same node n. The data attached to the 'Next' button in each of the branches is a link to the corresponding child node of n.

4   Conclusions

The essence of the method we presented is that we show, in each proof step, exactly the formulae that are *relevant* for the particular proof step and we put these formulae into our *focus*.

In the context of automated theorem proving, when proofs are naturally available as processable data objects (the "proof objects") this focusing operation can be described by an algorithm and can be made available for the users of automated theorem proving systems.

Note that the Focus Windows tool is *not* a prove method! The Focus Windows technique does not assert that each of the proof steps should be "easily" verifiable but, rather, it just gives a method to keep track of the relevant information used in each proof step a particular prover generates.

When comparing the linear proof presentation and the focus windows proof presentation of one and the same proof one may make the following observations:

• In short proofs, the focus windows presentation may generate presentation overhead that will distract the reader rather than help him.

• In proofs that are more than one or two pages long, the focus windows presentation may increase the possibility of verifying proofs drastically.

• Linear presentations are helpful for obtaining a quick overview on the overall flow of the proof whereas the focus windows presentation may drastically increase the process of thoroughly understanding proofs.

• Most probably, browsing a proof in linear representation and, then, studying the details of the proof by focus windows presentation style is the most reasonable and efficient way of understanding proofs.

After having implemented the Focus Windows technique in *Theorema*, we also made another, interesting and unexpected, experience: The tool can of course be applied to wrong proofs. In particular it can be used to check the proofs generated by theorem provers that are under construction and not yet fully tested. Here we noticed that checking the proofs by the Focus Windows technique makes it much easier to detect errors in the provers. Thus, the Focus Window tool may also be a useful research instrument for people working in the design and implementation of automated theorem provers.

References

1. B.Buchberger. *Focus Windows Presentation: A New Approach to Presenting Mathematical Proofs (in Automated Theorem Proving Systems)*. *Theorema* Technical Report, 2000–01–30, RISC,
   http://www.risc.uni–linz.ac.at/people/buchberg/downloads.html
2. B. Buchberger, C. Dupré, T. Jebelean, F. Kriftner, K. Nakagawa, D. Vasaru, W. Windsteiger. *The Theorema Project: A Progress Report.* In: Symbolic Computation and Automated Reasoning (Proceedings of CALCULEMUS 2000, Symposium on the Integration of Symbolic Computation and Mechanized Reasoning, August 6–7, 2000, St. Andrews, Scotland, M. Kerber and M. Kohlhase eds.), A.K. Peters, Natick, Massachusetts, pp. 98–113. ISBN 1–56881–145–4.
3. *The HOL System.* Developed at the University of Cambridge, directed by R. Milner. http://www.cl.cam.ac.uk/Research/HVG/HOL/.
4. *Mizar System.* Developed at the University of Warsaw, directed by A. Trybulec. http://mizar.uwb.edu.pl/system/.
5. *Omega System.* Developed at the University of Saarbrücken, directed by J. Siekmann. http://www.ags.uni–sb.de/ omega/intro.html.
6. D. Vasaru–Dupré, *Automated Theorem Proving by Integrating Proving, Solving and Computing.* RISC Institute, May 2000, RISC report 00–19. PhD Thesis.
7. F. Wiedijk, *The Fourteen Provers of the World.* 2001,
   http://www.cs.kun.nl/ freek/notes/index.html
8. W. Windsteiger, *A Set Theory Prover in Theorema: Implementation and Practical Applications*, RISC Institute, May 2001, RISC report 01–03. PhD Thesis.
9. S.Wolfram. *The Mathematica Book*, Wolfram Media and Cambridge University Press, 1996.

Author Index

Aiguier, Marc   51
Audemard, Gilles   231

Bahrami, Diane   51
Bai, Li   128
Beeson, Michael   246
Bernhaupt, Regina   168
Bertoli, Piergiorgio   231
Buchberger, Bruno   337

Campbell, J.A.   102
Castaing, Jacqueline   136
Chetty, Madhu   12
Cimatti, Alessandro   231
Colmerauer, Alain   2
Colton, Simon   259, 275
Corless, Robert M.   76
Crespo, José   38

Dennis, Louise A.   319
Deplagne, Eric   4
Dubois, Catherine   51

Freuder, Eugene C.   1

Golumbic, Martin Charles   196

Herrero, Begoña   38
Hunter, Andrew   117

Jeffrey, D.J.   76

Kirchner, Claude   4
Kitzelmann, Emanuel   26
Korniłowicz, Artur   231
Kutsia, Temur   290

Laita, Laura   38
Laita, Luis M.   38
Ledesma, Luis de   38
Liu, Yihui   128
Loriette-Rougegrez, S.   154

Maojo, Víctor   38
Meier, Andreas   275
Mühlpfordt, Martin   26

Nossum, Rolf   90

Pfalzgraf, Jochen   168
Pinkus, Ayal Z.   332
Piroi, Florina   337

Ranise, Silvio   305
Ratschan, Stefan   181
Roanes-Lozano, Eugenio   38

Schmid, Ute   26
Sebastiani, Roberto   231
Serafini, Luciano   90
Siani, Assaf   196
Smirnova, Elena   64
Sorge, Volker   275
Sturm, Thomas   7
Subramani, K.   217

Tounsi, Mohamed   208

Wiedijk, Freek   246
Winitzki, Serge   332
Wysotzki, Fritz   26

Zimmer, Jürgen   319

Lecture Notes in Artificial Intelligence (LNAI)

Vol. 2168: L. De Raedt, A. Siebes (Eds.), Principles of Data Mining and Knowledge Discovery. Proceedings, 2001. XVII, 510 pages. 2001.

Vol. 2173: T. Eiter, W. Faber, M. Truszczynski (Eds.), Logic Programming and Nonmonotonic Reasoning. Proceedings, 2001. XI, 444 pages. 2001.

Vol. 2174: F. Baader, G. Brewka, T. Eiter (Eds.), KI 2001: Advances in Artificial Intelligence. Proceedings, 2001. XIII, 471 pages. 2001.

Vol. 2175: F. Esposito (Ed.), AI*IA 2001: Advances in Artificial Intelligence. Proceedings, 2001. XII, 396 pages. 2001.

Vol. 2182: M. Klusch, F. Zambonelli (Eds.), Cooperative Information Agents V. Proceedings, 2001. XII, 288 pages. 2001.

Vol. 2190: A. de Antonio, R. Aylett, D. Ballin (Eds.), Intelligent Virtual Agents. Proceedings, 2001. VIII, 245 pages. 2001.

Vol. 2198: N. Zhong, Y. Yao, J. Liu, S. Ohsuga (Eds.), Web Intelligence: Research and Development. Proceedings, 2001. XVI, 615 pages. 2001.

Vol. 2203: A. Omicini, P. Petta, R. Tolksdorf (Eds.), Engineering Societies in the Agents World II. Proceedings, 2001. XI, 195 pages. 2001.

Vol. 2225: N. Abe, R. Khardon, T. Zeugmann (Eds.), Algorithmic Learning Theory. Proceedings, 2001. XI, 379 pages. 2001.

Vol. 2226: K.P. Jantke, A. Shinohara (Eds.), Discovery Science. Proceedings, 2001. XII, 494 pages. 2001.

Vol. 2246: R. Falcone, M. Singh, Y.-H. Tan (Eds.), Trust in Cyber-societies. VIII, 195 pages. 2001.

Vol. 2250: R. Nieuwenhuis, A. Voronkov (Eds.), Logic for Programming, Artificial Intelligence, and Reasoning. Proceedings, 2001. XV, 738 pages. 2001.

Vol. 2253: T. Terano, T. Nishida, A. Namatame, S. Tsumoto, Y. Ohsawa, T. Washio (Eds.), New Frontiers in Artificial Intelligence. Proceedings, 2001. XXVII, 553 pages. 2001.

Vol. 2256: M. Stumptner, D. Corbett, M. Brooks (Eds.), AI 2001: Advances in Artificial Intelligence. Proceedings, 2001. XII, 666 pages. 2001.

Vol. 2258: P. Brazdil, A. Jorge (Eds.), Progress in Artificial Intelligence. Proceedings, 2001. XII, 418 pages. 2001.

Vol. 2275: N.R. Pal, M. Sugeno (Eds.), Advances in Soft Computing – AFSS 2002. Proceedings, 2002. XVI, 536 pages. 2002.

Vol. 2281: S. Arikawa, A. Shinohara (Eds.), Progress in Discovery Science. XIV, 684 pages. 2002.

Vol. 2293: J. Renz, Qualitative Spatial Reasoning with Topological Information. XVI, 207 pages. 2002.

Vol. 2296: B. Dunin-Kęplicz, E. Nawarecki (Eds.), From Theory to Practice in Multi-Agent Systems. Proceedings, 2001. IX, 341 pages. 2002.

Vol. 2298: I. Wachsmuth, T. Sowa (Eds.), Gesture and Language in Human-Computer Interaction. Proceedings, 2001. XI, 323 pages.

Vol. 2302: C. Schulte, Programming Constraint Services. XII, 176 pages. 2002.

Vol. 2307: C. Zhang, S. Zhang, Association Rule Mining. XII, 238 pages. 2002.

Vol. 2308: I.P. Vlahavas, C.D. Spyropoulos (Eds.), Methods and Applications of Artificial Intelligence. Proceedings, 2002. XIV, 514 pages. 2002.

Vol. 2309: A. Armando (Ed.), Frontiers of Combining Systems. Proceedings, 2002. VIII, 255 pages. 2002.

Vol. 2313: C.A. Coello Coello, A. de Albornoz, L.E. Sucar, O.Cairó Battistutti (Eds.), MICAI 2002: Advances in Artificial Intelligence. Proceedings, 2002. XIII, 548 pages. 2002.

Vol. 2317: M. Hegarty, B. Meyer, N. Hari Narayanan (Eds.), Diagrammatic Representation and Inference. Proceedings, 2002. XIV, 362 pages. 2002.

Vol. 2321: P.L. Lanzi, W. Stolzmann, S.W. Wilson (Eds.), Advances in Learning Classifier Systems. Proceedings, 2002. VIII, 231 pages. 2002.

Vol. 2322: V. Mařík, O. Stěpánková, H. Krautwurmová, M. Luck (Eds.), Multi-Agent Systems and Applications II. Proceedings, 2001. XII, 377 pages. 2002.

Vol. 2336: M.-S. Chen, P.S. Yu, B. Liu (Eds.), Advances in Knowledge Discovery and Data Mining. Proceedings, 2002. XIII, 568 pages. 2002.

Vol. 2338: R. Cohen, B. Spencer (Eds.), Advances in Artificial Intelligence. Proceedings, 2002. X, 197 pages. 2002.

Vol. 2358: T. Hendtlass, M. Ali (Eds.), Developments in Applied Artificial Intelligence. Proceedings, 2002 XIII, 833 pages. 2002.

Vol. 2366: M.-S. Hacid, Z.W. Raś, D.A. Zighed, Y. Kodratoff (Eds.), Foundations of Intelligent Systems. Proceedings, 2002. XII, 614 pages. 2002.

Vol. 2385: J. Calmet, B. Benhamou, O. Caprotti, L. Henocque, V. Sorge (Eds.), Artificial Intelligence, Automated Reasoning, and Symbolic Computation. Proceedings, 2002. XI, 343 pages. 2002.

Vol. 2389: E. Ranchhod, N.J. Mamede (Eds.), Advances in Natural Language Processing. Proceedings, 2002. XII, 275 pages. 2002.

Lecture Notes in Computer Science

Vol. 2330: P.M.A. Sloot, C.J.K. Tan, J.J. Dongarra, A.G. Hoekstra (Eds.), Computational Science – ICCS 2002. Proceedings, Part II. XLI, 1115 pages. 2002.

Vol. 2331: P.M.A. Sloot, C.J.K. Tan, J.J. Dongarra, A.G. Hoekstra (Eds.), Computational Science – ICCS 2002. Proceedings, Part III. XLI, 1227 pages. 2002.

Vol. 2332: L. Knudsen (Ed.), Advances in Cryptology – EUROCRYPT 2002. Proceedings, 2002. XII, 547 pages. 2002.

Vol. 2334: G. Carle, M. Zitterbart (Eds.), Protocols for High Speed Networks. Proceedings, 2002. X, 267 pages. 2002.

Vol. 2335: M. Butler, L. Petre, K. Sere (Eds.), Integrated Formal Methods. Proceedings, 2002. X, 401 pages. 2002.

Vol. 2336: M.-S. Chen, P.S. Yu, B. Liu (Eds.), Advances in Knowledge Discovery and Data Mining. Proceedings, 2002. XIII, 568 pages. 2002. (Subseries LNAI).

Vol. 2337: W.J. Cook, A.S. Schulz (Eds.), Integer Programming and Combinatorial Optimization. Proceedings, 2002. XI, 487 pages. 2002.

Vol. 2338: R. Cohen, B. Spencer (Eds.), Advances in Artificial Intelligence. Proceedings, 2002. X, 197 pages. 2002. (Subseries LNAI).

Vol. 2340: N. Jonoska, N.C. Seeman (Eds.), DNA Computing. Proceedings, 2001. XI, 392 pages. 2002.

Vol. 2342: I. Horrocks, J. Hendler (Eds.), The Semantic Web – ISCW 2002. Proceedings, 2002. XVI, 476 pages. 2002.

Vol. 2345: E. Gregori, M. Conti, A.T. Campbell, G. Omidyar, M. Zukerman (Eds.), NETWORKING 2002. Proceedings, 2002. XXVI, 1256 pages. 2002.

Vol. 2346: H. Unger, T. Böhme, A. Mikler (Eds.), Innovative Internet Computing Systems. Proceedings, 2002. VIII, 251 pages. 2002.

Vol. 2347: P. De Bra, P. Brusilovsky, R. Conejo (Eds.), Adaptive Hypermedia and Adaptive Web-Based Systems. Proceedings, 2002. XV, 615 pages. 2002.

Vol. 2348: A. Banks Pidduck, J. Mylopoulos, C.C. Woo, M. Tamer Ozsu (Eds.), Advanced Information Systems Engineering. Proceedings, 2002. XIV, 799 pages. 2002.

Vol. 2349: J. Kontio, R. Conradi (Eds.), Software Quality – ECSQ 2002. Proceedings, 2002. XIV, 363 pages. 2002.

Vol. 2350: A. Heyden, G. Sparr, M. Nielsen, P. Johansen (Eds.), Computer Vision – ECCV 2002. Proceedings, Part I. XXVIII, 817 pages. 2002.

Vol. 2351: A. Heyden, G. Sparr, M. Nielsen, P. Johansen (Eds.), Computer Vision – ECCV 2002. Proceedings, Part II. XXVIII, 903 pages. 2002.

Vol. 2352: A. Heyden, G. Sparr, M. Nielsen, P. Johansen (Eds.), Computer Vision – ECCV 2002. Proceedings, Part III. XXVIII, 919 pages. 2002.

Vol. 2353: A. Heyden, G. Sparr, M. Nielsen, P. Johansen (Eds.), Computer Vision – ECCV 2002. Proceedings, Part IV. XXVIII, 841 pages. 2002.

Vol. 2358: T. Hendtlass, M. Ali (Eds.), Developments in Applied Artificial Intelligence. Proceedings, 2002 XIII, 833 pages. 2002. (Subseries LNAI).

Vol. 2359: M. Tistarelli, J. Bigun, A.K. Jain (Eds.), Biometric Authentication. Proceedings, 2002. XII, 373 pages. 2002.

Vol. 2360: J. Esparza, C. Lakos (Eds.), Application and Theory of Petri Nets 2002. Proceedings, 2002. X, 445 pages. 2002.

Vol. 2361: J. Blieberger, A. Strohmeier (Eds.), Reliable Software Technologies – Ada-Europe 2002. Proceedings, 2002 XIII, 367 pages. 2002.

Vol. 2363: S.A. Cerri, G. Gouardères, F. Paraguaçu (Eds.), Intelligent Tutoring Systems. Proceedings, 2002. XXVIII, 1016 pages. 2002.

Vol. 2364: F. Roli, J. Kittler (Eds.), Multiple Classifier Systems. Proceedings, 2002. XI, 337 pages. 2002.

Vol. 2366: M.-S. Hacid, Z.W. Raś, D.A. Zighed, Y. Kodratoff (Eds.), Foundations of Intelligent Systems. Proceedings, 2002. XII, 614 pages. 2002. (Subseries LNAI).

Vol. 2367: J. Fagerholm, J. Haataja, J. Järvinen, M. Lyly. P. Råback, V. Savolainen (Eds.), Applied Parallel Computing. Proceedings, 2002. XIV, 612 pages. 2002.

Vol. 2368: M. Penttonen, E. Meineche Schmidt (Eds.), Algorithm Theory – SWAT 2002. Proceedings, 2002. XIV, 450 pages. 2002.

Vol. 2370: J. Bishop (Ed.), Component Deployment. Proceedings, 2002. XII, 269 pages. 2002.

Vol. 2374: B. Magnusson (Ed.), ECOOP 2002 – Object-Oriented Programming. XI, 637 pages. 2002.

Vol. 2382: A. Halevy, A. Gal (Eds.), Next Generation Information Technologies and Systems. Proceedings, 2002. VIII, 169 pages. 2002.

Vol. 2385: J. Calmet, B. Benhamou, O. Caprotti, L. Henocque, V. Sorge (Eds.), Artificial Intelligence, Automated Reasoning, and Symbolic Computation. Proceedings, 2002. XI, 343 pages. 2002. (Subseries LNAI).

Vol. 2386: E.A. Boiten, B. Möller (Eds.), Mathematics of Program Construction. Proceedings, 2002. X, 263 pages. 2002.

Vol. 2389: E. Ranchhod, N.J. Mamede (Eds.), Advances in Natural Language Processing. Proceedings, 2002. XII, 275 pages. 2002. (Subseries LNAI).